LINEAR ALGEBRA

MICHAEL STECHER
Texas A & M University

1817
HARPER & ROW, PUBLISHERS, New York
Cambridge, Philadelphia, San Francisco, Washington,
London, Mexico City, São Paulo, Singapore, Sydney

Sponsoring Editor: Peter Coveney
Project Editor: Ellen MacElree
Text Design Adaptation: Barbara Bert, North 7 Atelier, LTD.
Cover Design: Edward H. Butler
Cover Painting: Edward H. Butler
Text Art: RDL Artset LTD.
Production Manager: Kewal Sharma
Compositor: Progressive Typographers
Printer and Binder: R. R. Donnelley & Sons Company
Cover Printer: New England Book Components

LINEAR ALGEBRA

Library of Congress Cataloging in Publication Data

Stecher, Michael.
Linear algebra.

Includes index.
1. Algebras, Linear. I. Title.
QA184.S74 1988 512′.5 87-17711
ISBN 0-06-046427-5

87 88 89 90 9 8 7 6 5 4 3 2 1

To
Jim,
Ann,
and
Fred

CONTENTS

7 NUMERICAL METHODS 230

8 APPLICATIONS 243

APPENDICES 269

PREFACE

This is a text for a first course in linear algebra. I believe that there is a need for a text that integrates the theory of vector spaces, matrices, and linear transformations with the computational aspects of the subject. All the crucial concepts and facts needed to understand linear algebra are here and, with a few exceptions, all theorems are proved. The ideas and computational techniques are discussed in full and illustrated with a range of examples.

Students often ask, "What is this stuff good for?" I have tried to answer this question by incorporating within the body of the text examples which illustrate the uses of linear algebra. I have also tried to explain not only how to achieve some result, but why it might be useful. Additional applications are included in Chapter 8 in order to provide instructors with added flexibility.

Each section of the text ends with numerous exercises of a routine nature to ensure that students understand the concepts and have learned how to do the necessary computations. In addition, supplementary exercises at the end of each chapter review, extend and amplify the material in the text. Although some exercises require calculus for their solution, students without this background need not become upset when encountering them, since they are included to further demonstate the range of applicability of linear algebra.

Chapter 4 is a discussion of the determinant of a matrix. I am primarily interested in using the determinant to define and compute the characteristic polynomial of a matrix. Because of this, Section 4.1 presents only the techniques needed to compute the determinant. The proofs of various theorems have been relegated to Appendix D. For those instructors who wish to cover this material more thoroughly, all of the relevant definitions and proofs are in Appendices C and D.

Chapter 7 is an elementary discussion of some numerical techniques that are employed in solving systems of equations and in computing eigenvalues and eigenvectors. I have found this material easily accessible to students and have presented it while covering the material it relates to. Thus, Sections 7.1 and 7.2 could be covered immediately following Chapter 1, and Section 7.3 after Section 5.2.

I have found the following sequence to be quite effective: Chapter 1, the first two sections of Chapter 7, Chapters 2 and 3, the first two sections of Chapter 4, and then Chapters 5 and 6. Depending on the amount of time left and personal inclinations, one may include any of the remaining material. Students unfamiliar with induction and summation notation will find Appendices A and B useful for self-study.

ACKNOWLEDGMENTS

I would like to thank the following individuals for their many helpful comments and suggestions during the reviewing process: Gerald Bradley, Claremont McKenna College; Sean Allen Broughton, University of Wisconsin; Martin Buntinas, Loyola University; Joaquin Bustoz, Arizona State University; John F. Cavalier, West Virginia Tech University; Carl Cowan, Purdue University; Lucy Dechéne, Fitchberg State College; Frank DeMeyer, Colorado State University; Bruce Edwards, University of Florida; Frederick Gass, Miami University; Jerry Goldman, DePaul University; Jerry Johnson, Oklahoma State University; Peter R. Jones, Marquette University; Kenneth Kalmanson, Montclair State College; Gail Kaplan, U. S. Naval Academy; Stanley Luckawecki, Clemson University; Roger Marty, Cleveland State University; Robert A. Moreland, Texas Tech University; Ancel C. Mewborn, University of North Carolina; Peter Tomas, University of Texas; William Trench, Trinity University.

A special thank you for careful problem checking and the preparation of the solutions Manual is due to Karen Zak of the U.S. Naval Academy. The assistance of Harper & Row in developing this manuscript was extremely helpful. I appreciate the work of Judy Rothman, Ann Trump, Peter Coveney, Ellen MacElree, Debra Bremer, and Jonathan Haber.

Michael Stecher

LIST OF SYMBOLS

1

MATRICES AND SYSTEMS OF LINEAR EQUATIONS

In Chapter 1 we discuss how to solve a system of linear equations. If there are not too many equations or unknowns our task is not very difficult; what we learned in high school will suffice. Fortunately our world is fairly complicated. We quite often meet problems that can be reduced to solving a system of several hundred or more equations with an equally large number of unknowns. Thus our method of solution should simplify the bookkeeping involved and be as direct and straightforward as possible. This simplification process leads in a natural manner to objects called matrices and various rules for manipulating them.

A great amount of time and effort will be spent on matrices, but we always need to keep in mind that we are discussing systems of linear equations.

1.1 SYSTEMS OF LINEAR EQUATIONS

Let's look at a simple example of a system of linear equations:

$$\begin{aligned} 2x_1 + 3x_2 &= 6 \\ -x_1 + 4x_2 &= 8 \end{aligned} \tag{1.1}$$

The problem is to determine all possible pairs of numbers, which we denote by x_1 and x_2, such that these numbers will satisfy both equations in (1.1). The trick (technique) in solving (1.1) is to first remove one of the unknowns from one of the equations. To accomplish this, multiply the second equation by 2 and then add the resulting equation to the first. This gives us the following system:

$$\begin{aligned} 11x_2 &= 22 \\ -x_1 + 4x_2 &= 8 \end{aligned} \tag{1.2}$$

This system of equations is easy to solve. Clearly $x_2 = \frac{22}{11} = 2$, and replacing x_2 with 2 in the second equation we have $-x_1 + 4(2) = 8$ or $-x_1 = 0$. Hence, $x_1 = 0$. We now check that we have actually found a solution to our original system (1.1). That is, we replace x_1 by 0 and x_2 by 2 and verify that both equations are satisfied.

Let's go back to the beginning and try to reduce some of the work involved. We started with (1.1), but now instead of writing the unknowns every time let's just write the coefficients. This leads us to

$$\begin{array}{lrrcr}
2x_1 + 3x_2 = 6 & & & & \\
-x_1 + 4x_2 = 8 & & & & \\
\text{List coefficients} & 2 & 3 & : & 6 \\
 & -1 & 4 & : & 8 \\[6pt]
2R_2 \text{ to } R_1 & 0 & 11 & : & 22 \\
 & -1 & 4 & : & 8 \\[6pt]
R_1 \sim R_2 & -1 & 4 & : & 8 \\
 & 0 & 11 & : & 22 \\[6pt]
-R_1 & 1 & -4 & : & -8 \\
\frac{1}{11}R_2 & 0 & 1 & : & 2
\end{array} \tag{1.3}$$

The notation is almost self-explanatory. For example, $2R_2$ to R_1 means multiply the second row (equation) by 2 and then add it to the first row, which gives us a new first row (equation). $R_1 \sim R_2$ means interchange the first and second rows. $-R_1$ and $\frac{1}{11}R_2$ mean multiply the first row by -1 and the second row by $\frac{1}{11}$. The colons are used to distinguish the right-hand side of equations (1.3) from the coefficients of the unknowns. Later we will not bother with the colons.

The next step in our procedure is to realize that the final array in (1.3) represents the equations

$$\begin{aligned} x_1 - 4x_2 &= -8 \\ x_2 &= 2 \end{aligned} \tag{1.4}$$

From (1.4) we again get the solution $x_2 = 2$, $x_1 = -8 + 8 = 0$. The form of this final array should be noted; namely, there are ones in the "main diagonal" and zeros below it, which means that we can easily solve for x_2 and then x_1.

Let's look at one more specific example before discussing linear systems in general.

EXAMPLE 1

$$\begin{aligned} 2x_1 + 3x_2 - x_3 &= 6 \\ -4x_1 + x_3 &= 7 \\ x_2 - 2x_3 &= 1 \end{aligned} \tag{1.5}$$

Remember that we're going to delete the variables and manipulate the rows in order to arrive at the following form if possible:

$$\begin{matrix} 1 & . & . & : & b_1 \\ 0 & 1 & . & : & b_2 \\ 0 & 0 & 1 & : & b_3 \end{matrix} \tag{1.6}$$

From (1.5) we have

$$\begin{bmatrix} 2 & 3 & -1 & : & 6 \\ -4 & 0 & 1 & : & 7 \\ 0 & 1 & -2 & : & 1 \end{bmatrix}$$

$$2R_1 \text{ to } R_2 \qquad \begin{bmatrix} 2 & 3 & -1 & : & 6 \\ 0 & 6 & -1 & : & 19 \\ 0 & 1 & -2 & : & 1 \end{bmatrix}$$

$$R_2 \sim R_3 \qquad \begin{bmatrix} 2 & 3 & -1 & : & 6 \\ 0 & 1 & -2 & : & 1 \\ 0 & 6 & -1 & : & 19 \end{bmatrix} \tag{1.7}$$

$$-6R_2 \text{ to } R_3 \qquad \begin{bmatrix} 2 & 3 & -1 & : & 6 \\ 0 & 1 & -2 & : & 1 \\ 0 & 0 & 11 & : & 13 \end{bmatrix}$$

$$\begin{matrix} \frac{1}{2}R_1 \\ \\ \frac{1}{11}R_3 \end{matrix} \qquad \begin{bmatrix} 1 & \frac{3}{2} & -\frac{1}{2} & : & 3 \\ 0 & 1 & -2 & : & 1 \\ 0 & 0 & 1 & : & \frac{13}{11} \end{bmatrix}$$

With this last array we associate the following system:

$$\begin{aligned} x_1 + \tfrac{3}{2}x_2 - \tfrac{1}{2}x_3 &= 3 \\ x_2 - 2x_3 &= 1 \\ x_3 &= \tfrac{13}{11} \end{aligned} \tag{1.8}$$

Since the leading coefficients are one, we can easily solve this system for $x_3 = \frac{13}{11}$, $x_2 = 1 + 2(\frac{13}{11}) = \frac{37}{11}$, and $x_1 = 3 + \frac{1}{2}x_3 - \frac{3}{2}x_2 = 3 + (\frac{1}{2})(\frac{13}{11}) -$

$(\frac{3}{2})(\frac{37}{11}) = -\frac{16}{11}$. The ease with which (1.8) is solved demonstrates why the form in (1.6) is desirable. ■

Question: The numbers x_1, x_2, and x_3 satisfy (1.8). Do they also satisfy (1.5)? They had better! The fact that they do can be demonstrated by either putting these numbers into equations (1.5) or observing that the row operations, which were performed, are all reversible. For example, the reverse (inverse) of $-6R_2$ to R_3 would be $6R_2$ to R_3. Starting with the final array of (1.7), if we execute the sequence of operations $11R_3, 2R_1, 6R_2$ to $R_3, R_3 \sim R_2, -2R_1$ to R_2, we will get the original array. Since (1.8) may be algebraically manipulated to get (1.5), every solution of (1.8) will also satisfy (1.5), and conversely every solution of (1.5) will also be a solution of (1.8).

We next discuss the form of a general system of m linear equations in n unknowns:

$$\begin{aligned} a_{11}x_1 + a_{12}x_2 + \cdots + a_{1n}x_n &= b_1 \\ a_{21}x_1 + a_{22}x_2 + \cdots + a_{2n}x_n &= b_2 \\ \cdots\cdots\cdots\cdots\cdots\cdots\cdots\cdots \\ a_{m1}x_1 + a_{m2}x_2 + \cdots + a_{mn}x_n &= b_m \end{aligned} \tag{1.9}$$

The symbols $x_1, x_2, \ldots, x_n$ represent the unknowns that we are trying to find. The a_{ij}, $1 \le i \le m$, $1 \le j \le n$, are called the coefficients of the system. It is assumed that the coefficients a_{ij} and the terms b_j, $1 \le j \le m$, are all known. A few words about the subscripts are in order. The first subscript, i, of a_{ij} refers to the equation in which this term appears, while the second subscript, j, tells us which of the n unknowns a_{ij} multiplies. Thus, if $a_{23} = 6$ we know that the variable x_3 in the second equation is multiplied by 6. To make sure this convention is understood, let's go back to (1.5):

$$\begin{aligned} 2x_1 + 3x_2 - x_3 &= 6 \\ -4x_1 + \qquad x_3 &= 7 \\ x_2 - 2x_3 &= 1 \end{aligned} \tag{1.5}$$

For the above system of equations we would have

$$\begin{array}{llll} a_{11} = 2 & a_{12} = 3 & a_{13} = -1 & b_1 = 6 \\ a_{21} = -4 & a_{22} = 0 & a_{23} = 1 & b_2 = 7 \\ a_{31} = 0 & a_{32} = 1 & a_{33} = -2 & b_3 = 1 \end{array}$$

One more bit of terminology. The procedure we've been using is referred to as Gaussian elimination. We summarize it below.

1. List the coefficients only.
2. Use row operations to obtain zeros below the "main diagonal."
3. Solve the new system.

The phrase "main diagonal" in 2 above refers to those a_{ij} for which $i = j$.

EXAMPLE 2

$$\begin{aligned} x_1 + x_2 - x_3 &= 1 \\ x_1 + \qquad x_3 &= 0 \end{aligned}$$

$$\begin{bmatrix} 1 & 1 & -1 & : & 1 \\ 1 & 0 & 1 & : & 0 \end{bmatrix}$$

$-R_1$ to R_2 $\quad \begin{bmatrix} 1 & 1 & -1 & : & 1 \\ 0 & -1 & 2 & : & -1 \end{bmatrix}$

$-R_2$ $\quad \begin{bmatrix} 1 & 1 & -1 & : & 1 \\ 0 & 1 & -2 & : & 1 \end{bmatrix}$

$-R_2$ to R_1 $\quad \begin{bmatrix} 1 & 0 & 1 & : & 0 \\ 0 & 1 & -2 & : & 1 \end{bmatrix}$

This last matrix implies that

$$x_1 = -x_3 \qquad x_2 = 1 + 2x_3$$ ■

Here we did not get numbers for an answer, but equations relating two of the unknowns to the third. This means that we have an infinite number of solutions, a different one for each choice of x_3. For example, if we set $x_3 = 0$, then $x_1 = 0$ and $x_2 = 1$ is a solution. Similarly the triple $x_3 = 1, x_2 = 3, x_1 = -1$ is also a solution.

EXAMPLE 3

$$\begin{aligned} x_1 + x_2 &= 1 \\ x_2 + 3x_3 &= 4 \\ x_1 + 2x_2 + 3x_3 &= 6 \end{aligned}$$

$$\begin{bmatrix} 1 & 1 & 0 & : & 1 \\ 0 & 1 & 3 & : & 4 \\ 1 & 2 & 3 & : & 6 \end{bmatrix}$$

$-R_1$ to R_3 $\quad \begin{bmatrix} 1 & 1 & 0 & : & 1 \\ 0 & 1 & 3 & : & 4 \\ 0 & 1 & 3 & : & 5 \end{bmatrix}$

$-R_2$ to R_3 $\quad \begin{bmatrix} 1 & 1 & 0 & : & 1 \\ 0 & 1 & 3 & : & 4 \\ 0 & 0 & 0 & : & 1 \end{bmatrix}$

Now what?

This last array represents the system of equations:

$$\begin{aligned} x_1 + x_2 \phantom{{}+3x_3} &= 1 \\ x_2 + 3x_3 &= 4 \\ 0x_1 + 0x_2 + 0x_3 &= 1 \end{aligned}$$

Clearly the third equation has no solution, which means that our original system can have no solution. ■

These examples indicate that for an arbitrary linear system we may have one of three possibilities:

1. No solution: Example 3
2. Exactly one solution: Example 1
3. An infinite number of solutions: Example 2

It is not clear that these are the only possibilities. For example, is there a system with exactly two solutions or some other finite number? We shall see later that this cannot happen and that the above three cases are the only possible ones.

Before concluding this first section we discuss in the next example the geometry of systems of equations, at least in two unknowns.

EXAMPLE 4 Sketch the solutions of each of the following systems of equations:

a. $2x_1 + x_2 = 1 \qquad 4x_1 + 2x_2 = 2$

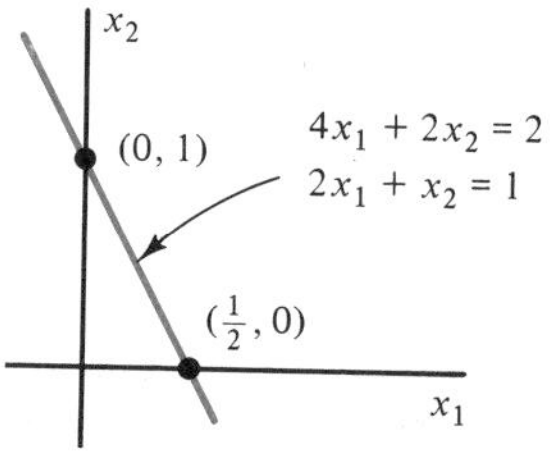

The locus of points that satisfy this system is a straight line as shown above.

b. $2x_1 + x_2 = 1$
$x_1 - x_2 = 0$

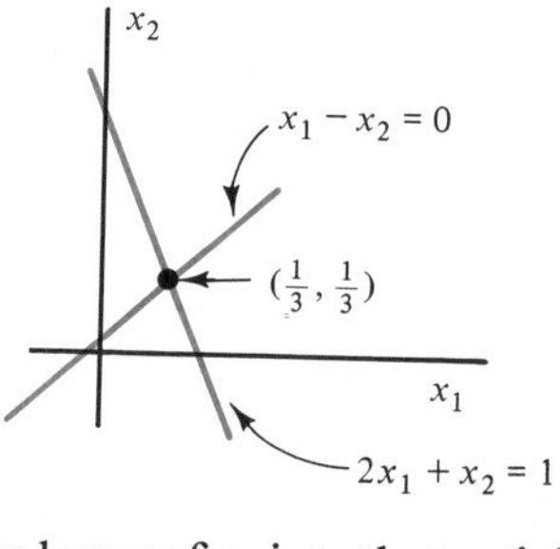

The locus of points that satisfy both equations is all points common to both straight lines. In this case there is only one such point, namely, $x_1 = \frac{1}{3} = x_2$.

c. $2x_1 + x_2 = 1$
$2x_1 + x_2 = 4$

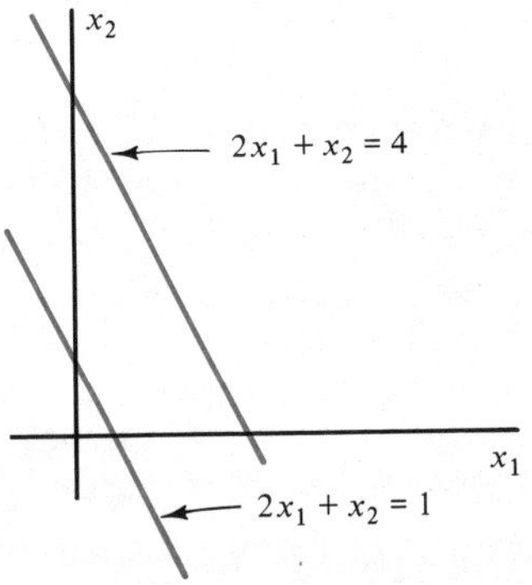

Each equation describes a straight line, and these lines are parallel. Since they do not intersect, the system of equations has no solution. ■

PROBLEM SET 1.1

1. Find all solutions, if any, of the following systems of linear equations:

a. $-3x_1 - 6x_2 = 0$
$2x_1 + 4x_2 = 0$

b. $-3x_1 - 6x_2 = 3$
$2x_1 + 4x_2 = 2$

2. Solve the following systems of linear equations:

a. $2x_1 + 3x_2 = 7$
$-6x_1 + x_2 = 1$

b. $2x_1 + 3x_2 = b_1$
$-6x_1 + x_2 = b_2$

3. Find all solutions, if any, of the following systems of linear equations:

a. $2x_1 + x_2 - x_3 = 1$
$x_1 + x_2 + x_3 = 2$

b. $2x_1 + x_2 - x_3 = 1$
$x_1 + x_2 + x_3 = 2$
$x_1 - x_2 = 2$

4. Find all numbers k such that the following system of equations has a solution and then solve that system.

$$\begin{aligned} 2x_1 + x_2 - x_3 &= 1 \\ x_1 + x_2 + x_3 &= 2 \\ 3x_1 + 2x_2 \phantom{{}+x_3} &= k \end{aligned}$$

5. Find all solutions of the following system of linear equations:

$$\begin{aligned} 2x_1 + 3x_2 - x_3 + x_4 &= 2 \\ x_1 + x_2 + x_3 - 2x_4 &= -6 \\ x_1 + 2x_3 + 4x_4 &= 13 \\ 2x_2 + x_3 - x_4 &= -5 \end{aligned}$$

6. Find all solutions of the following system:

$$\begin{aligned} x_1 \qquad\quad - x_3 + x_4 &= 0 \\ 4x_1 \qquad\quad + x_3 + 2x_4 &= 0 \\ x_1 + 2x_2 - x_3 - x_4 &= 0 \\ 2x_2 + 5x_3 - 4x_4 &= 0 \end{aligned}$$

7. Matt's, Dave's, and Abe's ages are not known but are related as follows: The sum of Dave's and Abe's ages is 3 more than Matt's. Matt's age plus Abe's age is 9 more than Dave's. If the sum of their ages is 35, how old is Abe?

8. Granny's favorite recipe for pecan pie calls for 1 cup of pecans, 3 cups of sugar, and 3 eggs. Her recipe for one batch of pecan puffs calls for 1 cup of sugar, 1 cup of pecans, and 2 eggs. If she has 2 pounds of pecans, 6 pounds of sugar, and a henhouse next door (i.e., as many eggs as she needs), how many pecan pies and batches of pecan puffs can she make while using up all her supplies? (1 pound = 2 cups)

9. Two cars are traveling toward each other. The first is moving at a speed of 45 mph and the second at 65 mph. If, at the start, they are 35 miles apart, how long before they meet and how far will each car have traveled?

10. Find all solutions of the following system:

$$\begin{aligned} 2x_1 + \qquad\quad x_3 + 6x_4 &= 0 \\ -x_1 + 3x_2 + 4x_3 + 2x_4 &= 0 \\ x_2 + 3x_3 + 4x_4 &= 1 \\ x_1 + x_2 - x_3 + x_4 &= 1 \end{aligned}$$

11. Find all solutions of the following system of equations:

$$\begin{aligned} 2x_1 + 2x_2 + \qquad\quad x_4 &= 0 \\ x_1 - x_2 + 2x_3 + x_4 &= 1 \\ x_1 + x_2 + \qquad\quad 4x_4 &= 1 \end{aligned}$$

12. If possible, solve the following systems of equations:

a. $3x_1 - 2x_2 + 3x_3 = 0 \quad x_1 + x_2 + x_4 = 0 \quad x_1 - x_3 - x_4 = 0$

b. $x_1 - x_2 = 1 \quad x_1 + x_2 - x_3 = -2 \quad x_1 - 5x_2 + 2x_3 = 1$

13. Dabs cement plant makes mortar and cement. Each ton of mortar requires $\frac{1}{2}$ ton of lime and $\frac{1}{5}$ ton of sand. Each ton of cement requies $\frac{1}{3}$ ton of lime and $\frac{3}{4}$ ton of sand. If Dabs has 9 tons of lime and 11 tons of sand available, how many tons of mortar and cement can the company make?

14. Graph the solution sets for each of the following systems of equations. Since there are two unknowns, the solution set is the set consisting of all pairs of numbers (x_1, x_2) that satisfy the equations

a. $x_1 - 2x_2 = 6$ **b.** $3x_1 + 4x_2 = 1$
$2x_1 + x_2 = 1$

15. Graph the solution sets of the following systems of equations. Here, the solution set consists of those triples of numbers (x_1, x_2, x_3) that satisfy the equations

a. $x_1 + x_2 + x_3 = 0$

b. $x_1 + x_2 + x_3 = 0 \quad x_1 + x_3 = 0$

c. $x_1 + x_2 + x_3 = 0 \quad x_1 + x_3 = 0 \quad x_1 - x_3 = 0$

16. Find an equation for the straight line that passes through the following points:

a. (1,6), (−1,4)

b. (−2,6), (4,3)

17. Is there a straight line that passes through the following sets of points?
 a. (1,6), (−1,4), (3,7)
 b. (−4,2) (3,7), (1,8)

18. Find an equation for the circle that passes through the three points (−4,2), (3,7), (1,8).

19. Find the solution set to the equations
 a. $x_1 + x_2 = 0$
 b. $x_1 + x_2 = 1$
 c. Graph the solutions sets of **a** and **b**. How are they related?

20. Suppose we have an infinite sequence of numbers x_k, $k = 1, 2, \ldots$, that are related to each other by the equations $x_{k+1} = 2x_k$.
 a. Find a formula for x_k in terms of x_1. For example, $x_2 = 2x_1$ and $x_3 = 4x_1$.
 b. Suppose the x_k are related by the equation $x_{k+2} = x_k + x_{k+1}$. Thus $x_3 = x_1 + x_2$ and $x_4 = x_2 + x_3 = x_1 + 2x_2$. Can you find a formula for x_k $(k > 3)$ in terms of x_1 and x_2?

21. Suppose we have two sequences of numbers x_k and y_k for $k = 1, 2, \ldots$, and that the sequences are related by the equations

$$\begin{aligned} x_{k+1} &= x_k + y_k \\ y_{k+1} &= x_k - y_k \end{aligned}$$

Find a formula for x_k and y_k in terms of x_1 and y_1 for all k if you can. If you cannot discover such a formula, compute x_3 and y_3 in terms of x_1 and y_1.

1.2 MATRICES

In the preceding section the first step taken in solving a system of equations was to remove the variables, which left us with a rectangular array of numbers. These arrays are given a special name.

Definition 1.1. An $m \times n$ matrix is a rectangular array of numbers arranged in m rows and n columns.

In most of this book we use real numbers, but from time to time, as the inclination or need arises, complex numbers will appear.

EXAMPLE 1

$$A = \begin{bmatrix} 1 & 2 \\ 3 & 4 \end{bmatrix} \text{ is a } 2 \times 2 \text{ matrix}$$

$$B = \begin{bmatrix} 1 \\ 3 \end{bmatrix} \text{ is a } 2 \times 1 \text{ matrix}$$

$$C = \begin{bmatrix} 1 & 0 & -1 & 6 \\ 4 & 0 & 1 & 0 \\ 1 & 0 & 0 & 0 \end{bmatrix} \text{ is a } 3 \times 4 \text{ matrix}$$

■

We will often denote an $m \times n$ matrix A by writing $A = [a_{ij}]$, $1 \leq i \leq m$, $1 \leq j \leq n$. The symbol a_{ij} is used to indicate the entry in the ith row and jth column. Thus, counting from the top left corner of the matrix, the first subscript tells us which row and the second subscript which column the entry appears in.

EXAMPLE 2

$$A = \begin{bmatrix} 1 & 2 \\ 3 & 4 \end{bmatrix} \qquad \begin{matrix} a_{11} = 1, a_{12} = 2 \\ a_{21} = 3, a_{22} = 4 \end{matrix}$$

$$B = \begin{bmatrix} 1 \\ 3 \end{bmatrix} \qquad \begin{matrix} b_{11} = 1 \\ b_{21} = 3 \end{matrix}$$

$$C = \begin{bmatrix} 1 & 5 & -1 & 6 \\ 4 & 0 & -2 & 8 \\ 1 & 2 & 3 & 4 \end{bmatrix} \qquad \begin{matrix} c_{11} = 1, c_{12} = 5, c_{13} = -1, c_{14} = 6 \\ c_{22} = 0, c_{23} = -2, c_{24} = 8 \\ c_{33} = 3, c_{34} = 4 \end{matrix}$$ ■

There are various operations that we may perform involving matrices, and we will spend the next few pages defining and exemplifying them. Before doing so, however, we need to state clearly what it means to say that two matrices are equal. Let A and B be two matrices. We say $A = B$ if they are of the same size and their corresponding entries are equal. For example, the matrices

$$\begin{bmatrix} 1 & 2 \\ 0 & 1 \end{bmatrix} \quad \text{and} \quad \begin{bmatrix} 1 & 2 \\ 1 & 1 \end{bmatrix}$$

are not equal since the entries in the second row and first column are not equal. The matrices

$$\begin{bmatrix} 1 & 2 \\ 0 & 1 \end{bmatrix} \quad \text{and} \quad [1 \quad 2]$$

are not equal since the first is a 2×2 matrix and the second is a 1×2 matrix. A precise definition of matrix equality is:

Definition 1.2. Let $A = [a_{ij}]$, $1 \leq i \leq m$, $1 \leq j \leq n$. Let $B = [b_{ij}]$, $1 \leq i \leq p$, $1 \leq j \leq q$, be two matrices. We say $A = B$ if and only if $m = p$, $n = q$, and $a_{ij} = b_{ij}$ for each i and j.

One algebraic operation involving matrices is that of addition. Given two matrices of the same size, we form their sum by adding together their respective entries. Thus

$$\begin{bmatrix} 1 & 2 \\ -1 & 3 \end{bmatrix} + \begin{bmatrix} 0 & 1 \\ 4 & 6 \end{bmatrix} = \begin{bmatrix} 1 & 3 \\ 3 & 9 \end{bmatrix}$$

and

$$[1 \quad 2 \quad 6] + [1 \quad -1 \quad 4] = [2 \quad 1 \quad 10]$$

A formal definition of matrix addition is:

Definition 1.3. Let $A = [a_{ij}]$, $1 \leq i \leq m$, $1 \leq j \leq n$. Let $B = [b_{ij}]$, $1 \leq i \leq m$, $1 \leq j \leq n$. The sum $S = A + B$ of these two $m \times n$ matrices is the $m \times n$ matrix whose i,j entry, s_{ij}, equals $a_{ij} + b_{ij}$.

EXAMPLE 3

a. $$\begin{bmatrix} 1 & -2 & 3 \\ -7 & 0 & 4 \end{bmatrix} + \begin{bmatrix} 6 & -2 & 6 \\ 1 & -8 & 0 \end{bmatrix} = \begin{bmatrix} 7 & -4 & 9 \\ -6 & -8 & 4 \end{bmatrix}$$

b. $$\begin{bmatrix} -1 & \frac{1}{2} \\ 3 & 5 \end{bmatrix} + \begin{bmatrix} \frac{1}{2} & 0 \\ 1 & -2 \end{bmatrix} = \begin{bmatrix} -\frac{1}{2} & \frac{1}{2} \\ 4 & 3 \end{bmatrix}$$ ■

There are also times when we wish to multiply a matrix by a number. To do this, just multiply every entry in the matrix by that number. For example,

$$2\begin{bmatrix} -1 & 1 \\ 4 & 3 \end{bmatrix} = \begin{bmatrix} -2 & 2 \\ 8 & 6 \end{bmatrix}$$

Definition 1.4. Let $A = [a_{ij}]$ be any $m \times n$ matrix and c any number. Then cA is the $m \times n$ matrix whose i,j entry is ca_{ij}.

EXAMPLE 4

a. $$2\begin{bmatrix} 1 & -1 \\ 4 & 3 \\ 1 & 2 \\ 0 & 8 \end{bmatrix} = \begin{bmatrix} 2 & -2 \\ 8 & 6 \\ 2 & 4 \\ 0 & 16 \end{bmatrix}$$

b. $$-1\begin{bmatrix} 1 & -2 \\ 3 & 4 \end{bmatrix} = \begin{bmatrix} -1 & 2 \\ -3 & -4 \end{bmatrix}$$ ■

As with real or complex numbers, these two operations satisfy certain laws, which we state in the following theorem:

Theorem 1.1. Let $A = [a_{ij}]$, $B = [b_{ij}]$, and $C = [c_{ij}]$ be any $m \times n$ matrices. Let a and b be any numbers. Then the following properties hold:

1. $A + B = B + A$
2. $(A + B) + C = A + (B + C)$
3. $a(A + B) = aA + aB$
4. $(a + b)A = aA + bA$
5. $a(bA) = (ab)A$

Proof. The proofs of these statements are straightforward, and we shall go through the details of properties 1 and 3 only.

1. $$\begin{aligned} A + B &= [a_{ij}] + [b_{ij}] \\ &= [a_{ij} + b_{ij}] = [b_{ij} + a_{ij}] \\ &= B + A \end{aligned}$$

3. $$\begin{aligned} a(A + B) &= a([a_{ij}] + [b_{ij}]) \\ &= a([a_{ij} + b_{ij}]) \\ &= [a(a_{ij} + b_{ij})] \\ &= [aa_{ij} + ab_{ij}] \\ &= [aa_{ij}] + [ab_{ij}] \\ &= a[a_{ij}] + a[b_{ij}] \\ &= aA + aB \end{aligned}$$

The next example repeats the above computations using particular matrices and numbers.

EXAMPLE 5

$$3\left(\begin{bmatrix} 1 & 2 & 0 \\ -1 & 6 & 7 \end{bmatrix} + \begin{bmatrix} -1 & -1 & 4 \\ 0 & 5 & 8 \end{bmatrix}\right)$$

$$= 3\begin{bmatrix} 0 & 1 & 4 \\ -1 & 11 & 15 \end{bmatrix} = \begin{bmatrix} 0 & 3 & 12 \\ -3 & 33 & 45 \end{bmatrix}$$

$$3\begin{bmatrix} 1 & 2 & 0 \\ -1 & 6 & 7 \end{bmatrix} + 3\begin{bmatrix} -1 & -1 & 4 \\ 0 & 5 & 8 \end{bmatrix}$$

$$= \begin{bmatrix} 3 & 6 & 0 \\ -3 & 18 & 21 \end{bmatrix} + \begin{bmatrix} -3 & -3 & 12 \\ 0 & 15 & 24 \end{bmatrix}$$

$$= \begin{bmatrix} 0 & 3 & 12 \\ -3 & 33 & 45 \end{bmatrix}$$

■

We see that our answer does not depend upon whether we first add and then multiply, or multiply and then add.

There is a special $m \times n$ matrix, called the zero matrix, which we denote by O_{mn} (when the size is clear we will drop the subscripts). This matrix has all its entries equal to zero. Thus,

$$O_{21} = \begin{bmatrix} 0 \\ 0 \end{bmatrix} \quad \text{and} \quad O_{22} = \begin{bmatrix} 0 & 0 \\ 0 & 0 \end{bmatrix}$$

It is clear that the following is true:

Theorem 1.2. Let A be any $m \times n$ matrix. Then

1. $A + O_{mn} = A$
2. $0A = O_{mn}$

Our next task is to learn how to multiply matrices. The rule at first seems unusual, but after we relate it in the next section to systems of equations, it will be clear that we are not being unreasonable. The first thing to remember is that two arbitrary sized matrices A and B cannot be multiplied together to form AB. The number of columns in the first factor A must equal the number of rows in the second factor B. To calculate the i,j entry of AB, we multiply the entries of the ith row of A by the corresponding entries of the jth column of B, and then sum these products. For example, if

$$A = \begin{bmatrix} 1 & 2 \\ -4 & 3 \end{bmatrix} \qquad B = \begin{bmatrix} -2 \\ 6 \end{bmatrix}$$

then AB will be the 2×1 matrix C, where

$$C = \begin{bmatrix} c_{11} \\ c_{21} \end{bmatrix} = \begin{bmatrix} \text{(first row of } A\text{) ``times'' (first column of } B\text{)} \\ \text{(second row of } A\text{) ``times'' (first column of } B\text{)} \end{bmatrix}$$

or

$$\underset{2 \times 2}{\begin{bmatrix} 1 & 2 \\ -4 & 3 \end{bmatrix}} \underset{2 \times 1}{\begin{bmatrix} -2 \\ 6 \end{bmatrix}} = \underset{2 \times 1}{\begin{bmatrix} (1)(-2) + (2)(6) \\ (-4)(-2) + (3)(6) \end{bmatrix}} = \begin{bmatrix} 10 \\ 26 \end{bmatrix}$$

The precise definition of matrix multiplication is:

Definition 1.5. Let $A = [a_{ij}]$ be an $m \times n$ matrix. Let $B = [b_{ij}]$ be an $n \times p$ matrix. Then $AB = C = [c_{ij}]$ is an $m \times p$ matrix where

$$\begin{aligned} c_{ij} &= a_{i1}b_{1j} + a_{i2}b_{2j} + \cdots + a_{in}b_{nj} \\ &= \sum_{k=1}^{n} a_{ik}b_{kj} \end{aligned}$$

Note the respective sizes of these matrices:

$$(m \times n)(n \times p) = m \times p$$

The number of columns in A must equal the number of rows in B. The number of rows in AB is the same as the number of rows in A and the number of columns in AB is the same as the number of columns in B. Readers unfamiliar with

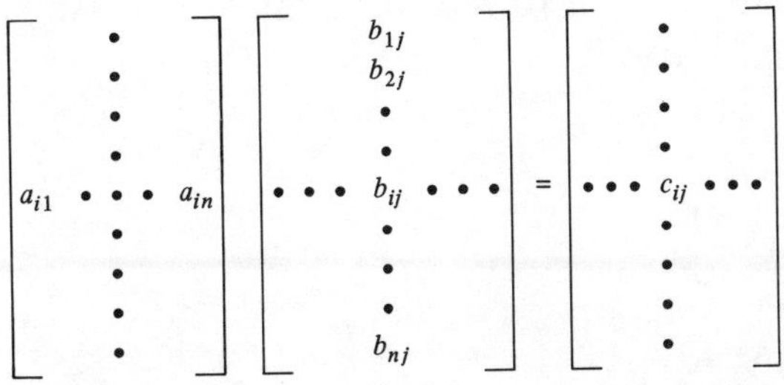

where

$$c_{ij} = a_{i1}b_{1j} + a_{i2}b_{2j} + \cdots + a_{in}b_{nj}$$

Figure 1.1

summation notation should refer to Appendix B. Figure 1.1 illustrates this definition.

EXAMPLE 6

a. $\begin{bmatrix} 1 & 1 & 0 \\ 4 & 3 & 6 \end{bmatrix} \begin{bmatrix} 1 & -1 & 6 \\ -2 & 0 & 3 \\ 8 & 1 & 2 \end{bmatrix}$

$$= \begin{bmatrix} 1-2+\ 0 & -1+0+0 & 6+3+\ 0 \\ 4-6+48 & -4+0+6 & 24+9+12 \end{bmatrix} = \begin{bmatrix} -1 & -1 & 9 \\ 46 & 2 & 45 \end{bmatrix}$$

Note that we cannot multiply these two matrices in opposite order since the number of columns (three) in the second matrix does not equal the number of rows (two) in the first matrix.

b. $\begin{bmatrix} 0 & 1 \\ 1 & 0 \end{bmatrix} \begin{bmatrix} 2 & 3 \\ 1 & 4 \end{bmatrix} = \begin{bmatrix} 0+1 & 0+4 \\ 2+0 & 3+0 \end{bmatrix} = \begin{bmatrix} 1 & 4 \\ 2 & 3 \end{bmatrix}$

Multiplying these matrices in the opposite order, we have

c. $\begin{bmatrix} 2 & 3 \\ 1 & 4 \end{bmatrix} \begin{bmatrix} 0 & 1 \\ 1 & 0 \end{bmatrix} = \begin{bmatrix} 0+3 & 2+0 \\ 0+4 & 1+0 \end{bmatrix} = \begin{bmatrix} 3 & 2 \\ 4 & 1 \end{bmatrix}$

Note that the products in b and c are not equal. Thus, even though AB and BA are both defined, they may not be equal. ■

Matrix multiplication does satisfy some algebraic laws, which are stated in the following theorem:

Theorem 1.3. Let A, B, and C denote matrices of appropriate size; that is, if we write AB, then A has the same number of columns as B has rows. Let a be any real number. Then the following formulas hold:

1. $(A + B)C = AC + BC$, $A(B + C) = AB + AC$
2. $(AB)C = A(BC)$
3. $a(AB) = aA(B) = A(aB)$

Proof. Let $A = [a_{ij}]$ be an $m \times n$ matrix, $B = [b_{ij}]$ be an $m \times n$ matrix, and $C = [c_{ij}]$ be an $n \times p$ matrix. Then

$$(A + B)C = [a_{ij} + b_{ij}][c_{ij}]$$

$$= \left[\sum_{k=1}^{n} (a_{ik} + b_{ik})c_{kj} \right]$$

$$= \left[\sum_{k=1}^{n} a_{ik}c_{kj} + \sum_{k=1}^{n} b_{ik}c_{kj} \right]$$

$$= \left[\sum_{k=1}^{n} a_{ik}c_{kj}\right] + \left[\sum_{k=1}^{n} b_{ik}c_{kj}\right]$$
$$= AC + BC$$

The proof that $A(B + C) = AB + AC$ is essentially the same as that for $(A + B)C = AC + BC$; so we omit it. Properties 2 and 3 may be proved in a similar manner, and the reader is asked to do so in the problems at the end of this section.

We've seen that there is a zero matrix O_{mn}. There is also an identity matrix I_n which plays the role of the number 1 in matrix multiplication. The $n \times n$ matrices, I_n, are defined as follows:

$$I_1 = [1] \qquad I_2 = \begin{bmatrix} 1 & 0 \\ 0 & 1 \end{bmatrix} \qquad I_3 = \begin{bmatrix} 1 & 0 & 0 \\ 0 & 1 & 0 \\ 0 & 0 & 1 \end{bmatrix}$$

For arbitrary n we define

$$I_n = [\delta_{ij}]$$

where $\delta_{ij} = \begin{cases} 1 & \text{if } i = j \\ 0 & \text{if } i \neq j \end{cases}$

That is, the symbol δ_{ij}, which is called the Kronecker delta, is equal to 1 if i equals j and zero if the subscripts are different.

EXAMPLE 7 Let $A = \begin{bmatrix} a & b \\ c & d \end{bmatrix}$ be any 2×2 matrix. Show that $I_2A = AI_2 = A$.

$$I_2A = \begin{bmatrix} 1 & 0 \\ 0 & 1 \end{bmatrix}\begin{bmatrix} a & b \\ c & d \end{bmatrix} = \begin{bmatrix} a+0 & b+0 \\ 0+c & 0+d \end{bmatrix} = \begin{bmatrix} a & b \\ c & d \end{bmatrix} = A$$

$$AI_2 = \begin{bmatrix} a & b \\ c & d \end{bmatrix}\begin{bmatrix} 1 & 0 \\ 0 & 1 \end{bmatrix} = \begin{bmatrix} a+0 & 0+b \\ c+0 & 0+d \end{bmatrix} = \begin{bmatrix} a & b \\ c & d \end{bmatrix} = A$$ ■

Theorem 1.4. Let A be an $m \times n$ matrix. Let I_n and I_m denote the $n \times n$ and $m \times m$ identity matrices, respectively. Then

$$I_mA = AI_n = A$$

Proof. Let $A = [a_{ij}]$. Then

$$I_mA = [\delta_{ij}][a_{ij}] = \left[\sum_{k=1}^{m} \delta_{ik}a_{kj}\right]$$
$$= [a_{ij}] = A$$

Since $\delta_{ik}=0$ unless $k=i$, the sum of the m terms $\delta_{ik}a_{kj}$ has only one nonzero summand, the one for which $k=i$. Similarly $AI_n=A$. Note that if A is a square $n\times n$ matrix, $I_nA=AI_n=A$.

By this time the reader should have the feeling that if a rule is true for numbers it is also true for matrices. This feeling should be maintained with two important exceptions:

1. $AB\neq BA$

This does not mean that AB always differs from BA but rather that equality need not hold.

2. $AB=AC$ does not imply that $B=C$

The reader will find examples of this in the problems at the end of this section.

There is one other operation that we need to discuss. It's called taking the transpose of a matrix. The operation is very simple to perform. We merely interchange rows and columns; that is, the ith row of A becomes the ith column of A^T (A transpose). For example, if

$$A=\begin{bmatrix}1&-1&0\\4&2&-6\end{bmatrix}$$

then

$$A^T=\begin{bmatrix}1&4\\-1&2\\0&-6\end{bmatrix}$$

Definition 1.6. Let $A=[a_{ij}]$ be an $m\times n$ matrix. The transpose of A, denoted by $A^T=[a_{ij}^T]$, is an $n\times m$ matrix where

$$a_{ij}^T=a_{ji}$$

In the future rather than write

$$\begin{bmatrix}1\\-1\\2\\4\end{bmatrix}$$

we will write $[1\quad -1\quad 2\quad 4]^T$.

EXAMPLE 8

1. $\begin{bmatrix}1&2\\3&4\end{bmatrix}^T=\begin{bmatrix}1&3\\2&4\end{bmatrix}$

2. $[1 \quad 2 \quad 3]^T = \begin{bmatrix} 1 \\ 2 \\ 3 \end{bmatrix}$

3. $\begin{bmatrix} 1 & 2 & 4 & 0 \\ -1 & -2 & 0 & 3 \\ 6 & 2 & 1 & 0 \end{bmatrix}^T = \begin{bmatrix} 1 & -1 & 6 \\ 2 & -2 & 2 \\ 4 & 0 & 1 \\ 0 & 3 & 0 \end{bmatrix}$ ■

The following theorem, whose proof we omit, is stated for future reference.

Theorem 1.5. Let A and B be two matrices. Then

$$(AB)^T = B^T A^T$$

EXAMPLE 9 Let $A = \begin{bmatrix} 1 & 2 \\ 0 & 3 \end{bmatrix}$ and $B = \begin{bmatrix} -1 & 0 \\ 3 & 4 \end{bmatrix}$. Verify Theorem 1.5 for these two matrices.

$$(AB)^T = \left(\begin{bmatrix} 1 & 2 \\ 0 & 3 \end{bmatrix}\begin{bmatrix} -1 & 0 \\ 3 & 4 \end{bmatrix}\right)^T = \begin{bmatrix} 5 & 8 \\ 9 & 12 \end{bmatrix}^T = \begin{bmatrix} 5 & 9 \\ 8 & 12 \end{bmatrix}$$

$$B^T A^T = \begin{bmatrix} -1 & 3 \\ 0 & 4 \end{bmatrix}\begin{bmatrix} 1 & 0 \\ 2 & 3 \end{bmatrix} = \begin{bmatrix} 5 & 9 \\ 8 & 12 \end{bmatrix}$$ ■

Definition 1.7. A square matrix A is said to be symmetric if $A = A^T$.

Thus a 2×2 matrix

$$A = \begin{bmatrix} a & b \\ c & d \end{bmatrix}$$

is symmetric if and only if $c = b$.

PROBLEM SET 1.2

1. Let $A = \begin{bmatrix} 1 & 2 & 3 \\ -1 & 0 & -2 \end{bmatrix}$ and let $B = \begin{bmatrix} -1 & 6 & -4 \\ -2 & -3 & 0 \end{bmatrix}$. Compute the following matrices:

a. $3A$ **b.** $A - B$ **c.** $2A + 3B$

2. Perform the indicated operations.

a. $2\begin{bmatrix} 6 & 7 & 1 & 1 \\ -1 & 3 & 1 & -1 \end{bmatrix}$

b. $-2\begin{bmatrix} 1 & 6 & 0 \\ 6 & -1 & 1 \end{bmatrix} + \begin{bmatrix} -2 & 4 & -1 \\ -1 & -2 & -1 \end{bmatrix}$

c. $\begin{bmatrix} 6 & -1 & 4 \\ -2 & 1 & -3 \\ 3 & 1 & 0 \\ 0 & 2 & 5 \end{bmatrix} - \begin{bmatrix} 2 & 3 & -2 \\ 0 & 1 & 1 \\ 1 & 0 & 1 \\ 6 & 2 & -1 \end{bmatrix}$

3. Let $A = \begin{bmatrix} 1 & 2 \\ 4 & -3 \end{bmatrix}$. Let $B = \begin{bmatrix} -1 & 0 \\ 0 & 3 \end{bmatrix}$. Compute the following:

a. AB **b.** BA **c.** $2A - B$

4. Let $A = \begin{bmatrix} 2 & 2 & 1 \\ -3 & 6 & 1 \end{bmatrix}$. Let $B = \begin{bmatrix} 1 & 0 & -1 \\ -2 & 1 & 4 \\ 6 & 0 & 3 \end{bmatrix}$. Compute if possible the following matrix products: AB, BA, $A^2 = AA$, and $B^2 = BB$.

5. Compute the following products:

a. $\begin{bmatrix} 1 & 0 \\ 2 & 1 \end{bmatrix}\begin{bmatrix} 3 & 6 \\ 4 & 9 \end{bmatrix}$ **b.** $\begin{bmatrix} 3 & 6 \\ 4 & 9 \end{bmatrix}\begin{bmatrix} 1 & 0 \\ 2 & 1 \end{bmatrix}$

6. Let $A = \begin{bmatrix} 1 & 6 \\ -1 & 2 \end{bmatrix}$. Are there any numbers x_1 and x_2 such that if $B = \begin{bmatrix} x_1 & 0 \\ 1 & x_2 \end{bmatrix}$, then $AB = BA$?

7. For each of the matrices below compute A^2 and A^3. Note $A^2 = AA$ and $A^3 = AA^2 = A^2A$.

a. $\begin{bmatrix} 1 & 0 \\ 0 & 2 \end{bmatrix}$ **b.** $\begin{bmatrix} 1 & 1 \\ 0 & 2 \end{bmatrix}$ **c.** $\begin{bmatrix} a & b \\ c & d \end{bmatrix}$

8. Find all numbers k such that the matrix $K = \begin{bmatrix} k & 1+k \\ 1-k & -k \end{bmatrix}$ satisfies the equation $K^2 = I_2$.

9. If $A = \begin{bmatrix} a & b \\ c & d \end{bmatrix}$, show that $A^2 = \begin{bmatrix} 0 & 1 \\ 0 & 0 \end{bmatrix}$ is not possible for any choice of a, b, c, or d.

10. Let $D = \begin{bmatrix} d & 0 \\ 0 & d \end{bmatrix}$, where d is an arbitrary number. Show that $DA = AD$ for any 2×2 matrix A. Conversely show that if D is a matrix for which $AD = DA$ for every 2×2 matrix A, then D has the above form for some number d. Such matrices are called scalar matrices.

11. Let $A = \begin{bmatrix} 2 & -1 \\ 4 & 1 \end{bmatrix}$. Let $B = \begin{bmatrix} 2 & 3 & 6 \\ 1 & -1 & 0 \end{bmatrix}$. Do there exist matrices X or Y such that the equations $AX = B$ or $YA = B$ have solutions?

12. Let $A = \begin{bmatrix} 1 & 0 & 0 \\ 0 & 2 & 0 \\ 0 & 0 & 3 \end{bmatrix}$ and let $B = \begin{bmatrix} 1 & 1 & 0 \\ 0 & -2 & 0 \\ 0 & 0 & 3 \end{bmatrix}$. Compute the following matrices:

a. A^2, B^2, AB, BA
b. A^T, B^T
c. A^3, B^3
d. A^4, B^4

13. Let A be an arbitrary $n \times n$ matrix. Define $A^0 = I_n$, $A^1 = A$, $A^2 = AA$, and $A^{n+1} = AA^n$, for $n = 1, 2, 3, \ldots$. Show that if p and q are arbitrary nonnegative integers then $A^pA^q = A^{p+q}$. Thus, we may conclude that $A^pA^q = A^qA^p$. (Hint: Use induction.)

14. Let $A = \begin{bmatrix} 1 & 1 \\ 2 & 4 \end{bmatrix}$, $B = \begin{bmatrix} 0 & 2 \\ 1 & 4 \\ -1 & -1 \end{bmatrix}$, $C = \begin{bmatrix} 6 & 0 & 1 \\ -4 & 1 & 0 \end{bmatrix}$, $D = \begin{bmatrix} -1 & 0 & 1 \\ -2 & -2 & 3 \end{bmatrix}$.

Calculate, when defined, each of the following matrices:
a. $2A + B$, $2C - D$, $B + D$ **b.** AB, BA **c.** A^2, B^2 **d.** A^T, B^T

15. Let $A = \begin{bmatrix} 2 & 1 & 3 \\ 4 & 2 & 6 \end{bmatrix}$, $B = \begin{bmatrix} 1 \\ 1 \\ 2 \end{bmatrix}$. Show $(AB)^T = B^TA^T$.

16. Let $A = \begin{bmatrix} -1 & 6 \\ 4 & 5 \end{bmatrix}$, $X = [1 \quad 2]$. Show that $(XA^T)^T = AX^T$.

17. Prove properties 2 and 3 of Theorem 1.3.

18. Let $A = \begin{bmatrix} -1 & 6 \\ 4 & 3 \end{bmatrix}$, $B = \begin{bmatrix} 7 & 5 \\ -2 & 1 \end{bmatrix}$. Verify that $A[7 \quad -2]^T$ is the first column of AB and that $A[5 \quad 1]^T$ is the second column of AB.

19. Let A and B be the matrices in problem 18. Verify that $[-1 \quad 6]B$ is the first row of AB and that $[4 \quad 3]B$ is the second row of AB.

20. Show that

a. $\begin{bmatrix} 2 & -1 \\ 6 & 7 \end{bmatrix}\begin{bmatrix} x_1 \\ x_2 \end{bmatrix} = x_1\begin{bmatrix} 2 \\ 6 \end{bmatrix} + x_2\begin{bmatrix} -1 \\ 7 \end{bmatrix}$

b. $[x_1 \quad x_2]\begin{bmatrix} 2 & -1 \\ 6 & 7 \end{bmatrix} = x_1[2 \quad -1] + x_2[6 \quad 7]$

21. Let A be any 3×3 matrix whose last row has only zeros. Let B be any 3×3 matrix. Show that the third row of AB has only zeros in it.

22. Generalize problem 21.

23. Which of the following matrices is symmetric?

a. $\begin{bmatrix} 1 & -1 \\ 1 & 2 \end{bmatrix}$ **b.** $\begin{bmatrix} 1 & 6 & 4 \\ 6 & 1 & -1 \\ 4 & -1 & 2 \end{bmatrix}$ **c.** $\begin{bmatrix} -1 & 0 & 4 \\ 0 & 4 & 6 \\ 4 & 6 & 2 \end{bmatrix}$

24. Let A and B be two $m \times n$ matrices. Show that $(A + B)^T = A^T + B^T$.

25. Show that $(A^T)^T = A$, for any matrix A.

26. Verify Theorem 1.5 for the matrices in problem 3.

27. Prove Theorem 1.5 for A, any 2×2 matrix, and B, any 2×3 matrix.

28. Show that if A and B are symmetric matrices of the same size, then $(AB)^T = BA$. Find two symmetric 2×2 matrices A and B such that AB is symmetric; then find two more 2×2 symmetric matrices such that their product is not symmetric.

29. Show that if A and B are symmetric, then so is $A + B$.

30. Prove Theorem 1.5 for A an arbitrary 2×3 matrix and B any 3×1 matrix.

31. An $n \times n$ matrix is upper triangular if $a_{ij} = 0$ when $i > j$, that is, all the terms below the main diagonal are zero. Show that the sum and product of two upper triangular matrices of the same size are also upper triangular.

32. A matrix is said to be lower triangular if its transpose is upper triangular. Show that a matrix is lower triangular if and only if $a_{ij} = 0$ when $i < j$. Show that the sum and product of two lower triangular matrices are also lower triangular.

33. Show that AA^T and A^TA are symmetric for any matrix A.

34. Prove Theorem 1.2.

35. Let

$$A = \begin{bmatrix} 2 & -1 \\ 3 & -\frac{3}{2} \end{bmatrix} \quad B = \begin{bmatrix} 0 & 1 \\ -5 & 3 \end{bmatrix} \quad C = \begin{bmatrix} -1 & -3 \\ -7 & -5 \end{bmatrix}$$

Show that $AB = AC$. Thus, it is possible for $AB = AC$ with A different from the zero matrix and B not equal to C.

1.3 ELEMENTARY ROW OPERATIONS

In this section we discuss in more detail the basic ideas involved in Gaussian elimination, and how one systematically reduces a matrix to the form (1.6).

The operations used in Gaussian elimination are called the elementary row operations. There are three of them which we list below.

Definition 1.8. The three elementary row operations are:

1. Interchange any two rows.
2. Multiply a row by a nonzero constant.
3. Add a multiple of one row to another.

The following example illustrates these three operations.

EXAMPLE 1 Let $A = \begin{bmatrix} 4 & 1 & 0 \\ -1 & 2 & 3 \\ 8 & -2 & 5 \end{bmatrix}$

a. Interchange rows 1 and 2 of A:

$$R_1 \sim R_2 \qquad \begin{bmatrix} 4 & 1 & 0 \\ -1 & 2 & 3 \\ 8 & -2 & 5 \end{bmatrix} \Rightarrow \begin{bmatrix} -1 & 2 & 3 \\ 4 & 1 & 0 \\ 8 & -2 & 5 \end{bmatrix}$$

b. Multiply row 3 of A by $\frac{1}{8}$:

$$\tfrac{1}{8}R_3 \qquad \begin{bmatrix} 4 & 1 & 0 \\ -1 & 2 & 3 \\ 8 & -2 & 5 \end{bmatrix} \Rightarrow \begin{bmatrix} 4 & 1 & 0 \\ -1 & 2 & 3 \\ 1 & -\frac{1}{4} & \frac{5}{8} \end{bmatrix}$$

c. Add four times the second row of A to the first row.

$$4R_2 \text{ to } R_1 \qquad \begin{bmatrix} 4 & 1 & 0 \\ -1 & 2 & 3 \\ 8 & -2 & 5 \end{bmatrix} \Rightarrow \begin{bmatrix} 0 & 9 & 12 \\ -1 & 2 & 3 \\ 8 & -2 & 5 \end{bmatrix}$$

■

Definition 1.9. Let A and B be two $m \times n$ matrices. We say that B is row equivalent to A if there is a sequence of elementary row operations that transforms A into B.

EXAMPLE 2a Let $A = \begin{bmatrix} 2 & 4 \\ 1 & 3 \end{bmatrix}$. Show that $I_2 = \begin{bmatrix} 1 & 0 \\ 0 & 1 \end{bmatrix}$ is row equivalent to A.

$$R_1 \sim R_2 \qquad \begin{bmatrix} 2 & 4 \\ 1 & 3 \end{bmatrix} \Rightarrow \begin{bmatrix} 1 & 3 \\ 2 & 4 \end{bmatrix}$$

$$-2R_1 \text{ to } R_2 \qquad \begin{bmatrix} 1 & 3 \\ 2 & 4 \end{bmatrix} \Rightarrow \begin{bmatrix} 1 & 3 \\ 0 & -2 \end{bmatrix}$$

$$-\tfrac{1}{2}R_2 \qquad \begin{bmatrix} 1 & 3 \\ 0 & -2 \end{bmatrix} \Rightarrow \begin{bmatrix} 1 & 3 \\ 0 & 1 \end{bmatrix}$$

$$-3R_2 \text{ to } R_1 \qquad \begin{bmatrix} 1 & 3 \\ 0 & 1 \end{bmatrix} \Rightarrow \begin{bmatrix} 1 & 0 \\ 0 & 1 \end{bmatrix}$$

■

Thus, the sequence of row operations $R_1 \sim R_2, -2R_1$ to $R_2, -\frac{1}{2}R_2$, and $-3R_2$ to R_1 performed in the given order transforms A into I_2.

EXAMPLE 2b Show that the matrix A above is row equivalent to I_2.

Solution. We know there is a sequence of elementary row operations that transforms A into I_2, and we expect that the row operations that undo the effect of these operations should then transform I_2 into A. The operations we used above and their "inverses" are

$R_1 \sim R_2$	$R_1 \sim R_2$
$-2R_1$ to R_2	$2R_1$ to R_2
$-\frac{1}{2}R_2$	$-2R_2$
$-3R_2$ to R_1	$3R_2$ to R_1

Performing the row operations in the second column in reverse order upon I_2 we have

$$3R_2 \text{ to } R_1 \qquad \begin{bmatrix} 1 & 0 \\ 0 & 1 \end{bmatrix} \Rightarrow \begin{bmatrix} 1 & 3 \\ 0 & 1 \end{bmatrix}$$

$$-2R_2 \qquad \begin{bmatrix} 1 & 3 \\ 0 & 1 \end{bmatrix} \Rightarrow \begin{bmatrix} 1 & 3 \\ 0 & -2 \end{bmatrix}$$

$$2R_1 \text{ to } R_2 \qquad \begin{bmatrix} 1 & 3 \\ 0 & -2 \end{bmatrix} \Rightarrow \begin{bmatrix} 1 & 3 \\ 2 & 4 \end{bmatrix}$$

$$R_1 \sim R_2 \qquad \begin{bmatrix} 1 & 3 \\ 2 & 4 \end{bmatrix} \Rightarrow \begin{bmatrix} 2 & 4 \\ 1 & 3 \end{bmatrix} = A$$

■

The previous example is a particular case of a general fact.

Lemma 1.1. Given any sequence of elementary row operations that transforms A into B, there is another sequence that transforms B into A. This is the same as saying that if B is row equivalent to A, then A is row equivalent to B.

Proof. We first show that each elementary row operation has a corresponding elementary row operation that cancels the effect of the first operation. Thus, let A be an arbitrary $m \times n$ matrix. Suppose

$$R_j \sim R_k \qquad A \Rightarrow B$$

That is, rows j and k of A are interchanged, and the resulting matrix is B. Clearly if we interchange the same two rows, then we transform B back into A. We also have:

$$\text{If } cR_j\text{: } A \Rightarrow B, \text{ then } (1/c)R_j\text{: } B \Rightarrow A$$
$$\text{If } cR_j \text{ to } R_k\text{: } A \Rightarrow B, \text{ then } -cR_j \text{ to } R_k\text{: } B \Rightarrow A$$

Now suppose B is row equivalent to A. Then there is a sequence of elementary row operations $E_1, \ldots, E_m$ which when performed in the given order, that is, E_m followed by $E_{m-1}, \ldots,$ followed by E_1, transforms A into B. Let F_j denote the "inverse" of E_j. Thus, if E_1 is $2R_3$, then F_1 is $\frac{1}{2}R_3$. Then $F_m, \ldots,$ F_1 when performed on B, in this order, will transform B into A; cf. Example 2.

The following theorem lists some properties of row equivalence:

Theorem 1.6. Let A, B, and C be $m \times n$ matrices. Then:

a. A is row equivalent to itself.
b. If A is row equivalent to B, then B is row equivalent to A.
c. If A is row equivalent to B and B is row equivalent to C, then A is row equivalent to C.

Proof. Part **a** is obvious, and **b** is Lemma 1.1. To see that **c** is true, let $E_1, \ldots, E_p$ denote a sequence of elementary row operations that transform A into B. Let $F_1, \ldots, F_q$ denote a sequence of elementary row operations that transform B into C. Clearly the following sequence of row operations, $E_1, \ldots, E_p, F_1, \ldots, F_q$, transforms A into C. Properties **a, b,** and **c** are called reflexivity, symmetry, and transitivity, respectively.

A more explicit description of what we meant by (1.6) is:

Definition 1.10. The matrix A is said to be in row echelon form if it satisfies the following conditions:

1. All rows consisting entirely of zeros must be at the bottom of the matrix; i.e., no row of zeros may precede any row that has a nonzero entry.
2. The first nonzero entry of any row must be the number 1.
3. All entries directly below the first nonzero entry of any row must be zero.
4. The number of the column containing the first nonzero entry of a row must increase as we move down the rows of the matrix.

EXAMPLE 3

a. The matrix $\begin{bmatrix} 1 & 0 & 1 \\ 0 & 3 & 0 \end{bmatrix}$ is not in row echelon form since the first nonzero entry in row 2 is not a 1. If row 2 is divided by 3, we get the row equivalent matrix

$$\begin{bmatrix} 1 & 0 & 1 \\ 0 & 1 & 0 \end{bmatrix}$$

which is in row echelon form.

b. $\begin{bmatrix} 0 & 0 & 0 \\ 0 & 1 & 0 \end{bmatrix}$ is not in row echelon form but the matrix

$$\begin{bmatrix} 0 & 1 & 0 \\ 0 & 0 & 0 \end{bmatrix}$$

which comes from the preceding matrix after a row interchange is in row echelon form.

c. $\begin{bmatrix} 1 & 1 & 0 & 1 \\ 0 & 1 & 1 & 1 \\ 0 & 1 & 0 & 0 \end{bmatrix}$ is not in row echelon form, since there is a nonzero entry in the 3,2 position, which is directly below the first nonzero entry of row 2.

d. $\begin{bmatrix} 1 & 0 & 0 \\ 0 & 0 & 1 \\ 0 & 1 & 0 \end{bmatrix}$ is not in row echelon form, since condition 4 of Definition 1.10 is not satisfied by this matrix. The first nonzero terms do not move to the right as we go down the rows. ■

Definition 1.11. A matrix is said to be in reduced row echleon form if it is in row echelon form and also satisfies:

5. All entries above the first nonzero entry of any row are zero.

EXAMPLE 4

a. $\begin{bmatrix} 1 & 1 & 0 \\ 0 & 1 & 0 \end{bmatrix}$ is in row echelon form, but not in reduced row echelon form.

$\begin{bmatrix} 1 & 0 & 0 \\ 0 & 1 & 0 \end{bmatrix}$ is in reduced row echelon form.

b. $\begin{bmatrix} 1 & 0 & 1 \\ 0 & 1 & 0 \\ 0 & 1 & 1 \end{bmatrix}$ is in neither row echelon form nor reduced row echelon form.

Condition 3 does not hold for this matrix.

c. $\begin{bmatrix} 1 & 0 & 2 \\ 0 & 1 & 1 \\ 0 & 0 & 0 \end{bmatrix}$ is in reduced row echelon form. ■

We next state two theorems but do not give their proofs.

Theorem 1.7. Every matrix is row equivalent to a unique matrix in reduced row echelon form.

Theorem 1.8. Two matrices are row equivalent if and only if they are row equivalent to the same matrix in reduced row echelon form.

In general if we wish to determine whether or not two matrices are row equivalent, we use row operations to find the reduced row echelon form they are row equivalent to, and then compare these two forms.

EXAMPLE 5 Show that $A = \begin{bmatrix} 2 & 3 & -1 \\ 4 & 0 & 1 \\ 0 & 0 & 1 \end{bmatrix}$ is row equivalent to the identity matrix. ■

To do this, one performs row operations on A until it is transformed into the 3×3 identity matrix.

$$\begin{bmatrix} 2 & 3 & -1 \\ 4 & 0 & 1 \\ 0 & 0 & 1 \end{bmatrix}$$

$$R_3 \text{ to } R_1 \qquad \begin{bmatrix} 2 & 3 & 0 \\ 4 & 0 & 1 \\ 0 & 0 & 1 \end{bmatrix}$$

$$-R_3 \text{ to } R_2 \qquad \begin{bmatrix} 2 & 3 & 0 \\ 4 & 0 & 0 \\ 0 & 0 & 1 \end{bmatrix}$$

$\frac{1}{4}R_2$ followed by $-2R_2$ to R_1 gives

$$\begin{bmatrix} 0 & 3 & 0 \\ 1 & 0 & 0 \\ 0 & 0 & 1 \end{bmatrix}$$

$\frac{1}{3}R_1$ $$\begin{bmatrix} 0 & 1 & 0 \\ 1 & 0 & 0 \\ 0 & 0 & 1 \end{bmatrix}$$

Interchanging R_1 and R_2, we have

$$\begin{bmatrix} 1 & 0 & 0 \\ 0 & 1 & 0 \\ 0 & 0 & 1 \end{bmatrix}$$

EXAMPLE 6 Show that $B = \begin{bmatrix} 2 & 0 & -1 \\ 0 & 0 & 1 \\ 0 & 0 & 1 \end{bmatrix}$ is not row equivalent to the matrix

A of Example 5. To do this, we show that B is row equivalent to a matrix that cannot be row equivalent to I_3. We deduce from this, via Theorem 1.8, that A and B cannot be row equivalent.

$$\begin{bmatrix} 2 & 0 & -1 \\ 0 & 0 & 1 \\ 0 & 0 & 1 \end{bmatrix}$$

R_3 to R_1 $$\begin{bmatrix} 2 & 0 & 0 \\ 0 & 0 & 1 \\ 0 & 0 & 1 \end{bmatrix}$$

$-R_2$ to R_3 $$\begin{bmatrix} 2 & 0 & 0 \\ 0 & 0 & 1 \\ 0 & 0 & 0 \end{bmatrix}$$

$\frac{1}{2}R_1$ $$\begin{bmatrix} 1 & 0 & 0 \\ 0 & 0 & 1 \\ 0 & 0 & 0 \end{bmatrix}$$ ■

Since the last matrix is in reduced row echelon form, and I_3 is also in reduced row echelon form, Theorem 1.7 implies B cannot be row equivalent to I_3, and hence B cannot be row equivalent to A. For if B were row equivalent to A, then by Theorem 1.8 and Example 3, B would be row equivalent to I_3. Actually there was no need to perform any row operations on B at all, since it is clear from the structure of B that no sequence of row operations can put a 1 in its second column.

For each of the elementary row operations there is a corresponding elementary column operation. For the most part, however, when solving systems of linear equations, only the elementary row operations are used.

The reader should also know that each of the elementary row (column) operations has an elementary row (column) matrix associated with it. For example, if we are dealing with $3 \times n$ matrices, then the elementary row matrix associated with a particular elementary row operation is that matrix which results when the given elementary row operation is performed upon I_3.

1. $R_1 \sim R_2$: $\begin{bmatrix} 1 & 0 & 0 \\ 0 & 1 & 0 \\ 0 & 0 & 1 \end{bmatrix} \Rightarrow \begin{bmatrix} 0 & 1 & 0 \\ 1 & 0 & 0 \\ 0 & 0 & 1 \end{bmatrix}$

2. cR_2: $\begin{bmatrix} 1 & 0 & 0 \\ 0 & 1 & 0 \\ 0 & 0 & 1 \end{bmatrix} \Rightarrow \begin{bmatrix} 1 & 0 & 0 \\ 0 & c & 0 \\ 0 & 0 & 1 \end{bmatrix}$

3. cR_2 to R_3: $\begin{bmatrix} 1 & 0 & 0 \\ 0 & 1 & 0 \\ 0 & 0 & 1 \end{bmatrix} \Rightarrow \begin{bmatrix} 1 & 0 & 0 \\ 0 & 1 & 0 \\ 0 & c & 1 \end{bmatrix}$

These examples demonstrate the relationship between the elementary row operation and its associated matrix. If we are in the class of $m \times n$ matrices (3×3 above), we would start out with I_m; i.e., the number of columns is irrelevant.

EXAMPLE 7 Let A be any 3×3 matrix. Let E be the elementary matrix associated with $2R_1$ to R_3. Show that EA equals the matrix obtained by transforming A with this row operation.

Solution. The matrix E, obtained by transforming I_3 with $2R_1$ to R_3, equals

$$\begin{bmatrix} 1 & 0 & 0 \\ 0 & 1 & 0 \\ 2 & 0 & 1 \end{bmatrix}$$

Computing EA for an arbitrary 3×3 matrix, we have

$$EA = \begin{bmatrix} 1 & 0 & 0 \\ 0 & 1 & 0 \\ 2 & 0 & 1 \end{bmatrix} \begin{bmatrix} a & b & c \\ d & e & f \\ g & h & i \end{bmatrix} = \begin{bmatrix} a & b & c \\ d & e & f \\ 2a+g & 2b+h & 2c+i \end{bmatrix}$$

and

$$2R_1 \text{ to } R_3\colon \begin{bmatrix} a & b & c \\ d & e & f \\ g & h & i \end{bmatrix} \Rightarrow \begin{bmatrix} a & b & c \\ d & e & f \\ 2a+g & 2b+h & 2c+i \end{bmatrix} = EA$$

■

The equality demonstrated in this example is not a coincidence. It is true in general, and the reader is asked to show this in the problems at the end of this section.

EXAMPLE 8 Let $A = \begin{bmatrix} 2 & 4 & 8 \\ 4 & 3 & 1 \\ 5 & 0 & 2 \end{bmatrix}$. Let $E = \begin{bmatrix} 1 & 0 & 0 \\ -2 & 1 & 0 \\ -\frac{5}{2} & 0 & 1 \end{bmatrix}$. Show that the first column of the matrix product EA equals $[2 \quad 0 \quad 0]^T$. Note that E is the matrix product

$$E = \begin{bmatrix} 1 & 0 & 0 \\ 0 & 1 & 0 \\ -\frac{5}{2} & 0 & 1 \end{bmatrix} \begin{bmatrix} 1 & 0 & 0 \\ -2 & 1 & 0 \\ 0 & 0 & 1 \end{bmatrix}$$

That is, E is the result of applying the two elementary row operations $-2R_1$ to R_2 and $-\frac{5}{2}R_1$ to R_3 consecutively to I_3.

$$EA = \begin{bmatrix} 1 & 0 & 0 \\ -2 & 1 & 0 \\ -\frac{5}{2} & 0 & 1 \end{bmatrix} \begin{bmatrix} 2 & 4 & 8 \\ 4 & 3 & 1 \\ 5 & 0 & 2 \end{bmatrix} = \begin{bmatrix} 2 & 4 & 8 \\ 0 & -5 & -15 \\ 0 & -10 & -18 \end{bmatrix}$$

Thus, EA equals the matrix that results from applying the above two row operations to A. ■

PROBLEM SET 1.3

1. Which of the following matrices are in row echelon form?

   a. $\begin{bmatrix} 0 & 1 \\ 1 & 2 \end{bmatrix}, \begin{bmatrix} 1 & -2 \\ 0 & 1 \end{bmatrix}, \begin{bmatrix} 0 & 0 \\ 1 & 0 \end{bmatrix}, \begin{bmatrix} 1 & 0 \\ 0 & 0 \end{bmatrix}$

   b. $\begin{bmatrix} 1 & 1 & 1 & 0 \\ 0 & 1 & 1 & 3 \\ 0 & 0 & 1 & 0 \end{bmatrix}, \begin{bmatrix} 0 & 1 & 2 & 4 \\ 0 & 0 & 0 & 1 \\ 0 & 0 & 1 & 6 \end{bmatrix}, \begin{bmatrix} 1 & 1 & 0 & -6 \\ 0 & 1 & 1 & 1 \\ 0 & 0 & 0 & 1 \end{bmatrix}$

2. Which of the following matrices are in reduced row echelon form?

a. $\begin{bmatrix} 0 & 1 \\ 1 & 2 \end{bmatrix}, \begin{bmatrix} 1 & -2 \\ 0 & 1 \end{bmatrix}, \begin{bmatrix} 0 & 0 \\ 1 & 0 \end{bmatrix}, \begin{bmatrix} 1 & 0 \\ 0 & 0 \end{bmatrix}$

b. $\begin{bmatrix} 1 & 1 & 1 & 0 \\ 0 & 1 & 1 & 3 \\ 0 & 0 & 1 & 0 \end{bmatrix}, \begin{bmatrix} 0 & 1 & 2 & 4 \\ 0 & 0 & 0 & 1 \\ 0 & 0 & 1 & 6 \end{bmatrix}, \begin{bmatrix} 1 & 1 & 0 & -6 \\ 0 & 1 & 1 & 1 \\ 0 & 0 & 0 & 1 \end{bmatrix}$

c. $\begin{bmatrix} 1 & 2 & 0 & 0 & 1 \\ 0 & 0 & 1 & 2 & 8 \\ 0 & 0 & 0 & 0 & 1 \end{bmatrix}, \begin{bmatrix} 1 & 2 & 0 & 0 & 0 \\ 0 & 0 & 1 & 2 & 0 \\ 0 & 0 & 0 & 0 & 1 \end{bmatrix}$

3. Find the reduced row echelon form to which each of the following matrices is row equivalent.

a. $\begin{bmatrix} 6 & 1 \\ 2 & 4 \end{bmatrix}, \begin{bmatrix} -1 & 2 \\ 3 & -6 \end{bmatrix}$

b. $\begin{bmatrix} 1 & -2 & 0 \\ 4 & 0 & 6 \\ 3 & 0 & 8 \end{bmatrix}, \begin{bmatrix} 6 & -5 & 1 \\ 3 & 1 & 4 \\ 3 & -13 & 10 \end{bmatrix}$

c. $\begin{bmatrix} 1 & 2 & 8 & -1 \\ 2 & 0 & 4 & -2 \\ 3 & 0 & 1 & 1 \end{bmatrix}$

4. Find the reduced row echelon form to which each of the following matrices is row equivalent.

a. $\begin{bmatrix} 6 & -2 \\ 1 & 4 \end{bmatrix}$ **b.** $\begin{bmatrix} 3 & 3 \\ 1 & 1 \end{bmatrix}$

5. Find the reduced row echelon form of each of the following matrices:

a. $\begin{bmatrix} 1 & 2 \\ -4 & -8 \\ 5 & 6 \\ 10 & 12 \end{bmatrix}$ **b.** $[-4 \quad 3 \quad 6 \quad -8]$ **c.** $\begin{bmatrix} -4 \\ 2 \\ 3 \\ 7 \end{bmatrix}$

d. $\begin{bmatrix} 9 & 2 & 8 & 4 & 5 \\ 6 & 1 & 0 & 3 & -2 \end{bmatrix}$

6. For each of the following elementary row operations, write the 3×3 elementary row matrix that corresponds to it.
a. $2R_2$ to R_3 **b.** $R_1 \sim R_3$ **c.** $-3R_2$

7. For each of the following sequences of row operations find a 2×2 matrix E such that, if A is any 2×3 matrix, EA is the matrix derived from A after the given sequence of row operations is performed on A.
a. $R_1 \sim R_2$, $2R_2$ to R_1
b. $-R_1$ to R_2, $2R_1$ to R_2, $3R_2$ to R_1

8. Let E be the matrix $\begin{bmatrix} 0 & 1 \\ 1 & 0 \end{bmatrix}$. We know that if A is any 2×2 matrix then EA is the matrix we get after interchanging the first two rows of A. Describe AE in terms of A.

9. Let $E = \begin{bmatrix} 1 & -2 \\ 0 & 1 \end{bmatrix}$. Let A be any 2×2 matrix. What column operations performed upon A will produce the product matrix AE?

10. If E is the $m \times m$ elementary matrix that corresponds to multiplying row i by c, then EA is the matrix we get when row i of A is multiplied by c. Show this for $m = 2$ or 3. How is AE related to A?

11. If E is the $m \times m$ elementary matrix that corresponds to interchanging row i with row j, then EA is the matrix we get when rows i and j of A are interchanged. Show this for $m = 2$ or 3. How is AE related to A?

12. An $n \times n$ matrix $D = [d_{ij}]$ is said to be a diagonal matrix if $d_{ij} = 0$ whenever $i \neq j$. That is, the only nonzero entries of D are on the main diagonal. Let D equal $\begin{bmatrix} -1 & 0 \\ 0 & 2 \end{bmatrix}$. For each of the following matrices, A, compute DA and AD.

 a. $\begin{bmatrix} 4 & 0 \\ 0 & 3 \end{bmatrix}$ **b.** $\begin{bmatrix} 2 & 0 \\ 1 & 0 \end{bmatrix}$ **c.** $\begin{bmatrix} 2 & 1 \\ 0 & 0 \end{bmatrix}$

13. Let $D = \begin{bmatrix} d_1 & 0 & 0 \\ 0 & d_2 & 0 \\ 0 & 0 & d_3 \end{bmatrix}$, $E = \begin{bmatrix} e_1 & 0 & 0 \\ 0 & e_2 & 0 \\ 0 & 0 & e_3 \end{bmatrix}$. Compute DE.

14. Let A be an arbitrary 3×3 matrix, and let D be any 3×3 diagonal matrix. Describe the rows of DA and the columns of AD.

15. Let $A = [a_{jk}]$ be any 3×3 matrix for which a_{11} is not zero. Let E be the matrix $\begin{bmatrix} 1 & 0 & 0 \\ -a_{21}/a_{11} & 1 & 0 \\ -a_{31}/a_{11} & 0 & 1 \end{bmatrix}$. Describe the first column of EA.

16. How many different 2×2 matrices in reduced row echelon form are there? Find at least four.

17. Let $A = \begin{bmatrix} a & b \\ c & d \end{bmatrix}$. Suppose E is a 2×2 matrix such that $EA = \begin{bmatrix} a+c & b+d \\ c & d \end{bmatrix}$. Clearly $E = \begin{bmatrix} 1 & 1 \\ 0 & 1 \end{bmatrix}$ is one such matrix. Are there any others?

18. For which 2×2 diagonal matrices D is there at least one nonzero matrix A such that DA is the 2×2 zero matrix?

19. Find all 2×2 matrices A such that AB is symmetric, where $B = \begin{bmatrix} 1 & 0 \\ 0 & 0 \end{bmatrix}$.

1.4 MATRICES AND LINEAR EQUATIONS

In this section we show how to write a system of linear equations as a matrix equation. We then show that the elementary row operations do not change the solution set.

Let's take a simple system of linear equations and rewrite it as an equation involving matrices.

$$\begin{aligned} 2x_1 + 6x_2 &= 5 \\ -x_1 + 3x_2 &= 4 \end{aligned} \tag{1.10}$$

Let $A = \begin{bmatrix} 2 & 6 \\ -1 & 3 \end{bmatrix}$ and let $X = \begin{bmatrix} x_1 \\ x_2 \end{bmatrix}$. Then we can write (1.10) as

$$AX = \begin{bmatrix} 2 & 6 \\ -1 & 3 \end{bmatrix}\begin{bmatrix} x_1 \\ x_2 \end{bmatrix} = \begin{bmatrix} 5 \\ 4 \end{bmatrix} \tag{1.11}$$

Now let's write the general linear system of m equations in n unknowns:

$$\begin{array}{l} a_{11}x_1 + a_{12}x_2 + \cdots + a_{1n}x_n = b_1 \\ a_{21}x_1 + a_{22}x_2 + \cdots + a_{2n}x_n = b_2 \\ \cdots\cdots\cdots\cdots\cdots\cdots\cdots\cdots \\ a_{m1}x_1 + a_{m2}x_2 + \cdots + a_{mn}x_n = b_m \end{array} \tag{1.12}$$

as a matrix equation.

Let $A = [a_{ij}]$, $X = [x_1, \ldots, x_n]^T$, and $B = [b_1, \ldots, b_m]^T$. The matrix A will be referred to as the coefficient matrix of (1.12) and X as the unknown solution. If some of the b_j are nonzero, B is called the nonhomogeneous term. We see, keeping in mind the definitions of matrix multiplication and equality, that (1.12) can be written as

$$AX = B \tag{1.13}$$

Another matrix associated with (1.12) is

$$\underline{A} = \begin{bmatrix} a_{11} & a_{12} & \cdots & a_{1n} & b_1 \\ a_{21} & a_{22} & \cdots & a_{2n} & b_2 \\ \cdots & \cdots & \cdots & \cdots & \cdots \\ a_{m1} & a_{m2} & \cdots & a_{mn} & b_m \end{bmatrix} \tag{1.14}$$

It is called the augmented matrix of (1.12). Notice that the augmented matrix $\underline{A}$ is just the coefficient matrix A with the right-hand side of (1.12) added on as an $(n + 1)$st column.

We want to show that if we have a system of equations and change it to a second system by performing some elementary row operations, we have neither lost nor gained any solutions. That is, the solution sets for the two systems are the same.

Theorem 1.9. Let A and B be the augmented matrices of two systems of linear equations. If A and B are row equivalent, the two systems have the same solution sets.

Proof. Since A and B are row equivalent only if there is a sequence of elementary row operations that transforms A into B, it suffices to prove this theorem in the case where only one of the three elementary row operations is used.

Case 1. Suppose two rows of A, when interchanged, yield B. Clearly all that we've done is to write the equations in a different order. Hence the solutions to both of these systems must coincide.

Case 2. Some row of A when multiplied by a nonzero constant yields B. Here we've multiplied one of the equations of the first system by a nonzero constant. This too leaves the solution set invariant.

Case 3. Suppose cR_j is added to R_k; that is, the kth row of B is c times the jth row of A added to the kth row of A. Suppose the numbers $x_1, x_2, \ldots, x_n$ satisfy the first system. In particular this means

$$\begin{aligned} a_{j1}x_1 + \cdots + a_{jn}x_n &= b_j \\ a_{k1}x_1 + \cdots + a_{kn}x_n &= b_k \end{aligned}$$

Adding c times the first equation to the second equation gives us the following pair of equations:

$$\begin{aligned} a_{j1}x_1 + \cdots + \qquad\qquad a_{jn}x_n &= b_j \\ (ca_{j1} + a_{k1})x_1 + \cdots + (ca_{jn} + a_{kn})x_n &= cb_j + b_k \end{aligned}$$

But the coefficients of these two equations are the jth and kth rows of B, respectively. Since the remaining $m-2$ rows of A and B are equal, we see that these n numbers x_i also satisfy the second system. Moreover since $-cR_j$ to R_k when performed upon B will yield A, a similar argument shows that every solution to the second system is also a solution to the first system.

EXAMPLE 1 Solve the following system of equations by finding a matrix in row echelon form which is row equivalent to the augmented matrix.

$$\begin{aligned} 2x_1 - x_2 + \qquad x_4 &= -1 \\ x_2 + x_3 + x_4 &= 0 \\ x_1 + x_2 + x_3 \qquad &= 6 \end{aligned}$$

The augmented matrix is

$$\begin{bmatrix} 2 & -1 & 0 & 1 & -1 \\ 0 & 1 & 1 & 1 & 0 \\ 1 & 1 & 1 & 0 & 6 \end{bmatrix}$$

and it is row equivalent to the following matrices:

$$R_1 \sim R_3 \qquad \begin{bmatrix} 1 & 1 & 1 & 0 & 6 \\ 0 & 1 & 1 & 1 & 0 \\ 2 & -1 & 0 & 1 & -1 \end{bmatrix}$$

$$-2R_1 \text{ to } R_3 \qquad \begin{bmatrix} 1 & 1 & 1 & 0 & 6 \\ 0 & 1 & 1 & 1 & 0 \\ 0 & -3 & -2 & 1 & -13 \end{bmatrix}$$

$$3R_2 \text{ to } R_3 \qquad \begin{bmatrix} 1 & 1 & 1 & 0 & 6 \\ 0 & 1 & 1 & 1 & 0 \\ 0 & 0 & 1 & 4 & -13 \end{bmatrix}$$

The above matrix is in row echelon form. The system of equations for which it is the augmented matrix has as solutions:

$$\begin{aligned} x_3 &= -13 - 4x_4 \\ x_2 &= 13 + 3x_4 \\ x_1 &= 6 + x_4 \end{aligned} \tag{1.15}$$

If a numerical value for x_4 is specified, setting x_1, x_2, and x_3 equal to those numbers which satisfy (1.15) gives us a solution to the original system of equations. Moreover, every solution may be generated in this manner. We obtain two explicit solutions by setting x_4 equal to 0 and then 1:

$$(6, 13, -13, 0) \qquad (7, 16, -17, 1)$$

■

As usual let $A = [a_{ij}]$ and $X = [x_1, \ldots, x_n]^T$ be $m \times n$ and $n \times 1$ matrices, respectively. Let $B = [b_1, \ldots, b_m]^T$ be an $m \times 1$ matrix.

Definition 1.12. A system of equations $AX = B$ is said to be homogeneous if $B = \mathrm{O}_{m1}$, i.e., $b_1 = 0$, $b_2 = 0$, . . . , $b_m = 0$. The zero solution to a homogeneous system is called the trivial solution. If $B \neq \mathrm{O}_{m1}$, the system is said to be nonhomogeneous.

The following result, which is almost obvious, is extremely useful.

Theorem 1.10. Let A and B be as above. If X_p is any solution to $AX = B$, every other solution to $AX = B$ can be written in the form $X = H + X_p$, where H is a solution to the associated homogeneous equation $AX = \mathrm{O}_{m1}$.

Proof. Suppose $AX = B$. Set $H = X - X_p$. Clearly $X = X_p + (X - X_p) = X_p + H$. Moreover $AH = A(X - X_p) = AX - AX_p = B - B = \mathrm{O}_{m1}$.

Note also that if $AH = \mathrm{O}_{m1}$, then $A(X_p + H) = AX_p + AH = B + \mathrm{O}_{m1} = B$.

EXAMPLE 2 Write the solutions to the system of equations in Example 1 in the form $H + X_p$.

Solution. From (1.15) one particular solution (obtained by setting x_4 equal to zero), is $(6,13,-13,0)$. Thus, we have

$$X = \begin{bmatrix} 6 \\ 13 \\ -13 \\ 0 \end{bmatrix} + x_4 \begin{bmatrix} 1 \\ 3 \\ -4 \\ 1 \end{bmatrix} = X_p + H$$

where H equals $x_4\,(1,3,-4,1)^T$. It is easy to show that, no matter what value x_4 is given, H is a solution to the homogeneous system. ■

Later we will need to know that every homogeneous system of equations with more unknowns than equations always has a nontrivial solution, that is, a solution for which not all the x_j's are zero. Before proving this, we look at a specific example.

EXAMPLE 3 Show that the following system has a nontrivial solution (two equations—three unknowns).

$$\begin{aligned} 3x_1 + 2x_2 \qquad &= 0 \\ -2x_1 + x_2 - x_3 &= 0 \end{aligned}$$

Solution. Writing the coefficient matrix, we have

$$\begin{bmatrix} 3 & 2 & 0 \\ -2 & 1 & -1 \end{bmatrix}$$

This matrix is row equivalent to

$$\begin{bmatrix} 1 & 0 & \frac{2}{7} \\ 0 & 1 & -\frac{3}{7} \end{bmatrix}$$

This matrix implies that $x_1 = -\frac{2}{7}x_3$ and $x_2 = \frac{3}{7}x_3$ are the only solutions to the system. Clearly there are many nontrivial solutions. One is obtained by setting x_3 equal to 1, another by setting x_3 equal to 2, etc. ■

Theorem 1.11. Let $A = [a_{ij}]$ be an $m \times n$ matrix with $m < n$ (more unknowns than equations). Then there is a nontrivial solution to the homogeneous equation $AX = O_{m1}$.

Proof. We prove this by induction on the number of equations. Suppose first that $m = 1$, i.e., only one equation.

$$a_1x_1 + a_2x_2 + \cdots + a_nx_n = 0 \tag{1.16}$$

Since we have an equation, at least one of the coefficients a_j is not zero. Suppose $a_1 \neq 0$. Set $x_2 = 1$ and $x_3 = \cdots = x_n = 0$. Setting $x_1 = -a_2/a_1$ gives us a nontrivial solution to (1.16). Now suppose that the theorem is true for p equations, $1 \leq p < m$. Using this we show that the theorem must be true for m equations. Let A_1 be a matrix in row echelon form that is row equivalent to A. To show that $AX = O_{m1}$ has a nontrivial solution it suffices to show that $A_1X = O_{m1}$ has a nontrivial solution.

Case 1. Assume A_1 has no row of zeros. Then there are exactly m columns with the number 1 as the only nonzero entry. Let $x_{i_1}, x_{i_2}, \ldots, x_{i_m}$, $1 \leq i_1 < i_2 < \cdots < i_m \leq n$ be the unknowns associated with these columns. Let k be any integer, $1 \leq k \leq n$, not equal to any of the i_j, $1 \leq j \leq m$. Set $x_k = 1$. Set all the other variables, excepting the x_{i_j}, equal to 0. Solve the resulting system for the remaining m unknowns x_{i_j}. Since $x_k = 1$, we have a nontrivial solution to our homogeneous system.

Case 2. A_1 has a row of zeros. This means that $A_1X = O_{m1}$ is actually a system of p equations with $p < m$. By our induction hypothesis this system has a nontrivial solution.

We conclude this section with two examples of how matrices may arise from different problems.

EXAMPLE 4 Suppose that a rabbit store starts out with 5 male rabbits and 4 female rabbits. Suppose also that every 6 months each female rabbit gives birth to 3 male rabbits and 3 female rabbits. How many male and female rabbits will there be (assuming no deaths) after 2 years?

Solution. We need to express how many males and females we have during any 6-month time period. Let m_k and f_k represent the number of males and females, respectively, during the kth time period. Then the following equations describe how the rabbit population grows from one time period to the next:

$$\begin{aligned} m_{k+1} &= m_k + 3f_k \\ f_{k+1} &= f_k + 3f_k \end{aligned}$$

Thus

$$\begin{bmatrix} m_{k+1} \\ f_{k+1} \end{bmatrix} = \begin{bmatrix} 1 & 3 \\ 0 & 1+3 \end{bmatrix} \begin{bmatrix} m_k \\ f_k \end{bmatrix}$$

Let A be the matrix $\begin{bmatrix} 1 & 3 \\ 0 & 4 \end{bmatrix}$. We have

$$\begin{aligned} \begin{bmatrix} m_{k+1} \\ f_{k+1} \end{bmatrix} &= A \begin{bmatrix} m_k \\ f_k \end{bmatrix} = A^2 \begin{bmatrix} m_{k-1} \\ f_{k-1} \end{bmatrix} = \cdots = A^{k+1} \begin{bmatrix} m_0 \\ f_0 \end{bmatrix} \\ &= A^{k+1} \begin{bmatrix} 5 \\ 4 \end{bmatrix} \end{aligned}$$

For a proof of this by induction see Appendix A. Thus, after 2 years or four time periods, we have

$$\begin{bmatrix} m_4 \\ f_4 \end{bmatrix} = A^4 \begin{bmatrix} 5 \\ 4 \end{bmatrix}$$

The reader is asked to complete this example by computing A^4. ■

EXAMPLE 5 Suppose we have some system that can be in any one of three different states, which are denoted by S_1, S_2, and S_3. The system is such that it can pass directly from some states to other states, as Figure 1.2 indicates. The diagram tells us that from S_1 the system can either stay in S_1 or enter S_2. From S_3

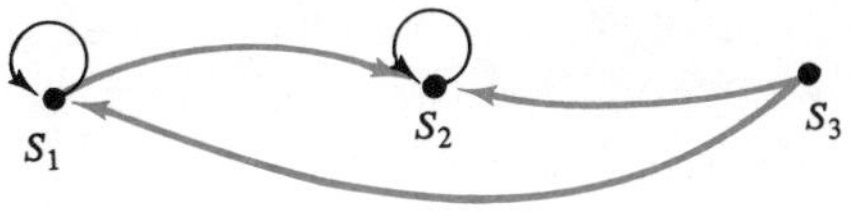

Figure 1.2

it must enter S_1 or S_2, and once the system is in S_2 it cannot leave. We catalog these data in a matrix $T = [t_{jk}]$. T is often referred to as a transition matrix. The t_{jk} are defined by

$$t_{jk} = 1, \text{ if the system can pass directly from state } j \text{ to } k.$$
$$t_{jk} = 0, \text{ if the system cannot pass directly from state } j \text{ to state } k.$$

Thus T equals

$$\begin{bmatrix} 1 & 1 & 0 \\ 0 & 1 & 0 \\ 1 & 1 & 0 \end{bmatrix}$$

Notice that the third column has all zeros; this is because the system can never enter the third state. We note too that each row must have at least one nonzero entry.

The positive powers of the transition matrix T contain information about the system. For example, $T^2 = [t_{2jk}]$ equals

$$\begin{bmatrix} 1 & 2 & 0 \\ 0 & 1 & 0 \\ 1 & 2 & 0 \end{bmatrix}$$

where

$$t_{2jk} = t_{j1}t_{1k} + t_{j2}t_{2k} + t_{j3}t_{3k}$$

Now $t_{jp}t_{pk} = 1$ or 0, and it equals 1 only if the system can pass from state S_j to state S_k by going through state S_p first. Thus t_{2jk} counts the number of ways that the system can pass from S_j to S_k with one intermediate state. In particular

$$\begin{aligned} t_{212} &= t_{11}t_{12} + t_{12}t_{22} + t_{13}t_{32} \\ &= (1)(1) + (1)(1) + (0)(0) = 2 \end{aligned}$$

Hence the system can pass from S_1 to S_2 with one intermediate state (both S_1 and S_2 are allowed to be used as intermediate states) in two different ways.

A similar analysis shows that the j, k entry of the matrix T^n counts the number of ways the system can pass from state S_j to state S_k with $n - 1$ intermediate states. We note that any of the n possible states can function as an intermediary state. For our particular T we have

$$T^n = \begin{bmatrix} 1 & n & 0 \\ 0 & 1 & 0 \\ 1 & n & 0 \end{bmatrix}$$

Thus, there must be three different ways in which the system can pass from state S_1 to S_2 with two intermediate stops, etc. ■

PROBLEM SET 1.4

1. For each of the following systems of equations, write the corresponding coefficient matrix and augmented matrix, and then determine the solution sets.
   a. $\begin{aligned} 10x_1 - x_2 + x_3 &= 9 \\ -x_1 + 6x_2 &= 4 \\ x_1 + 5x_2 - 7x_3 &= 1 \end{aligned}$
   b. $\begin{aligned} -3x_2 + 7x_3 + x_1 &= -1 \\ x_3 + x_1 - x_2 &= 4 \\ x_1 - x_2 - x_3 &= 5 \end{aligned}$
   c. $\begin{aligned} x_1 + x_2 &= 1 \\ x_2 + x_3 &= 2 \\ x_3 + x_4 &= 3 \\ x_4 + x_5 &= 4 \\ -x_1 + x_5 &= 5 \end{aligned}$
2. Assume each of the following matrices is the coefficient matrix for a system of homogeneous equations. Write the system and find all solutions.
   a. $\begin{bmatrix} 7 & 5 \\ 3 & 2 \end{bmatrix}$ b. $\begin{bmatrix} 4 & 6 & -2 \\ -3 & -2 & 1 \end{bmatrix}$ c. $\begin{bmatrix} -2 & 7 \\ 4 & -1 \\ 3 & 2 \end{bmatrix}$
3. Assume each of the matrices in problem 2 is the augmented matrix of a system of equations. Write out the system and solve if possible.
4. Assume each of the following matrices is the augmented matrix of a system of linear equations. Write the system and then find all solutions.
   a. $\begin{bmatrix} 4 & 6 & 0 \\ -3 & 0 & 1 \end{bmatrix}$ b. $\begin{bmatrix} 1 & -1 & 0 & 2 \\ 1 & 1 & 1 & 1 \\ 0 & 1 & 1 & -2 \end{bmatrix}$
5. Assume each of the matrices in problem 4 is the coefficient matrix of a homogeneous system. What are the systems? What are the solution sets?
6. Suppose that you have money in two separate investments, I_1 and I_2. Assume I_1 returns 10 percent per year and I_2 5 percent per year. At the end of each year half of the money you made in I_1 you reinvest in I_1, and the other half you invest in I_2, always leaving the original amount invested where it is. Suppose also that any monies made in I_2 are left there. If you invest \$1000 in I_1 and \$2000 in I_2, how much will be invested in each account during the second year? During the third year?

7. This is a continuation of problem 6. If a_{1k} and a_{2k} represent the amount of money invested in accounts I_1 and I_2, respectively, during the kth year, find a formula that gives $a_{1(k+1)}$ and $a_{2(k+1)}$ in terms of a_{1k} and a_{2k}. See if you can describe a_{1k} and a_{2k} in terms of a_{11} and a_{21}, i.e., in terms of the original investment.

8. Let $B = \begin{bmatrix} a & 0 \\ 1 & b \end{bmatrix}$.

 a. Show that $B^4 = \begin{bmatrix} a^4 & 0 \\ a^3 + a^2b + ab^2 + b^3 & b^4 \end{bmatrix}$.

 b. Show that $B^n = \begin{bmatrix} a^n & 0 \\ \sum_{k=0}^{n-1} a^{n-1-k}b^k & b^n \end{bmatrix}$.

 Notice that if $a = b$ we have

$$B^n = \begin{bmatrix} a^n & 0 \\ na^{n-1} & a^n \end{bmatrix}$$

9. The following diagrams describe the transition properties of three different systems. Determine the transition matrix for each system and then calculate the nth power of each matrix.

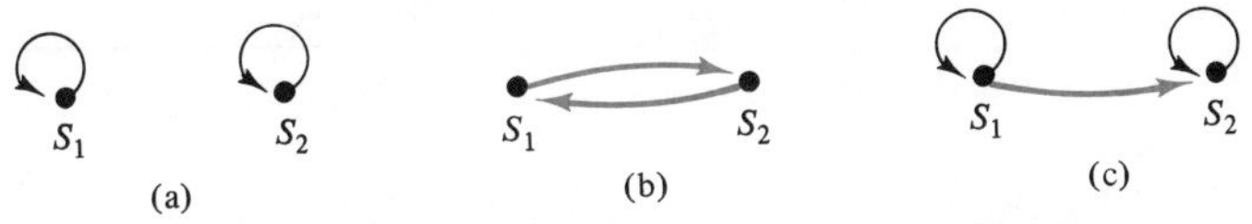

10. The following diagram describes the transition properties of a four-state system.
 a. Determine the transition matrix T for this system.

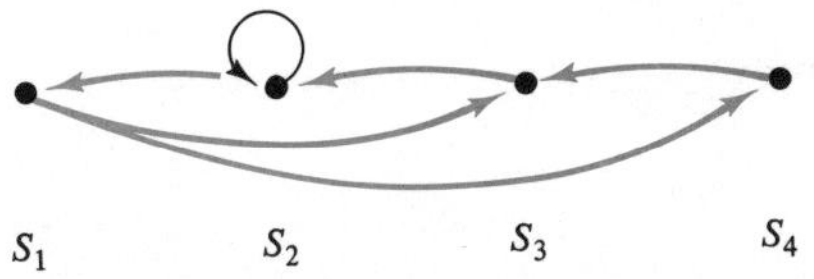

 b. In how many ways can the system go from S_1 to S_3 with one intermediate state?
 c. In how many ways can the system go from S_3 to S_4 with one intermediate state?
 d. $T^2 = ?$

11. Let T be the transition matrix of Example 5. Show that T^n equals

$$\begin{bmatrix} 1 & n & 0 \\ 0 & 1 & 0 \\ 1 & n & 0 \end{bmatrix}$$

12. Let T be a transition matrix. Suppose the 1, 2 entry of $(T^2 + T)$ is zero. Suppose also that the 1, 2 entry of T^3 is not zero. What does this say about going from state 1 to state 2?

13. Show that if x_1 and x_2 solve $Ax = 0$, then so does $c_1x_1 + c_2x_2$ for any constants c_1 and c_2.

14. Using Theorem 1.10 and the preceding problem, show that if the equation $Ax = b$ has at least two different solutions, then there are an infinite number of solutions; cf. the remark preceding Example 4 in Section 1.1.

15. Let $A = [a_{jk}]$ be an $m \times n$ matrix. Let C_k denote the kth column of A. That is,

$$C_k = [a_{1k} \quad a_{2k} \quad a_{3k} \quad \cdots \quad a_{mk}]^T \qquad 1 \le k \le n$$

Let $X = [x_1 \quad x_2 \quad \cdots \quad x_n]^T$. Show that $AX = x_1 C_1 + \cdots + x_n C_n$.

16. Solve the equation $AX + XA = B$ for X, where A, B, and X are 2×2 matrices. The matrix A equals

$$\begin{bmatrix} 1 & 2 \\ -1 & 4 \end{bmatrix}$$

while the matrix B is assumed known but arbitrary.

1.5 INVERSES OF MATRICES

If we were taking a beginning algebra course, one of the first things we would study would be how to solve one equation in one unknown. For example, to solve $2x = 3$, the technique is to divide the equation by 2 and get $x = \frac{3}{2}$. Instead of saying divide by 2, let's say multiply by $\frac{1}{2}$, and then, instead of writing $\frac{1}{2}$, let's write 2^{-1}, and our answer can be written $x = (2^{-1})\,(3)$. The reason for this discussion is that for some systems of linear equations we can do something analogous.

For an example of this approach we solve the following matrix equation:

$$\begin{bmatrix} 2 & 6 \\ -1 & 3 \end{bmatrix}\begin{bmatrix} x_1 \\ x_2 \end{bmatrix} = \begin{bmatrix} 5 \\ 4 \end{bmatrix} \tag{1.17}$$

Before solving this equation we note that

$$\begin{bmatrix} \frac{1}{4} & -\frac{1}{2} \\ \frac{1}{12} & \frac{1}{6} \end{bmatrix}\begin{bmatrix} 2 & 6 \\ -1 & 3 \end{bmatrix} = \begin{bmatrix} 1 & 0 \\ 0 & 1 \end{bmatrix} \tag{1.18}$$

Multiplying (1.17) on the left by the first matrix in (1.18) we have

$$\begin{bmatrix} \frac{1}{4} & -\frac{1}{2} \\ \frac{1}{12} & \frac{1}{6} \end{bmatrix}\begin{bmatrix} 2 & 6 \\ -1 & 3 \end{bmatrix}\begin{bmatrix} x_1 \\ x_2 \end{bmatrix} = \begin{bmatrix} \frac{1}{4} & -\frac{1}{2} \\ \frac{1}{12} & \frac{1}{6} \end{bmatrix}\begin{bmatrix} 5 \\ 4 \end{bmatrix}$$

Thus we have the equations

$$\begin{bmatrix} 1 & 0 \\ 0 & 1 \end{bmatrix}\begin{bmatrix} x_1 \\ x_2 \end{bmatrix} = \begin{bmatrix} \frac{1}{4} & -\frac{1}{2} \\ \frac{1}{12} & \frac{1}{6} \end{bmatrix}\begin{bmatrix} 5 \\ 4 \end{bmatrix} = \begin{bmatrix} -\frac{3}{4} \\ \frac{13}{12} \end{bmatrix}$$

These imply that

$$\begin{bmatrix} x_1 \\ x_2 \end{bmatrix} = \begin{bmatrix} -\frac{3}{4} \\ \frac{13}{12} \end{bmatrix}$$

Formally, given an equation of the form $AX = B$, we found a matrix A^{-1}, such that $A^{-1}A = I_2$, and then multiplied the equation (on the left) by A^{-1}. That is, $AX = B$ implies the following string of equalities:

$$\begin{aligned} A^{-1}(AX) &= A^{-1}B \\ (A^{-1}A)X &= A^{-1}B \\ (I_n)X &= A^{-1}B \\ X &= A^{-1}B \end{aligned}$$

Before proceeding any further we need:

Definition 1.13. A square matrix A is said to be nonsingular or invertible if there is a matrix, which we denote by A^{-1} (A inverse) such that

$$A^{-1}A = AA^{-1} = I = \text{identity matrix}$$

If no such matrix exists, A is said to be singular or noninvertible.

EXAMPLE 1 Show that the matrix $A = \begin{bmatrix} 0 & 1 \\ 0 & 0 \end{bmatrix}$ does not have an inverse.

Solution. Suppose that there is some matrix $B = \begin{bmatrix} a & b \\ c & d \end{bmatrix}$ such that $BA = I_2$. Then we must have

$$\begin{bmatrix} a & b \\ c & d \end{bmatrix}\begin{bmatrix} 0 & 1 \\ 0 & 0 \end{bmatrix} = \begin{bmatrix} 1 & 0 \\ 0 & 1 \end{bmatrix}$$

However, the first column of BA can only have zeros in it. Thus, the 1,1 entries of BA and I_2 are never equal. Hence A is a singular matrix. ■

The following lemmas are used to show that an $n \times n$ matrix is invertible if and only if it is row equivalent to the $n \times n$ identity matrix I_n.

Lemma 1.2. Let A be a nonsingular matrix. Let B_1 and B_2 be two inverses of A. That is, $B_jA = AB_j = I$ for $j = 1, 2$. Then $B_1 = B_2$.

Proof

$$B_1 = B_1 I = B_1(AB_2) = (B_1A)B_2 = IB_2 = B_2$$

Since each of the elementary row operations is reversible, we would expect each elementary row matrix to have an inverse. The next example illustrates this for a few 3×3 elementary row matrices.

EXAMPLE 2 Let $E_1 = \begin{bmatrix} 1 & 0 & 0 \\ 0 & 0 & 1 \\ 0 & 1 & 0 \end{bmatrix}$. Let $E_2 = \begin{bmatrix} 1 & 0 & 0 \\ 0 & 6 & 0 \\ 0 & 0 & 1 \end{bmatrix}$. Let

$$E_3 = \begin{bmatrix} 1 & 0 & 0 \\ 0 & 1 & -4 \\ 0 & 0 & 1 \end{bmatrix}.$$

Determine the inverses of these matrices.

a. E_1 corresponds to interchanging rows two and three. Hence E_1^{-1} should equal E_1. In fact we have:

$$E_1E_1 = \begin{bmatrix} 1 & 0 & 0 \\ 0 & 0 & 1 \\ 0 & 1 & 0 \end{bmatrix}\begin{bmatrix} 1 & 0 & 0 \\ 0 & 0 & 1 \\ 0 & 1 & 0 \end{bmatrix} = \begin{bmatrix} 1 & 0 & 0 \\ 0 & 1 & 0 \\ 0 & 0 & 1 \end{bmatrix}$$

b. E_2 corresponds to multiplying row two by 6. Thus we would expect E_2^{-1} to be the matrix that corresponds to multiplying row two by $\frac{1}{6}$. Indeed we have

$$\begin{bmatrix} 1 & 0 & 0 \\ 0 & \frac{1}{6} & 0 \\ 0 & 0 & 1 \end{bmatrix}\begin{bmatrix} 1 & 0 & 0 \\ 0 & 6 & 0 \\ 0 & 0 & 1 \end{bmatrix} = \begin{bmatrix} 1 & 0 & 0 \\ 0 & 1 & 0 \\ 0 & 0 & 1 \end{bmatrix}$$

c. E_3 is associated with the row operation $-4R_3$ to R_2 and we would expect to have its inverse associated with the row operation $4R_3$ to R_2. A quick computation shows that

$$\begin{bmatrix} 1 & 0 & 0 \\ 0 & 1 & 4 \\ 0 & 0 & 1 \end{bmatrix}\begin{bmatrix} 1 & 0 & 0 \\ 0 & 1 & -4 \\ 0 & 0 & 1 \end{bmatrix} = \begin{bmatrix} 1 & 0 & 0 \\ 0 & 1 & 0 \\ 0 & 0 & 1 \end{bmatrix}$$

To really verify that the above matrices are the inverses, we need to check that not only does $E^{-1}E = I$ but also that $EE^{-1} = I$. The reader may easily do this. It is also a fact that if we have a square $n \times n$ matrix A and another $n \times n$ matrix B such that either $AB = I_n$ or $BA = I_n$, then A is invertible with $A^{-1} = B$. ■

Lemma 1.3. Every elementary row matrix has an inverse.

Proof. We prove this for just one type of elementary row matrix, and ask the reader to supply the details for the other two cases. Let E be an $n \times n$ elementary row matrix obtained from I_n by adding c times row p to row q $(p \neq q)$. That is, $E = [e_{ij}]$ where $e_{ij} = \delta_{ij}$ if $i \neq q$ and $e_{qj} = c\delta_{pj} + \delta_{qj}$. Let F be the elementary row matrix obtained from I_n by adding $-c$ times row p to row q. So $F = [f_{ij}]$, where $f_{ij} = \delta_{ij}$ if $i \neq q$ and $f_{qj} = -c\delta_{pj} + \delta_{qj}$. To see that FE and EF equal I_n, one can actually compute each entry in the product of FE or just think about what is being done to the qth row. That is, first c times row p and then $-c$ times row p is added to row q. The end result is to add 0 times row p to row q; i.e., we wind up with the identity matrix. Hence $FE = I_n$. Similarly we have $EF = I_n$.

Lemma 1.4. If A and B are both invertible $n \times n$ matrices, then AB is invertible and $(AB)^{-1} = B^{-1}A^{-1}$.

Proof

$$(B^{-1}A^{-1})(AB) = B^{-1}(A^{-1}A)B = B^{-1}(I)B = B^{-1}B = I$$

Similarly $(AB)(B^{-1}A^{-1}) = I_n$. Thus, since we have exhibited a matrix that satisfies the condition of Definition 1.13, AB is invertible.

For a generalization of Lemma 1.4 to products of more than two invertible matrices, see the problems at the end of this section.

Lemma 1.5. If A is an invertible matrix, A^{-1} is also invertible, and its inverse is A. That is, $(A^{-1})^{-1} = A$.

Proof. From the definition of A^{-1} we have $AA^{-1} = A^{-1}A = I$. Thus the matrix A is such that when we multiply A^{-1} on either side by A we get the identity matrix. Thus A is the inverse of A^{-1}.

Lemmas 1.3 and 1.4 are the crucial facts needed to prove the following theorem.

Theorem 1.12. An $n \times n$ matrix is invertible if and only if it is row equivalent to the identity matrix.

Proof. Suppose A is row equivalent to the identity matrix. Then there is a matrix E, which is a product of elementary row matrices and hence invertible, such that $EA = I$. We then have

$$AE = (E^{-1}E)AE = E^{-1}(EA)E = E^{-1}(I)E = E^{-1}E = I$$

Thus A is invertible and its inverse is the product of the elementary row matrices that transform A into the identity matrix (this is our algorithm for computing A^{-1}). Suppose now that A is invertible. We want to show that A is row equivalent to the identity matrix. Suppose it isn't. Then, when we transform A to its reduced row echelon form B, at least one row of B must consist entirely of zeros. Thus, there is a matrix E such that $EA = B$, where E is a product of elementary row matrices and hence cannot have any row with only zeros in it. Since A has an inverse, we have the following:

$$E = E(AA^{-1}) = (EA)A^{-1} = BA^{-1}$$

A quick check of the rule for matrix multiplication tells us that since the last row of B is all zeros so too is the last row of BA^{-1}, and hence E must also have its last row filled with zeros, a contradiction. Thus A must be row equivalent to the identity matrix.

An easy consequence of Theorem 1.12 is that an $n \times n$ matrix $A = [a_{ij}]$ is nonsingular if and only if the homogeneous system of equations

$$\begin{array}{l} a_{11}x_1 + \cdots + a_{1n}x_n = 0 \\ \cdots\cdots\cdots\cdots\cdots \\ a_{n1}x_1 + \cdots + a_{nn}x_n = 0 \end{array}$$

has only the trivial solution $x_1 = x_2 = \cdots = x_n = 0$.

Before computing the inverses of a few matrices, we discuss how to keep track of the elementary row operations that are used to transform A into I (remember the product of the corresponding elementary row matrices will be A^{-1}). The technique is to write the identity matrix next to A and every time we perform a row operation on A perform the same operation on the matrix next to A. In this way we keep a running record of the cumulative effect of these elementary row operations. Moreover, when A has been transformed into I, the identity matrix will have been transformed into A^{-1}. The following examples illustrate this.

EXAMPLE 3 Determine whether the following matrices are nonsingular. If they are, compute their inverses.

a. $A = \begin{bmatrix} 1 & 2 & 3 \\ 0 & 1 & 2 \end{bmatrix}$ is singular, since A is not a square matrix

b. $A = \begin{bmatrix} 1 & 2 & 3 \\ 1 & 0 & 2 \\ 0 & 0 & 1 \end{bmatrix}$

Solution. Since A is a square matrix, A^{-1} might exist. To see if this is the case, we simultaneously transform A into its reduced row echelon form, and compute A^{-1}.

$$\begin{bmatrix} 1 & 2 & 3 & 1 & 0 & 0 \\ 1 & 0 & 2 & 0 & 1 & 0 \\ 0 & 0 & 1 & 0 & 0 & 1 \end{bmatrix}$$

$$-R_1 \text{ to } R_2 \qquad \begin{bmatrix} 1 & 2 & 3 & 1 & 0 & 0 \\ 0 & -2 & -1 & -1 & 1 & 0 \\ 0 & 0 & 1 & 0 & 0 & 1 \end{bmatrix}$$

$$R_3 \text{ to } R_2 \qquad \begin{bmatrix} 1 & 2 & 3 & 1 & 0 & 0 \\ 0 & -2 & 0 & -1 & 1 & 1 \\ 0 & 0 & 1 & 0 & 0 & 1 \end{bmatrix}$$

$$R_2 \text{ to } R_1 \qquad \begin{bmatrix} 1 & 0 & 3 & 0 & 1 & 1 \\ 0 & -2 & 0 & -1 & 1 & 1 \\ 0 & 0 & 1 & 0 & 0 & 1 \end{bmatrix}$$

$$-3R_3 \text{ to } R_1 \qquad \begin{bmatrix} 1 & 0 & 0 & 0 & 1 & -2 \\ 0 & -2 & 0 & -1 & 1 & 1 \\ 0 & 0 & 1 & 0 & 0 & 1 \end{bmatrix}$$

$$-\tfrac{1}{2}R_2 \qquad \begin{bmatrix} 1 & 0 & 0 & 0 & 1 & -2 \\ 0 & 1 & 0 & \frac{1}{2} & -\frac{1}{2} & -\frac{1}{2} \\ 0 & 0 & 1 & 0 & 0 & 1 \end{bmatrix}$$

Since A is row equivalent to the identity matrix, it is invertible, and its inverse is

$$A^{-1} = \begin{bmatrix} 0 & 1 & -2 \\ \frac{1}{2} & -\frac{1}{2} & -\frac{1}{2} \\ 0 & 0 & 1 \end{bmatrix}$$

The reader should check that

$$\begin{bmatrix} 0 & 1 & -2 \\ \frac{1}{2} & -\frac{1}{2} & -\frac{1}{2} \\ 0 & 0 & 1 \end{bmatrix}\begin{bmatrix} 1 & 2 & 3 \\ 1 & 0 & 2 \\ 0 & 0 & 1 \end{bmatrix} = \begin{bmatrix} 1 & 0 & 0 \\ 0 & 1 & 0 \\ 0 & 0 & 1 \end{bmatrix} = \begin{bmatrix} 1 & 2 & 3 \\ 1 & 0 & 2 \\ 0 & 0 & 1 \end{bmatrix}\begin{bmatrix} 0 & 1 & -2 \\ \frac{1}{2} & -\frac{1}{2} & -\frac{1}{2} \\ 0 & 0 & 1 \end{bmatrix}$$

c. $A = \begin{bmatrix} 1 & 0 & 2 \\ 0 & 1 & 1 \\ 1 & -1 & 1 \end{bmatrix}$

Solution. We again need to see if A is row equivalent to I_3.

$$\begin{bmatrix} 1 & 0 & 2 & 1 & 0 & 0 \\ 0 & 1 & 1 & 0 & 1 & 0 \\ 1 & -1 & 1 & 0 & 0 & 1 \end{bmatrix}$$

$-R_1$ to R_3 $\begin{bmatrix} 1 & 0 & 2 & 1 & 0 & 0 \\ 0 & 1 & 1 & 0 & 1 & 0 \\ 0 & -1 & -1 & -1 & 0 & 1 \end{bmatrix}$

R_2 to R_3 $\begin{bmatrix} 1 & 0 & 2 & 1 & 0 & 0 \\ 0 & 1 & 1 & 0 & 1 & 0 \\ 0 & 0 & 0 & -1 & 1 & 1 \end{bmatrix}$

Since A is not row equivalent to I_3, it is not invertible. ■

EXAMPLE 4 Solve the following systems of equations by finding the inverse of the coefficient matrix.

a.
$$\begin{aligned} x_1 + 2x_2 + 3x_3 &= 1 \\ x_1 + 2x_3 &= -2 \\ x_3 &= 6 \end{aligned}$$

Writing this as a matrix equation we have

$$\begin{bmatrix} 1 & 2 & 3 \\ 1 & 0 & 2 \\ 0 & 0 & 1 \end{bmatrix}\begin{bmatrix} x_1 \\ x_2 \\ x_3 \end{bmatrix} = \begin{bmatrix} 1 \\ -2 \\ 6 \end{bmatrix}$$

From Example 3b we know that the coefficient matrix A is invertible with

$$A^{-1} = \begin{bmatrix} 0 & 1 & -2 \\ \frac{1}{2} & -\frac{1}{2} & -\frac{1}{2} \\ 0 & 0 & 1 \end{bmatrix}$$

Thus $[x]^T = A^{-1}[1 \quad -2 \quad 6]^T$, or

$$\begin{bmatrix} x_1 \\ x_2 \\ x_3 \end{bmatrix} = \begin{bmatrix} 0 & 1 & -2 \\ \frac{1}{2} & -\frac{1}{2} & -\frac{1}{2} \\ 0 & 0 & 1 \end{bmatrix} \begin{bmatrix} 1 \\ -2 \\ 6 \end{bmatrix} = \begin{bmatrix} -14 \\ -\frac{3}{2} \\ 6 \end{bmatrix}$$

b.
$$\begin{aligned} 2x_1 + x_2 - x_3 &= -1 \\ x_1 + x_2 + 4x_4 &= 6 \\ x_3 - 2x_4 &= 0 \\ x_1 + x_3 + x_4 &= 2 \end{aligned}$$

Writing the above as a matrix equation, we have

$$\begin{bmatrix} 2 & 1 & -1 & 0 \\ 1 & 1 & 0 & 4 \\ 0 & 0 & 1 & -2 \\ 1 & 0 & 1 & 1 \end{bmatrix} \begin{bmatrix} x_1 \\ x_2 \\ x_3 \\ x_4 \end{bmatrix} = \begin{bmatrix} -1 \\ 6 \\ 0 \\ 2 \end{bmatrix}$$

Computing the inverse of the coefficient matrix, we have

$$\begin{bmatrix} 2 & 1 & -1 & 0 & 1 & 0 & 0 & 0 \\ 1 & 1 & 0 & 4 & 0 & 1 & 0 & 0 \\ 0 & 0 & 1 & -2 & 0 & 0 & 1 & 0 \\ 1 & 0 & 1 & 1 & 0 & 0 & 0 & 1 \end{bmatrix}$$

row equivalent to

$$\begin{bmatrix} 1 & 0 & 0 & 0 & \frac{1}{3} & -\frac{1}{3} & -\frac{1}{3} & \frac{2}{3} \\ 0 & 1 & 0 & 0 & \frac{1}{9} & \frac{8}{9} & \frac{11}{9} & -\frac{10}{9} \\ 0 & 0 & 1 & 0 & -\frac{2}{9} & \frac{2}{9} & \frac{5}{9} & \frac{2}{9} \\ 0 & 0 & 0 & 1 & -\frac{1}{9} & \frac{1}{9} & -\frac{2}{9} & \frac{1}{9} \end{bmatrix}$$

Thus the inverse of the coefficient matrix is

$$A^{-1} = \begin{bmatrix} \frac{1}{3} & -\frac{1}{3} & -\frac{1}{3} & \frac{2}{3} \\ \frac{1}{9} & \frac{8}{9} & \frac{11}{9} & -\frac{10}{9} \\ -\frac{2}{9} & \frac{2}{9} & \frac{5}{9} & \frac{2}{9} \\ -\frac{1}{9} & \frac{1}{9} & -\frac{2}{9} & \frac{1}{9} \end{bmatrix}$$

Using this inverse we have

$$\begin{bmatrix} x_1 \\ x_2 \\ x_3 \\ x_4 \end{bmatrix} = A^{-1}\begin{bmatrix} -1 \\ 6 \\ 0 \\ 2 \end{bmatrix} = (\tfrac{1}{9})\begin{bmatrix} 3 & -3 & -3 & 6 \\ 1 & 8 & 11 & -10 \\ -2 & 2 & 5 & 2 \\ -1 & 1 & -2 & 1 \end{bmatrix}\begin{bmatrix} -1 \\ 6 \\ 0 \\ 2 \end{bmatrix}$$

$$= (\tfrac{1}{9})\begin{bmatrix} -9 \\ 27 \\ 18 \\ 9 \end{bmatrix} = \begin{bmatrix} -1 \\ 3 \\ 2 \\ 1 \end{bmatrix}$$

■

We now have two methods for solving systems of equations when the coefficient matrix A is nonsingular: Gaussian elimination or computing A^{-1}. The most efficient method is, in general, Gaussian elimination. This is true even when we wish to solve $Ax = b_k$, for different b_k. One just keeps track of the elementary row operations that transform A into a matrix that is in row echelon form. Then these operations are performed upon each b_k. This new system, which has an upper triangular coefficient matrix, is then solved by back substitution. It turns out that usually fewer arithmetical operations are involved in this procedure than in computing $A^{-1}b_k$, for each individual b_k.

The main use of matrix inversions is of a theoretical nature. For example, what are the consequences if a particular matrix is invertible? For us, one such consequence is that if the coefficient matrix of a system of equations is nonsingular, not only does the system have a solution, but the solution is unique.

We next illustrate how matrices and their inverses can be used to model and then solve problems that arise in actual applications.

W. Leontief, who received a Nobel prize for economics in 1974, developed a method called input-output analysis in his study of economic systems and their component sectors. For example, a manufacturing plant might be the system and the various sectors could represent the different products made in that plant. Or our system could be the entire economic complex of this country with some of the sectors being energy, agriculture, sports, manufacturing, etc.

To avoid unnecessary complications we describe this analysis for a two-sector economic system. Thus, let S_1 and S_2 represent the two sectors involved. We assume that part of each sector's output is used within each of the sectors. That is, part of S_j's output is used by both S_1 and S_2 to produce their products. For example, suppose each unit of S_1's product requires 0.01 unit from S_1 and 0.4 unit from S_2 while each unit of S_2's product requires 0.1 unit from S_1 and 0.3 unit from S_2. We construct a matrix A using these data.

$$A = [a_{ij}] = \begin{bmatrix} 0.01 & 0.1 \\ 0.4 & 0.3 \end{bmatrix}$$

is called the input-output matrix of the system. We need to make sure that the convention used to construct A is understood. Thus,

$a_{11} = 0.01$ = amount of x_1 needed to produce 1 unit of x_1.

$a_{12} = 0.1$ = amount of x_1 needed to produce 1 unit of x_2.

$a_{21} = 0.4$ = amount of x_2 needed to produce 1 unit of x_1.

$a_{22} = 0.3$ = amount of x_2 needed to produce 1 unit of x_2.

More generally,

a_{jk} = amount of sector j's output needed to produce 1 unit of sector k's goods.

Thus, the jth row of A tells us how much is needed from sector j by each of the other sectors, and the kth column of A tells how much the kth sector needs from each of the sectors.

Let x_1 and x_2 represent the amounts produced by S_1 and S_2, respectively. Let u_1 and u_2 represent the amount of S_1's and S_2's output consumed internally; that is, u_1 represents the amount of S_1's product needed to produce x_1 units from S_1 and x_2 units from S_2. Thus, we have

$$u_1 = a_{11}x_1 + a_{12}x_2 = 0.01x_1 + 0.1x_2$$
$$u_2 = a_{21}x_1 + a_{22}x_2 = 0.4x_1 + 0.3x_2$$

Writing this as a matrix equation, we have

$$U = AX$$

where $U = [u_1 \quad u_2]^T$ and $X = [x_1 \quad x_2]^T$. Clearly $x_1 - u_1$ and $x_2 - u_2$ represent the amount of goods that would be available for use outside the economic system. In terms of these matrices this amount can be written as

$$X - AX = (I - A)X$$

Suppose that the system is required to produce for use outside itself d_1 and d_2 units, respectively, of S_1's and S_2's goods. Setting $D = [d_1 \quad d_2]^T$, our problem is to solve the following system of equations for x_1 and x_2:

$$d_1 = x_1 - u_1 = x_1 - (a_{11}x_1 + a_{12}x_2)$$
$$d_2 = x_2 - u_2 = x_2 - (a_{21}x_1 + a_{22}x_2)$$

or in matrix form,

$$D = X - AX = (I - A)X$$

The matrices D and X are commonly called the demand vector and the intensity vector, respectively. Moreover, the solution, if the matrix $(I - A)$ is nonsingular, is given by

$$X = (I - A)^{-1}D$$

We note that if the input-output matrix A equals O_{nn}, this means that no sector's goods are used by itself or any other sector. If $A = I_n$, each sector uses up all its product while making it.

EXAMPLE 5 Let $A = \begin{bmatrix} 0.3 & 0.2 \\ 0.1 & 0.4 \end{bmatrix}$ be an input-output matrix for a two-sector economic system. How much of each sector's output will be needed to produce 6 and 5 units from the first and second sectors, respectively?

Solution. The first sector must supply $6(0.3) = 1.8$ units to make 6 units of its goods and must also supply $5(0.2) = 1$ unit to make 5 units of the second sector's product. Thus 2.8 units of the first sector's output will be used. Similarly the second sector will need to supply $6(0.1) + 5(0.4) = 2.6$ units. ■

EXAMPLE 6 Let $A = \begin{bmatrix} 0.2 & 0.01 \\ 0.4 & 0.1 \end{bmatrix}$ be the input-output matrix for a two-sector economic system.

a. How much of the second sector's product will be used in producing 6 units of the first sector's goods and 3 units of the second sector's goods?

Solution. In terms of the above notation, we are asked to find u_2.

$$u_2 = 6(0.4) + 3(0.1) = 2.4 + 0.3 = 2.7$$

Thus 2.7 units of the second sector's output is needed to produce 6 and 3 units of the first and second sector's products, respectively.

b. If $D = \begin{bmatrix} 10 \\ 3 \end{bmatrix}$ is a demand vector, how much should the economic system produce in order to supply this demand from outside the system?

Solution. We need to find $X = [x_1 \quad x_2]^T$ such that

$$(I - A)X = D = \begin{bmatrix} 10 \\ 3 \end{bmatrix}$$

Thus we should calculate $(I - A)^{-1}$ if possible, for then $X = (I - A)^{-1}D$.

$$I - A = \begin{bmatrix} 1 & 0 \\ 0 & 1 \end{bmatrix} - \begin{bmatrix} 0.2 & 0.01 \\ 0.4 & 0.1 \end{bmatrix} = \begin{bmatrix} 0.8 & -0.01 \\ -0.4 & 0.9 \end{bmatrix}$$

We easily see that $(I - A)$ is invertible and its inverse is approximately equal to

$$(I - A)^{-1} \approx \begin{bmatrix} 1.256 & 0.0139 \\ 0.5587 & 1.1173 \end{bmatrix}$$

Thus, since $X = (I - A)^{-1}[10,3]^T$, we have

$$x_1 \approx 12.6017 \qquad x_2 \approx 8.9389$$

■

PROBLEM SET 1.5

1. Determine which of the following matrices are invertible and then calculate their inverses.

   a. $\begin{bmatrix} 2 & 3 \\ 1 & 4 \end{bmatrix}$ **b.** $\begin{bmatrix} -2 & -1 \\ 6 & 3 \end{bmatrix}$

   c. $\begin{bmatrix} 1 & -2 & 3 \\ -1 & 0 & 4 \end{bmatrix}$ **d.** $\begin{bmatrix} -1 & 4 & -1 \\ 0 & 2 & 1 \\ 1 & 0 & 3 \end{bmatrix}$

2. Find the inverses of the following matrices if they exist.

   a. $\begin{bmatrix} 1 & 2 & 3 & 4 \\ 2 & 3 & 4 & 5 \\ 3 & 4 & 5 & 6 \\ 4 & 5 & 6 & 7 \end{bmatrix}$ **b.** $\begin{bmatrix} 1 & 0 & 1 & 0 \\ 0 & 1 & 0 & 1 \\ 1 & 1 & 0 & 0 \\ 0 & 1 & 1 & 0 \end{bmatrix}$

   c. $\begin{bmatrix} 2 & 3 & 1 & -1 \\ 3 & -2 & 0 & 2 \\ 1 & -1 & -1 & 1 \\ 2 & 3 & 1 & 2 \end{bmatrix}$ **d.** $\begin{bmatrix} 6 & -4 & 3 \\ 4 & 2 & 8 \\ 1 & 2 & 3 \end{bmatrix}$

3. Suppose A is an $n \times n$ invertible matrix. If k is any positive integer, let $A^{-k} = (A^{-1})^k$. Show that $A^pA^q = A^{p+q}$ where p and q are any integers, positive, negative, or zero.

4. Let $A = \begin{bmatrix} 0.3 & 0.1 \\ 0 & 0.2 \end{bmatrix}$ be an input-output matrix for some two-sector economic system. Suppose 10 units of the second sector's product are produced.
   a. How many units of the first sector's product are needed by the second sector?
   b. How many units of the second sector's product are needed by the second sector?
   c. If the first sector produces exactly 10 units, what is the largest number of units that sector 2 can make?

5. A biologist has been studying two different species of plants. Let x_k and y_k represent the population of each species in the kth year. Suppose she has observed the following relationships between the populations:

   $$\begin{aligned} x_{k+1} &= 2x_k + y_k \\ y_{k+1} &= 3y_k + x_k \end{aligned}$$

   a. Find a matrix A such that

   $$\begin{bmatrix} x_{k+1} \\ y_{k+1} \end{bmatrix} = A \begin{bmatrix} x_k \\ y_k \end{bmatrix}$$

b. Show that

$$\begin{bmatrix} x_{k+1} \\ y_{k+1} \end{bmatrix} = A^2 \begin{bmatrix} x_{k-1} \\ y_{k-1} \end{bmatrix}$$

c. Show that

$$\begin{bmatrix} x_k \\ y_k \end{bmatrix} = A^k \begin{bmatrix} x_0 \\ y_0 \end{bmatrix} \qquad \text{for } k = 1, 2, \ldots$$

d. If the population sizes in the third year are known, how can the initial populations be determined; i.e., what population size at time zero leads to the known populations?

e. Compute A^3 and A^{-3}.

Questions 6 and 7 refer to the input-output matrix A given below:

$$A = \begin{bmatrix} 0.1 & 0.2 & 0.1 \\ 0.2 & 0.1 & 0.4 \\ 0.0 & 0.2 & 0.3 \end{bmatrix}$$

6. How much of sector 2's output is needed to produce 6 units of sector 3's goods?

7. How much of the first sector's product is needed to produce 5, 8, and 9 units of each of the first, second, and third sector's products, respectively?

8. The matrix

$$\begin{bmatrix} \frac{1}{2} & \frac{1}{3} & 0 \\ \frac{1}{2} & \frac{1}{2} & \frac{1}{3} \\ \frac{1}{4} & \frac{1}{5} & \frac{1}{2} \end{bmatrix}$$

is the input-output matrix for a three-sector economic system. How many units of the third sector's product will be needed to simultaneously produce 8 units from sector 1, 30 units from sector 2, and 14 units from sector 3?

9. The matrix

$$\begin{bmatrix} \frac{1}{2} & \frac{1}{2} & 0 \\ \frac{1}{2} & 0 & \frac{1}{2} \\ 0 & \frac{1}{2} & \frac{1}{3} \end{bmatrix}$$

is the input-output matrix for a three-sector economic system. If some outside agency wants to purchase 2 units of sector 1's product, 4 units of sector 2's product, and 1 unit of the third sector's, how much of the second product must be made?

10. Let E be the $m \times m$ elementary row matrix that corresponds to interchanging row i and row j. Show that $E^{-1} = E$.

11. Let E be the $m \times m$ elementary row matrix that corresponds to multiplying row i by $c (c \neq 0)$. Find E^{-1}.

12. Let $A_1, A_2, \ldots, A_k$ be a collection of $n \times n$ invertible matrices, $k \geq 3$. Show $(A_1 \cdots A_k)^{-1} = (A_k)^{-1} \cdots (A_1)^{-1}$.

13. An invertible matrix P such that each row has exactly one entry equal to one and all the other entries in this row equal to zero is called a permutation matrix. There are exactly two 2×2 permutation matrices, namely,

$$\begin{bmatrix} 1 & 0 \\ 0 & 1 \end{bmatrix} \quad \text{and} \quad \begin{bmatrix} 0 & 1 \\ 1 & 0 \end{bmatrix}$$

Notice that not only the rows of these matrices but also the columns contain exactly one nonzero entry.

a. List all the 3×3 permutation matrices. There are $3! = 3(2) = 6$ of them.

b. Show there are $n! = n(n-1)(n-2) \cdots (1)$ $n \times n$ permutation matrices.

c. If P is a permutation matrix, show that it can be obtained from the identity matrix by a sequence of row interchanges.

14. Let P be an $n \times n$ permutation matrix (see problem 13). Show that P^T is also a permutation matrix and that $P^T = P^{-1}$.

15. Let D be any 2×2 diagonal matrix. Give a necessary and sufficient condition on the diagonal entries so that D has an inverse. Compute the inverse of any such matrix.

16. Generalize problem 15 to $n \times n$ diagonal matrices.

17. Suppose A is a 3×3 upper triangular matrix. Show that A has an inverse if and only if the diagonal elements of A are nonzero.

18. Suppose that A is an invertible matrix. Show that A^T is also invertible and that $(A^T)^{-1} = (A^{-1})^T$. (Hint: Use Theorem 1.5.)

19. Show that if A is symmetric and invertible, then A^{-1} is also symmetric.

20. Let $A = \begin{bmatrix} \frac{1}{2} & -1 \\ -2 & \frac{1}{3} \end{bmatrix}$. Show that $I - A$ is invertible and that all its entries are positive.

SUPPLEMENTARY PROBLEMS

1. Define and give examples of each of the following:

a. The coefficient matrix of a nonhomogeneous system of equations

b. An elementary row matrix

c. An elementary row operation

d. A singular matrix

2. For each of the following find a system of equations for which the given statement is correct:

a. The solution set equals $\{(1,-2,0)\}$.

b. The solution set equals $\{(x_1,x_2) : x_1 = 2\}$.

c. The system has no solution.

3. Find the solution set for the system of equations:

$$\begin{aligned} 6x_1 \qquad - 4x_3 \qquad &= 1 \\ 2x_1 + 3x_2 \qquad\qquad &= 2 \\ + \; x_3 + x_4 &= -1 \end{aligned}$$

4. If A is a finite set of numbers, let $\text{av}(A)$ denote the numerical average of the numbers in A. Thus, if $A = \{1,2,-4\}$, then $\text{av}(A) = -\frac{1}{3}$. Let $A = \{a,b,c\}$ and $B = \{a,b,c,d\}$.

a. Show that, in general, it is not true that $\text{av}(B) = [\text{av}(A) + d]/2$.

b. Let A_1 and A_2 be two sets of numbers. Find a condition that will guarantee

$$\text{av}(A_1 \cup A_2) = \frac{\text{av}(A_1) + \text{av}(A_2)}{2}$$

5. Find the solution sets for each of the following systems of equations:

a. $$\begin{aligned} x_1 + x_2 - x_3 &= 4 \\ 2x_1 + 5x_2 - 2x_3 &= 10 \end{aligned}$$

b. $$\begin{aligned} x_1 - x_2 + 2x_3 &= 1 \\ x_1 + x_2 + x_3 &= 2 \\ 3x_1 + x_2 + 4x_3 &= -4 \end{aligned}$$

c. $$\begin{aligned} x_1 - x_2 + 2x_3 &= 1 \\ x_1 + x_2 + x_3 &= 2 \\ 3x_1 + x_2 + 4x_3 &= 5 \end{aligned}$$

6. Suppose $A = \begin{bmatrix} a & b \\ c & d \end{bmatrix}$ is invertible with $A^{-1} = \begin{bmatrix} e & f \\ g & h \end{bmatrix}$. Let $B = \begin{bmatrix} c & d \\ a & b \end{bmatrix}$; that is, B is obtained from A by a row interchange. Show that $B^{-1} = \begin{bmatrix} f & e \\ h & g \end{bmatrix}$.

7. Let A be any $n \times n$ invertible matrix. Let B be the matrix obtained by interchanging rows i and j of A. Show that B^{-1} equals the matrix obtained by interchanging columns i and j of A^{-1}. (Hint: $B = EA$, where E is an elementary row matrix. What happens if we interchange two columns of A?)

8. Let $M = \left\{ \begin{bmatrix} a & b \\ c & d \end{bmatrix} : ad - bc = 1 \right\}$.

a. Show that M is closed under the operations of inversion and transposition; that is, if A is in M, so are A^{-1} and A^T.

b. If A and B are in M, must $A + B$ be in M? What about their product AB?

9. The matrix $A = \begin{bmatrix} a & b \\ c & d \end{bmatrix}$ is invertible if and only if there is a matrix $B = \begin{bmatrix} x_1 & x_2 \\ x_3 & x_4 \end{bmatrix}$ such that $AB = I_2 = BA$. Write out the system of four equations in the unknowns x_j that arises from the matrix equation $AB = I$. Show that there is a solution if and only if $ad - bc \neq 0$. Assuming this is true, solve for the unknowns x_j; then show that not only does $AB = I$, but so too does BA.

10. Let B be any fixed $n \times n$ matrix. Let $C(B)$ be the set of $n \times n$ matrices that commute with B. That is, A is in $C(B)$ if and only if $AB = BA$. Let A_1 and A_2 be any two matrices that are in $C(B)$.

a. Show that $aA_1 + bA_2$ is in $C(B)$ for any choice of constants a and b.

b. Show that A_1A_2 is in $C(B)$.

11. Let $A = [a_{jk}]$ be an $n \times n$ invertible matrix. Suppose y_j equals $\sum_{k=1}^{n} a_{jk}x_k$. Show that $x_j = \sum_{k=1}^{n} b_{jk}y_k$, where $[b_{jk}] = A^{-1}$.

12. Let A be an $n \times n$ invertible matrix. Let E_j be the $n \times 1$ matrix whose only nonzero entry is a 1 in the jth row. Show that the solution to the equation $AX = E_j$ is the jth column of A^{-1}.

13. Show that an $n \times n$ matrix A is invertible if and only if the equation $AX = O_{n1}$ has exactly one solution, namely, O_{n1}.

14. Let A be an $n \times n$ matrix that has been partitioned into four 2×2 submatrices. That is,

$$A = \begin{bmatrix} A_{11} & A_{12} \\ A_{21} & A_{22} \end{bmatrix}$$

where each A_{jk} is a 2×2 matrix. Let $B = [B_{jk}]$ be a 4×4 matrix partitioned in the same manner as A. Thus, each B_{jk} is also a 2×2 matrix. Show that AB equals

$$\begin{bmatrix} A_{11}B_{11} + A_{21}B_{21} & A_{11}B_{12} + A_{12}B_{22} \\ A_{21}B_{11} + A_{22}B_{21} & A_{21}B_{12} + A_{22}B_{22} \end{bmatrix}$$

15. Let $A(t) = [a_{jk}(t)]$ be an $m \times n$ matrix with its entries $a_{jk}(t)$ differentiable functions of t. Define $A'(t)$ by $A'(t) = [a'_{jk}(t)]$. Thus, if

$$A(t) = \begin{bmatrix} \cos t & \sin t \\ -\sin t & \cos t \end{bmatrix} \quad \text{then} \quad A'(t) = \begin{bmatrix} -\sin t & \cos t \\ -\cos t & -\sin t \end{bmatrix}$$

Let $A(t)$ and $B(t)$ be two matrices whose entries are differentiable functions. Show that $(A + B)' = A' + B'$ and that $(AB)' = A'B + AB'$ whenever it is possible to add or multiply the two given matrices.

16. Prove Theorem 1.5.

2
VECTOR SPACES

One of my favorite dictionaries (the one from Oxford) defines a vector as "A quantity having direction as well as magnitude, denoted by a line drawn from its original to its final position." What is useful about this definition is that we can draw pictures and use our spatial intuition. An objection to this being used as the definition of a vector is: It's not very precise, and thus will be hard to compute with to any degree of accuracy.

The first section of this chapter makes the phrase "a quantity having direction and magnitude" more precise and at the same time develops an algebraic structure. That is, the addition of one vector to another and the multiplication of vectors by numbers will be defined, and various properties of these operations will be stated and proved.

In order to avoid confusing vectors with numbers, all vectors will be in boldface type.

2.1 $\mathbb{R}^2$ THROUGH $\mathbb{R}^n$

Fix some point (think of it as the origin in the Euclidean plane), draw two short rays from this point, and put arrowheads at the tips of the rays (see Figure 2.1).

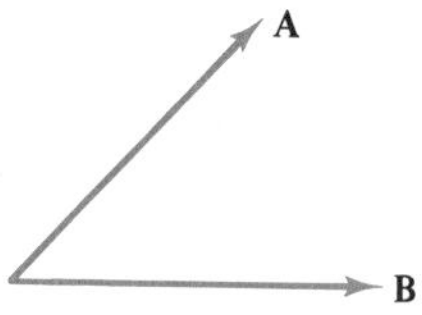

Figure 2.1

You've just drawn two vectors. Now label one of them **A** and the other **B**. The magnitudes of these vectors are their lengths.

In many applications, vectors are used to represent forces. Since several forces may act on an object, and the result of this will be the same as if a single force, called the resultant or sum, acted on the object, we define the sum of two vectors in order to model how forces combine. Thus, if we want the vector **A** + **B**, we draw a dashed line segment starting at the tip of **A**, parallel to **B**, with the same length as **B**; cf. Figure 2.2a and b. Label the end of this dashed line c and now draw the vector **C**; i.e., draw a line segment from the origin to c and put an arrowhead there; cf. Figure 2.2a. The vector **C** is called **A** + **B**, and this way of combining **A** and **B** is referred to as the parallelogram law of vector addition. Note that the construction of **B** + **A** will be different from that of **A** + **B**; cf. Figure 2.2b. However, a little geometry convinces us that we get the same two vectors. We repeat this. The fact that **A** + **B** = **B** + **A** is something that has to be proved. Its truth is not self-evident.

There is another operation (scalar multiplication) that we can perform on a vector, and that is to multiply it by a number. If **A** is a vector, 2**A**'s meaning is clear. It is the same as **A** + **A**; i.e., 2**A** is a vector twice as long, and with the same direction, as **A**. Thus, we define c**A** ($c \geq 0$) to be the vector pointing in the same direction as **A** but whose magnitude is c times the magnitude of **A**. If c is negative, c**A** points in the direction opposite to **A**.

Since we will be doing a lot of computing with vectors, we need a method for doing so other than using a ruler, compass, and protractor. Following in the footsteps of Descartes and others, we assign a pair of numbers to each vector and then see how these pairs should be combined in order to model vector addition and scalar multiplication.

Let's go back to Figure 2.1. This time we label our initial point (0,0), the origin in the Euclidean plane. We also draw two perpendicular lines that inter-

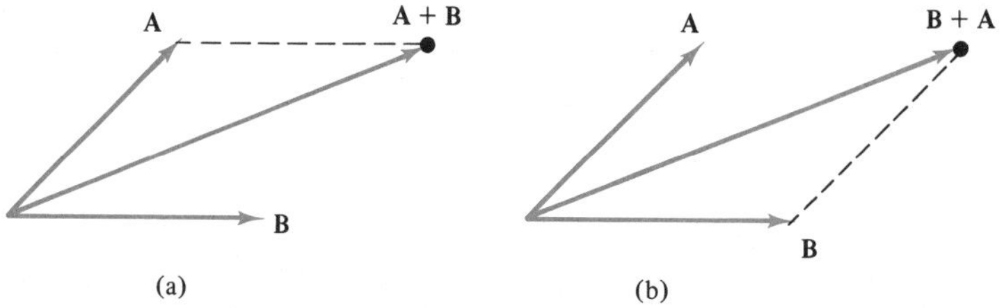

Figure 2.2

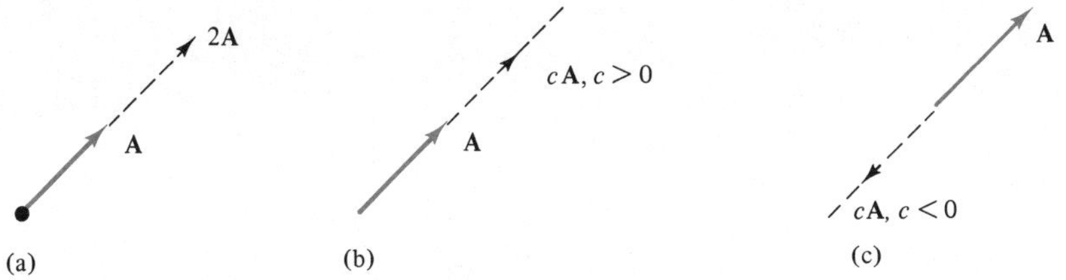

Figure 2.3

sect at (0,0). We call the horizontal line the x_1 axis and the vertical line the x_2 axis. We next associate a number with each point on these axes. This number indicates the directed distance of the point from the origin; that is, the point on the x_1 axis labeled 2 is 2 units to the right of the origin while the point labeled -2 is 2 units to the left. How large a distance the number 1 represents is arbitrary. For the x_2 axis, a positive number indicates that the point lies above the origin while a negative number means that the point lies below the origin. We now associate an ordered pair of numbers with each point in the plane. The first number tells us the directed distance of the point from the x_2 axis along a line parallel to the x_1 axis and the second number gives us the directed distance of the point from the x_1 axis along a line parallel to the x_2 axis; cf. Figure 2.5. Every vector, which we picture as an arrow emanating from the origin, is uniquely determined once we know where its tip is located. This means that every vector can be uniquely associated with an ordered pair of numbers. In other words, two vectors are equal if and only if their respective coordinates are the same.

The preceding discussion may lead to some confusion as to what an ordered pair of numbers represents: a point in the plane or a vector. In practice this will not cause any problem, since one can just as easily think of a vector as a point in the plane, that point where the tip of the vector is located.

EXAMPLE 1 Sketch the following vectors:

a. (1,1)

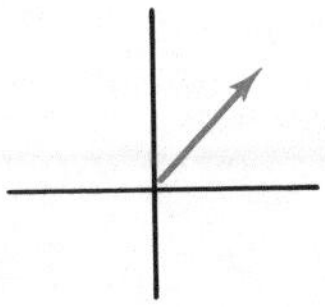

b. $(-1,3)$

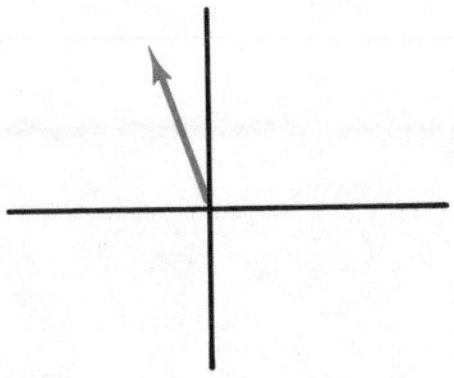

c. (2,−3)

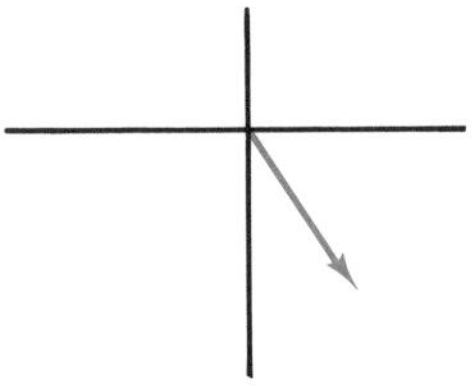

■

Our next task is to determine how the number pairs, i.e., the coordinates, of two vectors should be combined to give their sum. Let $\mathbf{A}=(a_1,a_2)$ and $\mathbf{B}=(b_1,b_2)$ be two vectors. Then a simple proof using congruent triangles yields that $\mathbf{A}+\mathbf{B}=(a_1+b_1,a_2+b_2)$; cf. Figure 2.6a. Similar arguments show us that $c\mathbf{A}=(ca_1,ca_2)$ if c is a rational number; cf. Figure 2.6b. We haven't talked about subtracting one vector from another yet, but $\mathbf{A}-\mathbf{B}$ should be equal to a vector such that $(\mathbf{A}-\mathbf{B})+\mathbf{B}=\mathbf{A}$. Thus if $\mathbf{A}=(a_1,a_2)$ and $\mathbf{B}=(b_1,b_2)$, then $\mathbf{A}-\mathbf{B}=(a_1-b_1,a_2-b_2)$.

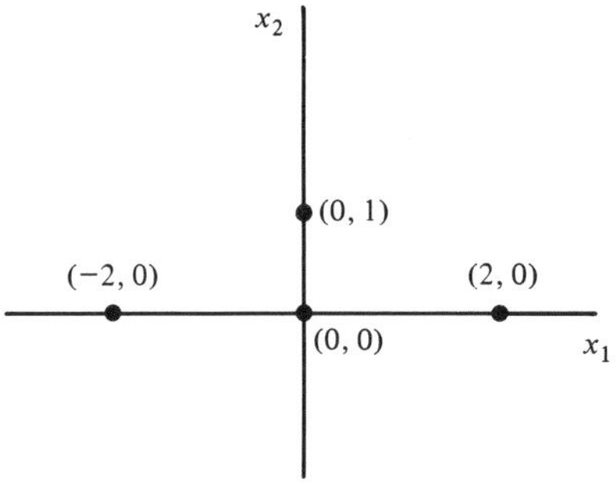

Figure 2.4

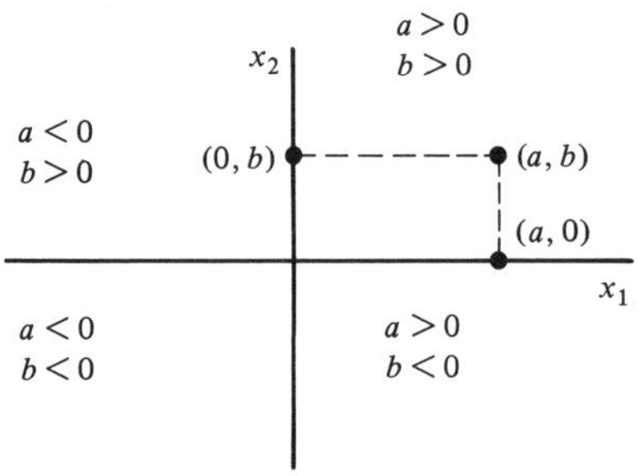

Figure 2.5

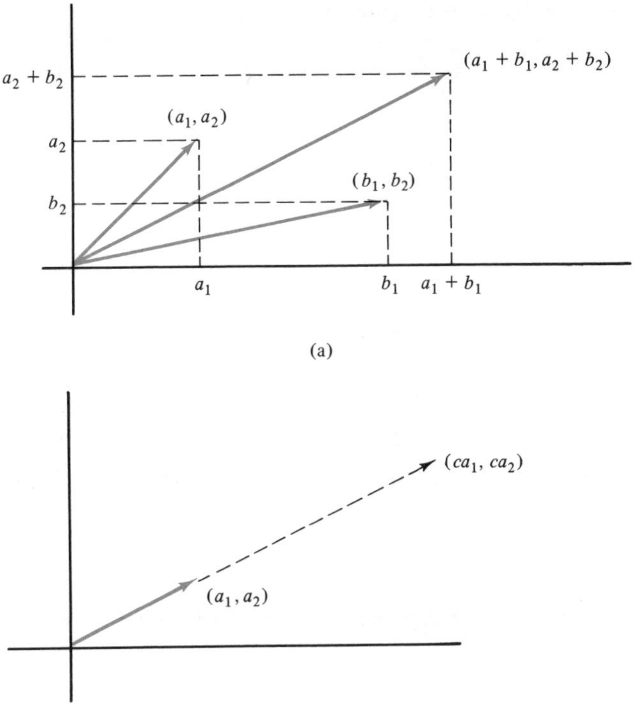

Figure 2.6

We now formally define $\mathbb{R}^2$ along with vector addition and scalar multiplication for these particular vectors. $\mathbb{R}^2$ is the classical example of a two-dimensional vector space.

Definition 2.1. $\mathbb{R}^2 = \{(x_1, x_2)$: x_1 and x_2 are arbitrary real numbers$\}$. If $\mathbf{A} = (a_1, a_2)$, $\mathbf{B} = (b_1, b_2)$, and c is any real number, then

1. $\mathbf{A} + \mathbf{B} = (a_1 + b_1, a_2 + b_2)$
2. $c\mathbf{A} = (ca_1, ca_2)$

Note that $\mathbb{R}^2$ can also be thought of as all possible 1×2 matrices, with vector addition being matrix addition and scalar multiplication the same as multiplying a matrix by a scalar; cf. Definitions 1.3 and 1.4.

A few words about the notation $\{:\}$. It is used to describe a set of elements. The phrase after the colon describes the property or attributes that something must possess in order for it to be in the set. For example, $\{x : x = 1, 2\}$ is the set $\{1,2\}$, and $\{x\colon 1 \le x < 100\}$ is the set of real numbers between 1 and 100 including 1 but excluding 100. Thus (1,23), (3.14159,53), and (0,−7) are in $\mathbb{R}^2$, but (0,\$), (#,f), and !e, are not, since the last three objects are not pairs of real numbers.

EXAMPLE 2 Solve the following vector equation:

$$\begin{aligned} 2(6,-1) + 3\mathbf{X} &= (3,-4) \\ 3\mathbf{X} &= (3,-4) - 2(6,-1) \\ &= (3,-4) - (12,-2) \\ &= (3-12,-4+2) = (-9,-2) \end{aligned}$$

Thus

$$\mathbf{X} = \tfrac{1}{3}(-9,-2) = (-3,-\tfrac{2}{3})$$

■

EXAMPLE 3 Given any vector in $\mathbb{R}^2$ write it as the sum of two vectors that are parallel to the coordinate axes.

Solution. Let $\mathbf{A} = (a_1, a_2)$ be any vector in $\mathbb{R}^2$. To say that a vector $\mathbf{X}$ is parallel to the x_1 axis is to say that the second component of $\mathbf{X}$'s representation as an ordered pair of numbers is zero. A similar comment applies to vectors parallel to the x_2 axis. Thus,

$$\mathbf{A} = (a_1, a_2) = (a_1, 0) + (0, a_2)$$

The first vector, $(a_1, 0)$, is parallel to the x_1 axis and the second vector, $(0, a_2)$, is parallel to the x_2 axis. We could also write

$$\mathbf{A} = (a_1, 0) + (0, a_2) = a_1(1,0) + a_2(0,1)$$

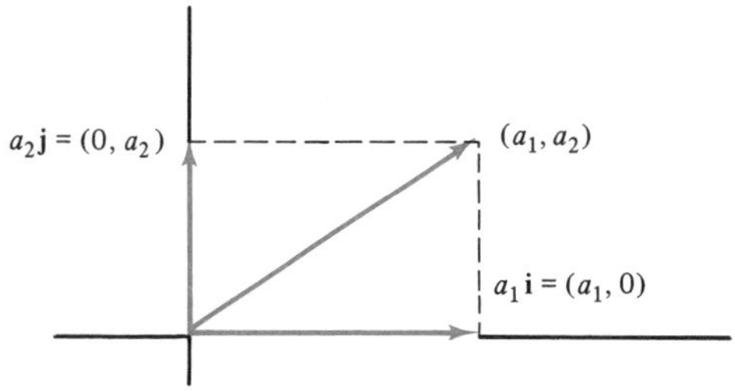

Figure 2.7

The vectors $(1,0)$ and $(0,1)$ are often denoted by **i** and **j**, respectively, and in this notation $(a_1, a_2) = a_1\mathbf{i} + a_2\mathbf{j}$. ■

The theorem below lists some of the algebraic properties that vector addition and scalar multiplication in $\mathbb{R}^2$ satisfy.

Theorem 2.1. Let $\mathbf{A} = (a_1, a_2)$, $\mathbf{B} = (b_1, b_2)$, and $\mathbf{C} = (c_1, c_2)$ be three arbitrary vectors in $\mathbb{R}^2$. Let a and b be any two numbers. Then the following equations are true.

1. $\mathbf{A} + \mathbf{B} = \mathbf{B} + \mathbf{A}$
2. $(\mathbf{A} + \mathbf{B}) + \mathbf{C} = \mathbf{A} + (\mathbf{B} + \mathbf{C})$
3. Let $\mathbf{0} = (0,0)$, then $\mathbf{A} + \mathbf{0} = \mathbf{A}$ [zero vector]

4. For every $\mathbf{A}$ there is a $(-\mathbf{A})$ such that $\mathbf{A} + (-\mathbf{A}) = \mathbf{0}$ $[-\mathbf{A} = (-a_1,-a_2)]$
5. $a(\mathbf{A} + \mathbf{B}) = a\mathbf{A} + a\mathbf{B}$
6. $(a + b)\mathbf{A} = a\mathbf{A} + b\mathbf{A}$
7. $(ab)\mathbf{A} = a(b\mathbf{A})$
8. $1\mathbf{A} = \mathbf{A}$

Proof. We verify equations 1, 4, 5, and 8, leaving the others for the reader.

1. $\mathbf{A} + \mathbf{B} = (a_1,a_2) + (b_1,b_2) = (a_1 + b_1, a_2 + b_2) = (b_1 + a_1, b_2 + a_2)$
$= (b_1,b_2) + (a_1,a_2) = \mathbf{B} + \mathbf{A}$
4. $\mathbf{A} + (-\mathbf{A}) = (a_1,a_2) + (-a_1,-a_2) = (a_1 - a_1, a_2 - a_2) = (0,0) = \mathbf{0}$
5. $a(\mathbf{A} + \mathbf{B}) = a[(a_1,a_2) + (b_1,b_2)] = a(a_1 + b_1, a_2 + b_2)$
$= (a(a_1 + b_1), a(a_2 + b_2)) = (aa_1 + ab_1, aa_2 + ab_2)$
$= (aa_1, aa_2) + (ab_1, ab_2) = a(a_1,a_2) + a(b_1,b_2)$
$= a\mathbf{A} + a\mathbf{B}$
8. $1\mathbf{A} = 1(a_1,a_2) = (1a_1, 1a_2) = (a_1,a_2) = \mathbf{A}$

We next discuss the standard three-dimensional space $\mathbb{R}^3$. As with $\mathbb{R}^2$, we first picture arrows starting at some fixed point and ending at any other point in three-dimensional space. We draw three mutually perpendicular coordinate axes x_1, x_2, and x_3 and impose a distance scale on each of these axes. To each point P in three space we associate an ordered triple of numbers (a,b,c), where a denotes the directed distance from P to the x_2, x_3 plane, b the directed distance from P to the x_1, x_3 plane, and c the directed distance from P to the x_1, x_2 plane. Just as we did for $\mathbb{R}^2$, we now think of vectors in three space both as arrows and as ordered triples of numbers.

EXAMPLE 4 For each of the triples of numbers below sketch the vector they represent.

a. (1,0,0), (0,1,0), (0,0,1): these three vectors are commonly denoted by **i**, **j**, and **k**, respectively. This notation is somewhat ambiguous; e.g., does **i** represent

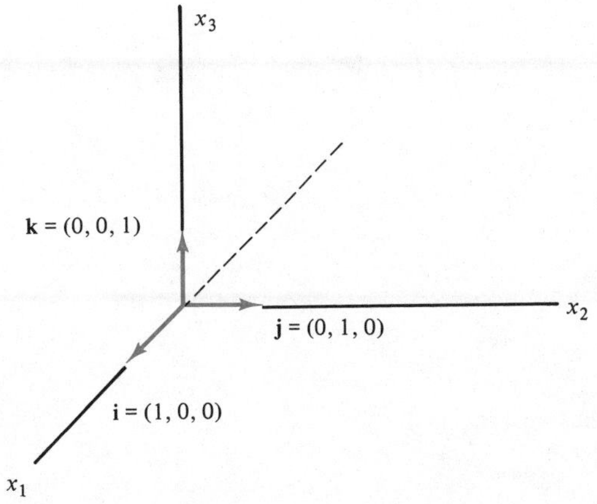

(1,0) or (1,0,0)? There should, however, be no confusion, since which vector space we are talking about will be clear from the discussion, and this will determine the meaning of **i**.

b. $\mathbf{A} = (-1,2,1)$ $\mathbf{B} = (-1,-1,1)$ $\mathbf{C} = (1,2,-1)$

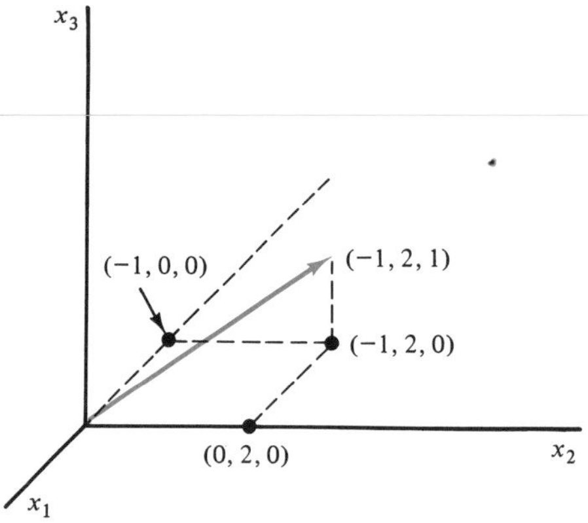

■

Definition 2.2 defines $\mathbb{R}^3$ and the algebraic operations of vector addition and scalar multiplication for this vector space.

Definition 2.2. $\mathbb{R}^3 = \{(x_1,x_2,x_3): x_1, x_2, \text{ and } x_3 \text{ are any real numbers}\}$. Vector addition and scalar multiplication are defined as follows:

1. $\mathbf{A} \pm \mathbf{B} = (a_1,a_2,a_3) \pm (b_1,b_2,b_3) = (a_1 \pm b_1, a_2 \pm b_2, a_3 \pm b_3)$, for any vectors **A** and **B** in $\mathbb{R}^3$.
2. $c\mathbf{A} = c(a_1,a_2,a_3) = (ca_1,ca_2,ca_3)$, for any vector **A** and any number c.

EXAMPLE 5 Let $\mathbf{A} = (1,-1,0)$, $\mathbf{B} = (0,1,2)$. Compute the following vectors:

a. $\mathbf{A} + \mathbf{B} = (1,-1,0) + (0,1,2) = (1,0,2)$
b. $2\mathbf{A} = 2(1,-1,0) = (2,-2,0)$ ■

Having defined $\mathbb{R}^2$ and $\mathbb{R}^3$, we now define $\mathbb{R}^n$, i.e., the set of ordered n-tuples of real numbers.

Definition 2.3. $\mathbb{R}^n = \{(x_1,x_2, \ldots ,x_n): x_1, x_2, \ldots , x_n \text{ are arbitrary real numbers}\}$. If **A** and **B** are any two vectors in $\mathbb{R}^n$ and a is any real number, we define vector addition and scalar multiplication in $\mathbb{R}^n$ as follows:

1. $\mathbf{A} \pm \mathbf{B} = (a_1,a_2, \ldots ,a_n) \pm (b_1,b_2, \ldots ,b_n) =$
   $(a_1 \pm b_1, a_2 \pm b_2, \ldots , a_n \pm b_n)$
2. $c\mathbf{A} = c(a_1,a_2, \ldots ,a_n) = (ca_1,ca_2, \ldots ,ca_n)$

We note that $\mathbb{R}^n$ may also be thought of as the set of $1 \times n$ matrices. Sometimes we will want to think of $\mathbb{R}^n$ as the set of $n \times 1$ matrices, i.e., $\mathbb{R}^n$ consists of columns rather than rows. One reason for this is so that $A\mathbf{x}$ will be defined for A an $m \times n$ matrix and $\mathbf{x}$ in $\mathbb{R}^n$.

There are times when complex numbers are used for scalars. We then define $\mathbb{C}^n = \{(z_1, z_2, \ldots, z_n): z_k$ are arbitrary complex numbers$\}$. Vector addition and scalar multiplication are defined as in $\mathbb{R}^n$, except that the scalars may now be complex numbers instead of just real numbers.

EXAMPLE 6 Write the vector $(-1,6,4)$ as a sum of three vectors each of which is parallel to one of the coordinate axes.

$$(-1,6,4) = (-1,0,0) + (0,6,0) + (0,0,4)$$

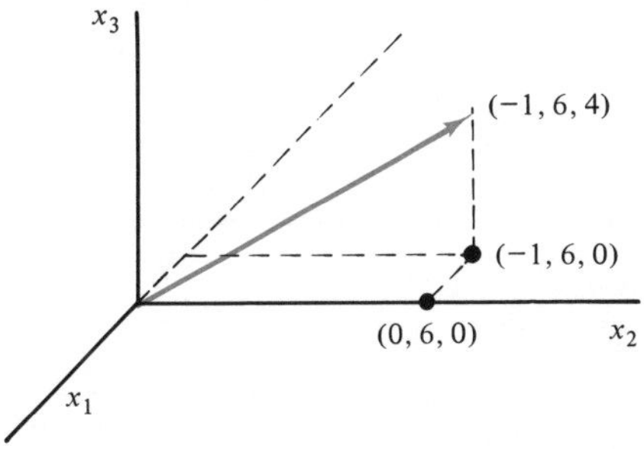

We can also write $(-1,6,4)$ as a linear combination of the three vectors $\mathbf{i}$, $\mathbf{j}$, and $\mathbf{k}$.

$$(-1,6,4) = -(1,0,0) + 6(0,1,0) + 4(0,0,1) = -\mathbf{i} + 6\mathbf{j} + 4\mathbf{k}$$ ■

A common mistake is to think that $\mathbb{R}^2$ is a subset of $\mathbb{R}^3$; that is, $W = \{(x_1, x_2, 0): x_1$ and x_2 arbitrary$\}$ is equated with $\mathbb{R}^2$. Clearly W is not $\mathbb{R}^2$, since W consists of triples of numbers, while $\mathbb{R}^2$ consists of pairs of numbers.

EXAMPLE 7 Solve the following vector equation in $\mathbb{R}^5$:

$$2(-1,4,2,0,1) + 6\mathbf{X} = 3(2,0,6,1,-1)$$
$$6\mathbf{X} = (6,0,18,3,-3) - 2(-1,4,2,0,1) = (8,-8,14,3,-5)$$

Thus,

$$\mathbf{X} = \tfrac{1}{6}(8,-8,14,3,-5) = (\tfrac{4}{3}, -\tfrac{4}{3}, \tfrac{7}{3}, \tfrac{1}{2}, -\tfrac{5}{6})$$ ■

The algebraic operations we have defined in $\mathbb{R}^n$ have the same properties as those in $\mathbb{R}^2$, and we list them in the following theorem.

Theorem 2.2. Let **A**, **B**, and **C** be three arbitrary vectors in $\mathbb{R}^n$. Let a and b be any two numbers. Then the following are true:

1. $\mathbf{A} + \mathbf{B} = \mathbf{B} + \mathbf{A}$
2. $(\mathbf{A} + \mathbf{B}) + \mathbf{C} = \mathbf{A} + (\mathbf{B} + \mathbf{C})$
3. Let $\mathbf{0} = (0,0, \ldots ,0)$, then $\mathbf{0} + \mathbf{A} = \mathbf{A}$
4. For every **A** in $\mathbb{R}^n$ there is a $(-\mathbf{A})$ in $\mathbb{R}^n$ such that $\mathbf{A} + (-\mathbf{A}) = \mathbf{0}$. If $\mathbf{A} = (a_1, \ldots ,a_n)$, $-\mathbf{A} = (-a_1, \ldots ,-a_n)$
5. $a(\mathbf{A} + \mathbf{B}) = a\mathbf{A} + a\mathbf{B}$
6. $(a + b)\mathbf{A} = a\mathbf{A} + b\mathbf{A}$
7. $(ab)\mathbf{A} = a(b\mathbf{A})$
8. $1\mathbf{A} = \mathbf{A}$

We ask the reader to check that these identities are valid.

PROBLEM SET 2.1

1. Sketch the following vectors in $\mathbb{R}^2$:
   a. $(-1,2)$, $(0,6)$, $-2(1,4)$
   b. $(1,1) + (-1,1)$, $3(-2,1) - 2(1,1)$
2. If $\mathbf{A} = (-1,6)$ and $\mathbf{B} = (3,-2)$, sketch the vectors
   a. $\mathbf{A}, -\mathbf{A}$
   b. $\mathbf{A} + \mathbf{B}, -\mathbf{A} + \mathbf{B}, \mathbf{A} - \mathbf{B}$
   c. $2\mathbf{A} - 3\mathbf{B}$
3. Let $\mathbf{A} = (1,-1)$. Sketch all vectors of the form $t\mathbf{A}$, where t is an arbitrary real number.
4. Let $\mathbf{A} = (1,-2)$. Let $\mathbf{B} = (-2,6)$. Find **X** such that $6\mathbf{X} + 2\mathbf{A} = 3\mathbf{B}$.
5. Let **A** and **B** be the same vectors as in problem 4. Sketch vectors of the form $\mathbf{X} = c_1\mathbf{A} + c_2\mathbf{B}$ for various values of c_1 and c_2. Which vectors in $\mathbb{R}^2$ can be written in this manner?
6. Prove Theorem 2.2 for $n = 3$.
7. Let $\mathbf{A} = (1,-1,2)$. Find all vectors **x** in $\mathbb{R}^3$ such that $\mathbf{x} = t\mathbf{A}$ for some number t.
8. Let $\mathbf{A} = (1,-1,2)$, $\mathbf{B} = (0,1,0)$.
   a. Find all vectors **x** in $\mathbb{R}^3$ such that $\mathbf{x} = t_1\mathbf{A} + t_2\mathbf{B}$ for some constants t_1 and t_2.
   b. Is the vector $(0,0,1)$ one of the vectors you found in part a?
9. Let $\mathbf{x} = (1,0)$, $\mathbf{y} = (0,1)$. Give a geometrical description of the following sets of vectors.
   a. $\{t\mathbf{x} + (1 - t)\mathbf{y}: 0 \le t \le 1\}$
   b. $\{t_1\mathbf{x} + t_2\mathbf{y}: 0 \le t_1 \le 1, 0 \le t_2 \le 1\}$
   c. $\{t\mathbf{x} + (1 - t)\mathbf{y}: t \text{ any real number}\}$
   d. $\{t_1\mathbf{x} + t_2\mathbf{y}: 0 \le t_1, 1 \le t_2\}$
   e. $\{t_1\mathbf{x} + t_2\mathbf{y}: t_1 \text{ and } t_2 \text{ arbitrary}\}$
10. Let $\mathbf{x} = (1,2)$, $\mathbf{y} = (-1,1)$. Describe the following sets of vectors.
    a. $\{t\mathbf{x} + (1 - t)\mathbf{y}: 0 \le t \le 1\}$

b. $\{t_1\mathbf{x} + t_2\mathbf{y}: 0 \le t_1 \le 1, 0 \le t_2 \le 1\}$
c. $\{t\mathbf{x} + (1-t)\mathbf{y}: t \text{ any real number}\}$
d. $\{t_1\mathbf{x} + t_2\mathbf{y}: 0 \le t_1, 1 \le t_2\}$
e. $\{t_1\mathbf{x} + t_2\mathbf{y}: t_1 \text{ and } t_2 \text{ arbitrary}\}$

11. Let $\mathbf{x} = (1,1,0)$ and let $\mathbf{y} = (1,1,1)$. Describe the following sets of vectors:
a. $\{t\mathbf{x} + (1-t)\mathbf{y}: 0 \le t \le 1\}$
b. $\{t_1\mathbf{x} + t_2\mathbf{y}: 0 \le t_1 \le 1, 0 \le t_2 \le 1\}$
c. $\{t\mathbf{x} + (1-t)\mathbf{y}: t \text{ any real number}\}$

12. Given two sets A and B, their intersection, $A \cap B$, is defined as everything they have in common.

$$A \cap B = \{x: x \in A \text{ and } x \in B\}$$

Let

$$W_1 = \{(x_1,x_2,x_3): x_1 = x_2, x_3 = 0\}$$
$$W_2 = \{(x_1,x_2,x_3): x_1 = x_2\}$$
$$W_3 = \{(x_1,x_2,x_3): x_1 + x_2 = 0\}$$

Graph each of the following sets:
a. $W_1 \cap W_2$ **b.** $W_2 \cap W_3$ **c.** $W_1 \cap W_2 \cap W_3$

2.2 DEFINITION OF A VECTOR SPACE

In the last section we defined the classical vector spaces $\mathbb{R}^n$, for $n = 2, 3, \ldots$. In this section we extract from these sets their crucial properties, and then define a vector space to be anything that has these properties. Thus, let V denote a set or collection of objects. The elements of V will be called vectors and we assume that there are two operations defined. The first (vector addition) will be denoted by a "+" sign, and the second (scalar multiplication) by a "·" or by juxtaposition. Specifically if $\mathbf{x}$ and $\mathbf{y}$ are any two vectors, i.e., elements of V, then $\mathbf{x} + \mathbf{y}$ is <u>in V</u>. If a is any real number and $\mathbf{x}$ any vector, then $a\mathbf{x}$ is <u>in V</u>. The condition that $\mathbf{x} + \mathbf{y}$ and $a\mathbf{x}$ be in V is expressed by saying that V is closed under vector addition and scalar multiplication. The term "<u>in V</u>" has been underlined to underscore the idea that there may be times when an addition or multiplication has been defined such that the result will not be back in V. In such cases V is not a vector space.

In order for V to be a vector space, it is not enough that these two operations have been defined. They must also satisfy certain laws: those needed so that a reasonable arithmetic is possible. The requirements are listed in the following definition.

Definition 2.4. Let V be a set on which two operations, vector addition and scalar multiplication, have been defined. V will be called a real vector space if the following properties are true:

1. For any $\mathbf{x}$ and $\mathbf{y}$ in V, $\mathbf{x}+\mathbf{y}$ is in V.
2. For any $\mathbf{x}$ in V, and any real number a in R, $a\mathbf{x}$ is in V.
3. $\mathbf{x}+\mathbf{y}=\mathbf{y}+\mathbf{x}$, for any $\mathbf{x}$ and $\mathbf{y}$ in V.
4. $\mathbf{x}+(\mathbf{y}+\mathbf{z})=(\mathbf{x}+\mathbf{y})+\mathbf{z}$, for any $\mathbf{x}$, $\mathbf{y}$, and $\mathbf{z}$ in V.
5. There is a vector in V denoted by $\mathbf{0}$ such that $\mathbf{x}+\mathbf{0}=\mathbf{x}$, for every vector $\mathbf{x}$ in V.
6. For any vector $\mathbf{x}$ in V, there is a vector $(-\mathbf{x})$ in V such that $\mathbf{x}+(-\mathbf{x})=\mathbf{0}$.
7. $a(\mathbf{x}+\mathbf{y})=a\mathbf{x}+a\mathbf{y}$, for any two vectors $\mathbf{x}$ and $\mathbf{y}$ in V and any real number a.
8. $(a+b)\mathbf{x}=a\mathbf{x}+b\mathbf{x}$, for any two real numbers a and b and any vector $\mathbf{x}$.
9. $(ab)\mathbf{x}=a(b\mathbf{x})$, for any two real numbers a and b and any vector $\mathbf{x}$.
10. $1\mathbf{x}=\mathbf{x}$, for every vector $\mathbf{x}$ in V.

What the above properties mean practically is that all the algebraic manipulations you've learned to do with real numbers and vectors in $\mathbb{R}^n$ can also be done in this more abstract setting.

Remember, whenever someone says, "V is a real vector space" they are merely stating that the elements of V along with V's version of vector addition and scalar multiplication satisfy rules 1 through 10.

One more comment: The scalars are required to be real numbers and V is called a real vector space. There will be times when the scalars are complex numbers, in which case we call V a complex vector space. In the following we will not explicitly use the adjective real, since all our vector spaces will be real, unless the contrary is stated.

EXAMPLE 1 $\mathbb{R}^n (n \geq 2)$ is a vector space. Definition 2.3 defines $\mathbb{R}^n$ as the set of n tuples of real numbers as well as what vector addition and scalar multiplication mean in this setting. Theorem 2.2 shows that 1 through 10 of Definition 2.4 hold. ■

EXAMPLE 2 $\mathbb{R}^1 = \{x\colon x$ is a real number$\}$ with vector addition and scalar multiplication being ordinary addition and multiplication of real numbers. ■

EXAMPLE 3 M_{mn}, the set of all $m \times n$ matrices with real entries, with the usual way of adding matrices and multiplying matrices by numbers, is a vector space. ■

EXAMPLE 4 Let $V = \{(x_1,0,x_3)$: such that x_1 and x_3 are real numbers$\}$. Define addition and scalar multiplication by

$$\begin{aligned}\mathbf{x}+\mathbf{y} &= (x_1,0,x_3)+(y_1,0,y_3)=(x_1+y_1,0,x_3+y_3)\\ a\mathbf{x} &= a(x_1,0,x_3)=(ax_1,0,x_3)!!\end{aligned}$$

Clearly V is closed under these versions of vector addition and scalar multiplication; i.e., 1 and 2 hold. As a matter of fact, one can quickly check that 3 through 7 hold, but 8 does not. For

$$\begin{aligned} a\mathbf{x} + b\mathbf{x} &= a(x_1,0,x_3) + b(x_1,0,x_3) = (ax_1,0,x_3) + (bx_1,0,x_3) \\ &= [(a+b)x_1,0,2x_3] \end{aligned}$$

while

$$(a+b)\mathbf{x} = (a+b)(x_1,0,x_3) = [(a+b)x_1,0,x_3]$$

Since these two expressions will be equal only if $2x_3 = x_3$ or $x_3 = 0$ it is clear that 8 is not true for all the elements of V. Thus V is not a vector space. ■

EXAMPLE 5 Let $V = \{x: x > 0\}$. That is, V is the set of all positive real numbers. In this example $\oplus$ will denote vector addition and $\odot$ will denote scalar multiplication. Define these operations as follows:

$\mathbf{x} \oplus \mathbf{y} = xy$; that is, vector addition will be ordinary multiplication

$a \odot \mathbf{x} = x^a$

Since the product of two positive numbers is positive, and a positive number raised to any power is still positive, V is closed under these two operations; i.e., 1 and 2 in the definition of a vector space hold. We now proceed to check, as we must, each of the other required properties.

3. $\mathbf{x} \oplus \mathbf{y} = xy = yx = \mathbf{y} \oplus \mathbf{x}$
4. $x \oplus (\mathbf{y} \oplus \mathbf{z}) = x(yz) = (xy)z = (\mathbf{x} \oplus \mathbf{y}) \oplus \mathbf{z}$
5. For the $\mathbf{0}$ vector we try the number 1: $\mathbf{1} \oplus \mathbf{x} = 1x = x = \mathbf{x}$
6. For $(-\mathbf{x})$ we try x^{-1}: $(-\mathbf{x}) \oplus \mathbf{x} = x^{-1}x = 1 = \mathbf{0}$. Note that if $x > 0$, so is x^{-1}.
7. $a \odot (\mathbf{x} \oplus \mathbf{y}) = (xy)^a = (x^a)(y^a) = (a \odot \mathbf{x}) \oplus (a \odot \mathbf{y})$
8. $(a+b) \odot \mathbf{x} = x^{a+b} = x^a x^b = (a \odot \mathbf{x}) \oplus (b \odot \mathbf{x})$
9. $(ab) \cdot \mathbf{x} = x^{ab} = (x^b)^a = a \odot (b \odot \mathbf{x})$
10. $1 \odot \mathbf{x} = x^1 = \mathbf{x}$

Notice that in 5, the number 1 is treated as a vector in V, while in 10, it is treated as a scalar. Since V, with these definitions of vector addition and scalar multiplication, satisfies 1 through 10 of Definition 2.4, V is a vector space. ■

The moral of this example, if there is one, is that neither "addition" nor "multiplication" need always be what we're used to. Notice that the zero vector in the above example is not the number zero but the number 1. What is important about the zero vector is not that it be zero (the number) but that it satisfy the algebraic property 5 of Definition 2.4.

EXAMPLE 6 $V = \{(x_1, x_2, 0)$: x_1, x_2 are arbitrary real numbers$\}$. The usual definitions of addition and multiplication of vectors in $\mathbb{R}^3$ will be used. ■

The reader is asked to check that all 10 properties do indeed hold and conclude that V is a vector space.

EXAMPLE 7 $V = \{\mathbf{f}$: $\mathbf{f}$ is a real-valued function defined on $[0,1]\}$. For $\mathbf{f}$ and $\mathbf{g}$ in V, $\mathbf{f} + \mathbf{g}$ must be in V; that is, it must be a function defined on $[0,1]$. Thus for each t, $0 \le t \le 1$, define $(\mathbf{f} + \mathbf{g})(t) = \mathbf{f}(t) + \mathbf{g}(t)$. We similarly define $a\mathbf{g}$, by $(a\mathbf{g})(t) = a\mathbf{g}(t)$. With these definitions of vector addition and scalar multiplication, 1 and 2 are satisfied. The reader is asked to check that the other eight properties are also true. ■

EXAMPLE 8 Let $P_n = \{$polynomials in t of degree $\le n\} =$

$$\{a_n t^n + a_{n-1} t^{n-1} + \cdots + a_1 t + a_0\text{: where the } a_j \text{ are real numbers}\}$$

Thus

$$P_0 = \{a\text{: } a \text{ is any real number}\} = \mathbb{R}^1$$

$$P_1 = \{a_0 + a_1 t\text{: } a_0 \text{ and } a_1 \text{ are arbitrary real numbers}\}$$

If $\mathbf{f} = f_0 + \cdots + f_n t^n$ and $\mathbf{g} = g_0 + \cdots + g_n t^n$ are any two polynomials in P_n, we define $(\mathbf{f} + \mathbf{g})(t)$ by

$$[\mathbf{f} + \mathbf{g}](t) = (f_0 + g_0) + (f_1 + g_1)t + \cdots + (f_n + g_n)t^n$$

Thus $\mathbf{f} + \mathbf{g}$ is in P_n. We define $a\mathbf{f}$ by

$$a\mathbf{f}(t) = (af_0) + (af_1)t + \cdots + (af_n)t^n$$

which is also in P_n. To see that P_n satisfies the other conditions in Definition 2.4 is routine and is left to the reader. We note that P_n, for each value of n, is a subset of the vector space in Example 7. ■

We remind the reader that a vector space is not just a set, but a set with two operations: vector addition and scalar multiplication. If we change any one of these three, all ten properties must again be checked before we can say that we still have a vector space.

In the future when we deal with any of the standard vector spaces $\mathbb{R}^n$, P_n, M_{mn}, or the vector space in Example 7, the operations of vector addition and scalar multiplication will not be explicitly stated. The reader, unless told otherwise, should assume the standard operations in these spaces.

PROBLEM SET 2.2

1. Verify the details of Example 6.
2. Let $V = \{(x_1, x_2)\colon x_1 \text{ and } x_2 \text{ are real numbers}\}$. Define addition by

$$(x_1,x_2) + (y_1,y_2) = (x_1 + y_1, x_2 - y_2)$$

and scalar multiplication by

$$a(x_1,x_2) = (ax_1, ax_2)$$

Is V a vector space? If not, list all the axioms that V fails to satisfy.

3. Let $V = \{(x_1, x_2)\colon x_1 \text{ and } x_2 \text{ are real numbers}\}$. Define addition and scalar multiplication by

$$(x_1,x_2) + (y_1,y_2) = (x_2 + y_2, x_1 + y_1)$$
$$a(x_1,x_2) = (ax_1, ax_2)$$

Is V a vector space? If not, list all the axioms that V fails to satisfy. What happens if we define $a(x_1,x_2) = (ax_2, ax_1)$?

4. Verify that M_{mn} (Example 3) is a vector space.
5. Verify that the set in Example 4 satisfies all the properties of Definition 2.4 except number 8.
6. Let $C[0,1]$ be the set of real-valued continuous functions defined on the interval $[0,1]$. This set is a subset of the vector space defined in Example 7. Define addition and multiplication as in that example, and show that $C[0,1]$ is also a vector space.
7. Show that V in Example 7 is a vector space.
8. Show that P_n for $n = 1, 2, \ldots,$ is a vector space.
9. Let V_0 be all 2×2 matrices for which the 1,1 entry is zero. Let V_1 be all 2×2 matrices for which the 1,1 entry is nonzero. Determine if either of these two sets is a vector space. Define addition and multiplication as we did in M_{mn}.
10. Let $V = \{A, B, \ldots, Z\}$; that is, V is the set of capital letters. Define $A + B = C$, $B + C = D$, $A + E = F$; i.e., the sum of two letters is the first letter larger than either of the summands, unless Z is one of the letters to be added. In that case the sum will always be A. Can you define a way to multiply the elements in V by real numbers in such a way that V becomes a vector space?
11. Let $V = \{(x,0,y)\colon x \text{ and } y \text{ are arbitrary real numbers}\}$. Define addition and scalar multiplication as follows:

$$(x_1,0,y_1) + (x_2,0,y_2) = (x_1 + x_2, y_1 + y_2)$$
$$c(x,0,y) = (cx, cy)$$

Is V a vector space?

12. Let $V_c = \{(x_1,x_2)\colon x_1 + 2x_2 = c\}$.
    a. For each value of c sketch V_c.
    b. For what values of c, if any, is V_c a vector space?
13. Let $V_c = \left\{\begin{bmatrix} a_1 & a_2 & a_3 \\ a_4 & a_5 & a_6 \end{bmatrix}\colon a_1 + a_2 + a_3 = c\right\}$. For what values of c is V_c a vector space?

14. Let $V_c = \{\mathbf{p}: \mathbf{p}$ is in P_4 and $\mathbf{p}(c) = 0\}$. For what values of c is V_c a vector space?

15. Let $V = \{A: A$ is in M_{23}, and $\Sigma_{i,j}a_{ij} = 0\}$. That is, A is in V if A is a 2×3 matrix whose entries sum to zero. Is V a vector space?

16. Let $V_1 = \{(x_1,x_2): x_1^2 + x_2^2 = 1\}$, $V_2 = \{(x_1,x_2): x_1^2 + x_2^2 \leq 1\}$, $V_3 = \{(x_1,x_2): x_1 \geq 0, x_2 \geq 0\}$, and $V_4 = \{(x_1,x_2): x_1 + x_2 \geq 0\}$. Which of these subsets of $\mathbb{R}^2$ is a vector space?

17. Let $V = \{\mathbf{p}: \mathbf{p}$ is in P_{17} and $\mathbf{p}(1) + \mathbf{p}(6) - \mathbf{p}(2) = 0\}$. Is V a vector space? Does the subscript 17 have any bearing on the matter? What if the constraint is $\mathbf{p}(1) + \mathbf{p}(6) - \mathbf{p}(2) = 1$?

2.3 SPANNING SETS AND SUBSPACES

The material covered so far has been relatively concrete and easy to absorb, but we now have to start thinking about an abstract concept, vector spaces. The only rules that we may use are those listed in Definition 2.4 and any others we are able to deduce from them. This, especially for the novice, is not easy and is somewhat tedious, but well worth the effort.

In the definition of a vector space, we have as axioms the existence of a zero vector (axiom 5) and an additive inverse (axiom 6) for every vector. One question that occurs is, how many zero vectors and inverses are there? Let's see if we can find an answer to this, at least in $\mathbb{R}^2$. Suppose $\mathbf{k} = (k_1,k_2)$ is another zero vector in $\mathbb{R}^2$. Thus $\mathbf{x} = \mathbf{x} + \mathbf{k} = (x_1,x_2) + (k_1,k_2) = (x_1 + k_1, x_2 + k_2)$. This implies that $x_1 = x_1 + k_1$ and $x_2 = x_2 + k_2$, from which $k_1 = k_2 = 0$. Hence $\mathbf{k} = (0,0) = \mathbf{0}$, which shows that $\mathbb{R}^2$ has a unique zero vector. Suppose now that $\mathbf{y}$ is an additive inverse for $\mathbf{x}$ in $\mathbb{R}^2$. Then $\mathbf{0} = \mathbf{x} + \mathbf{y} = (x_1,x_2) + (y_1,y_2) = (x_1 + y_1, x_2 + y_2)$. Hence $x_1 + y_1 = 0 = x_2 + y_2$. Thus $y_1 = -x_1$ and $y_2 = -x_2$, and we have uniqueness of inverses, at least in $\mathbb{R}^2$. What about other vector spaces?

Theorem 2.3. Let V be a vector space. Then

1. If $\mathbf{0}$ and $\mathbf{k}$ are both zero vectors (i.e., both satisfy axiom 5), then $\mathbf{0} = \mathbf{k}$.
2. If $-\mathbf{x}$ and $\mathbf{y}$ are both additive inverses of $\mathbf{x}$, then $-\mathbf{x} = \mathbf{y}$.

Thus, in any vector space the zero vector is unique, and the additive inverse $-\mathbf{x}$ of any vector $\mathbf{x}$ is uniquely determined by $\mathbf{x}$.

Proof

1. Since $\mathbf{0}$ satisfies 5, we have $\mathbf{k} + \mathbf{0} = \mathbf{k}$ but $\mathbf{k}$ also satisfies 5; thus $\mathbf{0} + \mathbf{k} = \mathbf{0}$. These two equations plus commutativity of addition imply $\mathbf{k} = \mathbf{0}$. Thus, there is only one zero vector.
2. $\mathbf{y} = \mathbf{y} + \mathbf{0} = \mathbf{y} + [\mathbf{x} + (-\mathbf{x})] = (\mathbf{y} + \mathbf{x}) + (-\mathbf{x}) = \mathbf{0} + (-\mathbf{x}) = -\mathbf{x}$. We remind the reader that $\mathbf{y} + \mathbf{x} = \mathbf{0}$ since $\mathbf{y}$ is assumed to be an additive inverse for $\mathbf{x}$.

Some additional properties of vector spaces are listed in the following theorem. The reader should try to prove them for V equal to $\mathbb{R}^2$; then read the proof of Theorem 2.4 and compare his or her version with ours.

Theorem 2.4. Let V be a vector space. Let $\mathbf{X}$ be any vector in V and a any number. Then

1. $0\mathbf{X} = \mathbf{0}$
2. $a\mathbf{0} = \mathbf{0}$
3. $(-1)\mathbf{X} = -\mathbf{X}$
4. $a\mathbf{X} = \mathbf{0}$ only if $a = 0$ or $\mathbf{X} = \mathbf{0}$

Proof

1. Let $\mathbf{x}$ be any vector in V. Then,

$$\mathbf{x} + 0\mathbf{x} = 1\mathbf{x} + 0\mathbf{x} = (1 + 0)\mathbf{x} = 1\mathbf{x} = \mathbf{x}$$

Adding $-\mathbf{x}$ to both sides of this equation, we get

$$\mathbf{0} = -\mathbf{x} + \mathbf{x} = -\mathbf{x} + (\mathbf{x} + 0\mathbf{x}) = (-\mathbf{x} + \mathbf{x}) + 0\mathbf{x} = \mathbf{0} + 0\mathbf{x} = 0\mathbf{x}$$

2. If $a = 0$, then from 1 above we have $a\mathbf{0} = \mathbf{0}$. Thus we may suppose $a \neq 0$. Then,

$$a\mathbf{0} + \mathbf{x} = a\mathbf{0} + (aa^{-1})\mathbf{x} = a(\mathbf{0} + a^{-1}\mathbf{x}) = a(a^{-1}\mathbf{x}) = 1\mathbf{x} = \mathbf{x}$$

Hence from Theorem 2.3 $a\mathbf{0} = \mathbf{0}$.

3. $\mathbf{x} + (-1)\mathbf{x} = 1\mathbf{x} + (-1)\mathbf{x} = [1 + (-1)]\mathbf{x} = 0\mathbf{x} = \mathbf{0}$. Theorem 2.3 now implies that $(-1)\mathbf{x} = -\mathbf{x}$.
4. Suppose $a\mathbf{x} = \mathbf{0}$ and $a \neq 0$. Then

$$\mathbf{0} = a^{-1}\mathbf{0} = a^{-1}(a\mathbf{x}) = (aa^{-1})\mathbf{x} = 1\mathbf{x} = \mathbf{x}$$

Thus $a\mathbf{x} = \mathbf{0}$ only when $a = 0$ or $\mathbf{x} = \mathbf{0}$.

Quite often when the statement of a theorem or definition is read for the first time, its meaning is lost in a maze of strange words and concepts. The reader is strongly urged to always go back to $\mathbb{R}^2$ or $\mathbb{R}^3$ and try to understand what the statement means in this perhaps friendlier setting. For example, before proving Theorem 2.3 we analyzed the special case $V = \mathbb{R}^2$.

Given a collection of vectors $\{\mathbf{x}_k: k = 1, \ldots, n\}$, we often wish to form arbitrary sums of these vectors; that is, we wish to look at all vectors $\mathbf{x}$ of the form

$$\mathbf{x} = c_1\mathbf{x}_1 + c_2\mathbf{x}_2 + \cdots + c_n\mathbf{x}_n = \sum_{k=1}^{n} c_k\mathbf{x}_k \tag{2.1}$$

where the c_k are arbitrary real numbers. Such sums are called *linear combinations* of the vectors $\mathbf{x}_k$.

EXAMPLE 1 Let $\mathbf{x}_1 = (1,2,0)$, $\mathbf{x}_2 = (-1,1,0)$. Determine all linear combinations of these two vectors.

Solution. Any linear combination of these two vectors must be of the form

$$\begin{aligned}\mathbf{x} &= c_1\mathbf{x}_1 + c_2\mathbf{x}_2 \\ &= c_1(1,2,0) + c_2(-1,1,0) = (c_1,2c_1,0) + (-c_2,c_2,0) \\ &= (c_1 - c_2, 2c_1 + c_2, 0)\end{aligned}$$

Thus, no matter what values the constants c_1 and c_2 have the third component of $\mathbf{x}$ will always be zero, and it seems likely that as c_1 and c_2 vary over all pairs of real numbers, so too will the terms $c_1 - c_2$ and $2c_1 + c_2$. We conjecture, then, that the set of all linear combinations of these two vectors will be the set of vectors S in $\mathbb{R}^3$ whose third component is zero. To prove this, we only have to show that if $\mathbf{x}$ is in S, then there are constants c_1 and c_2 such that $\mathbf{x} = c_1\mathbf{x}_1 + c_2\mathbf{x}_2$. Thus suppose $\mathbf{x}$ is in S. Then $\mathbf{x} = (a,b,0)$ for some numbers a and b. We want to find constants c_1 and c_2 such that

$$(a,b,0) = c_1(1,2,0) + c_2(-1,1,0) = (c_1 - c_2, 2c_1 + c_2, 0)$$

or

$$c_1 - c_2 = a \quad \text{and} \quad 2c_1 + c_2 = b$$

The solution is

$$c_1 = \frac{a+b}{3} \qquad c_2 = \frac{-2a+b}{3}$$

Thus $(a,b,0)$ is a linear combination of $(1,2,0)$ and $(-1,1,0)$. ■

The set of all linear combinations of a set of vectors will occur frequently enough that we give it a special name and notation.

Definition 2.5. Given a set of vectors $A = \{\mathbf{x}_k\colon k = 1, 2, \ldots, n\}$ in some vector space V, the set of all linear combinations of these vectors will be called their linear span, or span, and denoted by $S[A]$.

The result of the previous example restated in this notation is $S[(1,2,0),(-1,1,0)] = \{(x_1,x_2,0)\colon x_1 \text{ and } x_2 \text{ are arbitrary real numbers}\}$.

EXAMPLE 2 Find the span of the vector $(-1,1)$.

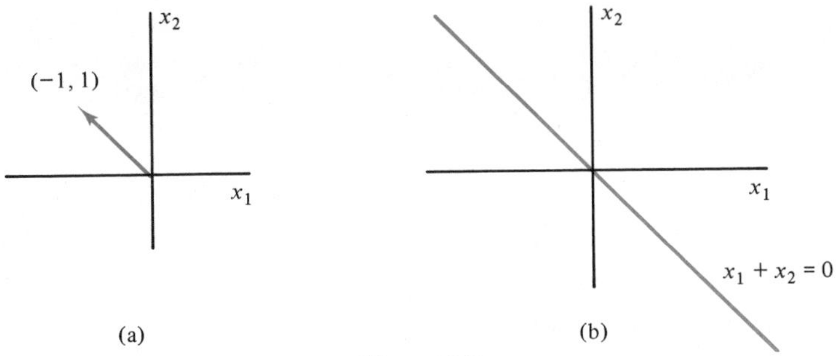

Figure 2.8

$$S[(-1,1)] = \{c(-1,1)\colon c \text{ is any real number}\}$$
$$= \{(-c,c)\colon c \text{ is any real number}\}$$

Thus the span of $(-1,1)$ may be thought of as the straight line $x_1 + x_2 = 0$. ■

EXAMPLE 3 Find the span of $A = \{(1,1,0), (-1,0,0)\}$.

$$S[A] = \{a(1,1,0) + b(-1,0,0)\colon a \text{ and } b \text{ arbitrary real numbers}\}$$
$$= \{(a-b,a,0)\colon a,\ b \text{ arbitrary numbers}\}$$
$$= \{(x,y,0)\colon x,\ y \text{ arbitrary numbers}\}$$

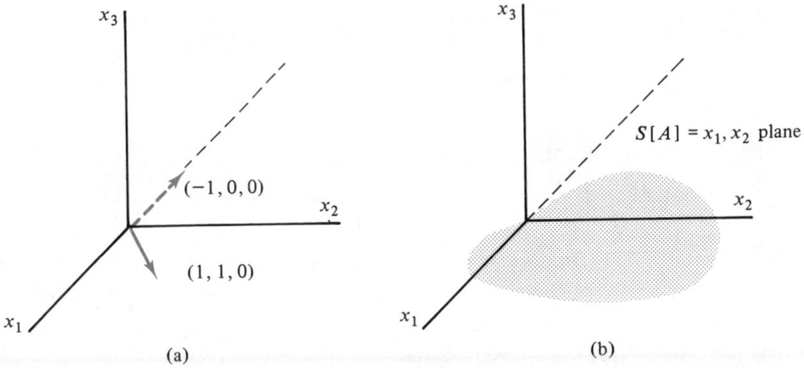

Figure 2.9

■

The span of a nonempty set of vectors has some properties that we state and prove in the next theorem.

Theorem 2.5. Let $S[A]$ be the linear span of the set $A = \{\mathbf{x}_k\colon k = 1, 2, \ldots, n\}$. Then $S[A]$ satisfies axioms 1 through 10 of the definition of a vector space. That is, $S[A]$ is a vector space.

Proof

1. Let $\mathbf{y}$ and $\mathbf{z}$ be in $S[A]$. Then there are constants a_j and b_j, $j = 1, 2, \ldots, n$ such that

$$\mathbf{y} = \sum_{j=1}^{n} a_j \mathbf{x}_j \qquad \mathbf{z} = \sum_{j=1}^{n} b_j \mathbf{x}_j$$

Thus

$$\begin{aligned}\mathbf{y} + \mathbf{z} &= \sum_{j=1}^{n} a_j \mathbf{x}_j + \sum_{j=1}^{n} b_j \mathbf{x}_j = \sum_{j=1}^{n} (a_j \mathbf{x}_j + b_j \mathbf{x}_j) \\ &= \sum_{j=1}^{n} (a_j + b_j) \mathbf{x}_j\end{aligned}$$

and we see that $\mathbf{y} + \mathbf{z}$ is also in $S[A]$.

2. Let $\mathbf{x}$ be in $S[A]$ and k a real number. Then there are constants c_j, $j = 1, 2, \ldots, n$ such that $\mathbf{x} = \Sigma_{j=1}^{n} c_j \mathbf{x}_j$. Thus

$$k\mathbf{x} = k\left(\sum_{j=1}^{n} c_j \mathbf{x}_j\right) = \sum_{j=1}^{n} k c_j \mathbf{x}_j$$

and we have that $k\mathbf{x}$ is in $S[A]$.

3 and **4**. Commutativity and associativity of vector addition are true because we are in a vector space to begin with.

5. Clearly each of the constants in a linear combination of the vectors $\mathbf{x}_j$ can be set equal to zero. Thus we have

$$\sum_{j=1}^{n} 0\mathbf{x}_j = \sum_{j=1}^{n} \mathbf{0} = \mathbf{0}$$

and $S[A]$ contains the zero vector.

6. Let $\mathbf{x}$ be in $S[A]$. Then $\mathbf{x} = \Sigma_{j=1}^{n} c_j \mathbf{x}_j$ and $-\mathbf{x} = -(\Sigma_{j=1}^{n} c_j \mathbf{x}_j) = \Sigma_{j=1}^{n} (-c_j) \mathbf{x}_j$. Thus, $-\mathbf{x}$ is also in $S[A]$.

7, 8, 9, and **10** are true because they are true for all the vectors in V, and hence these properties are true for the vectors in $S[A]$. Remember everything in $S[A]$ is automatically in V, since V is a vector space and is closed under linear combinations.

This fact, that the span of a set of vectors is itself a vector space contained in the original vector space, is expressed by saying that the span is a subspace of V.

Definition 2.6. Let V be a vector space. Let W be a nonempty subset of V. Then if W (using the same operations of vector addition and scalar multiplication as in V) is also a vector space, we say that W is a subspace of V.

EXAMPLE 4 $V = \mathbb{R}^3$. Let $W = \{(r,0,0)\colon r \text{ is any real number}\}$. Show that W is a subspace of V.

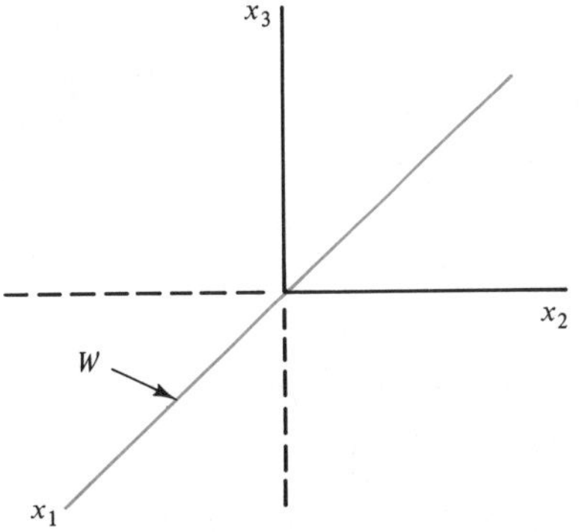

Figure 2.10

Solution. Since the operations of vector addition and scalar multiplication will be unchanged, we know that W satisfies properties 3, 4, 7, 8, 9, and 10. Hence, we merely need to verify that W satisfies 1, 2, 5, and 6 of Definition 2.4. Thus suppose that $\mathbf{x}$ and $\mathbf{y}$ are in W. Then $\mathbf{x} = (r,0,0)$ and $\mathbf{y} = (s,0,0)$ for some numbers r and s, and

$$\mathbf{x} + \mathbf{y} = (r,0,0) + (s,0,0) = (r + s,0,0)$$

Thus $\mathbf{x} + \mathbf{y}$ is in W. Now let a be any real number, then

$$a\mathbf{x} = a(r,0,0) = (ar,0,0)$$

is also in W. To see that $\mathbf{0}$ and $-\mathbf{x}$ are in W we note that

$$\mathbf{0} = 0\mathbf{x} = (0,0,0) \quad \text{and} \quad -\mathbf{x} = -(r,0,0) = (-r,0,0)$$

Thus W is a subspace of $V = \mathbb{R}^3$. ■

EXAMPLE 5 Let V be any vector space and let A be any nonempty subset of V. Then $S[A]$ is a subspace of V. This is merely Theorem 2.5 restated. ■

EXAMPLE 6 Let $\mathbf{x} = (a,b,c)$ be any nonzero vector in $\mathbb{R}^3$. Then $S[\mathbf{x}]$, which is a subspace of $\mathbb{R}^3$, is just the straight line passing through the origin and the point with coordinates (a,b,c). See Figure 2.11. The reader should verify the details of this example. ■

To verify that a subset of a vector space is a subspace can be tedious. The following theorem shows that it is sufficient to verify axioms 1 and 2 of Definition 2.4.

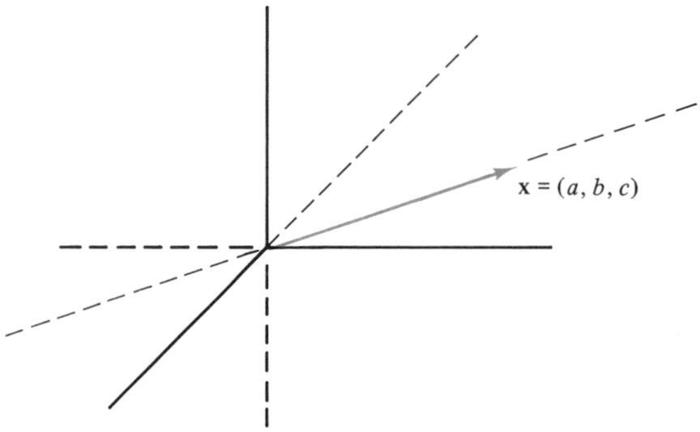

Figure 2.11

Theorem 2.6. A nonempty subset W of a vector space V is a subspace of V if and only if

1. For any $\mathbf{x}$ and $\mathbf{y}$ in W, $\mathbf{x} + \mathbf{y}$ is also in W.
2. For any scalar c and any vector $\mathbf{x}$ in W, $c\mathbf{x}$ is in W.

Proof. If W is a subspace of V, clearly 1 and 2 must be true. Thus, suppose 1 and 2 are true. Then we need to verify that axioms 3 through 10 in the definition of a vector space are also true. Since we have not changed how vectors are added or multiplied by scalars, axioms 3, 4, and 7 through 10 are automatically true. Thus, only axioms 5 and 6 need to be verified.

5. To show that zero is in W, we use the fact that W is assumed to be nonempty. Let $\mathbf{x}$ be any vector in W. Then since W is closed under scalar multiplication we have $\mathbf{0} = 0\mathbf{x}$ and $\mathbf{0}$ must be in W.
6. Let $\mathbf{x}$ be any vector in W, then $(-1)\mathbf{x} = -\mathbf{x}$ must also be in W.

In the following, whenever we wish to indicate that a subspace W equals $S[A]$, for some set A, we will say W is spanned by A or A is a spanning set of W.

EXAMPLE 7 Let W be that subspace of $\mathbb{R}^4$ spanned by the vectors $(1,-1,0,2)$, $(3,0,1,6)$, and $(0,1,1,-1)$. Is the vector $\mathbf{x} = (2,-3,4,0)$ in W?

Solution. If $\mathbf{x}$ is in W, it must be a linear combination of the three vectors whose span is W. Thus, there should exist constants c_1, c_2, and c_3 such that

$$(2,-3,4,0) = c_1(1,-1,0,2) + c_2(3,0,1,6) + c_3(0,1,1,-1)$$

This vector equation has a solution if and only if the following system has a solution.

$$\begin{aligned} c_1 + 3c_2 \qquad &= 2 \\ -c_1 \qquad + c_3 &= -3 \\ c_2 + c_3 &= 4 \\ 2c_1 + 6c_2 - c_3 &= 0 \end{aligned}$$

A few row operations on the augmented matrix of this system show that it is row equivalent to the matrix

$$\begin{bmatrix} 1 & 3 & 0 & 2 \\ 0 & 1 & 0 & 0 \\ 0 & 0 & 1 & 4 \\ 0 & 0 & 0 & -5 \end{bmatrix}$$

The nonhomogeneous system, which has the above matrix as its augmented matrix, has no solution. This means that W does not contain the given vector. ■

EXAMPLE 8 Let $V = \mathbb{R}^2$. Let $W = \{(x, \sin x): x \text{ is a real number}\}$. Show that while W contains **0** and the additive inverse $-\mathbf{x}$ of every vector $\mathbf{x}$ in W, it still is not a subspace.

Solution. To see that $\mathbf{0} = (0,0)$ is in W, we only need observe that $\sin 0 = 0$. Thus $(0, \sin 0) = (0,0)$ is in W. If $\mathbf{x}$ is in W, then $\mathbf{x} = (r, \sin r)$ for some number r. But then

$$-\mathbf{x} = -(r, \sin r) = (-r, -\sin r) = (-r, \sin(-r))$$

must also be in W. To see that W is not a subspace, we note that if W were a subspace, $c(1, \sin 1)$ must be in W for any choice of the scalar c. Since $\sin 1 > 0$, by making c large enough we will have $c \sin 1 > 1$. Since $-1 \leq \sin x \leq 1$ for every x, we cannot have $\sin x = c \sin 1$ for any x. Thus, $c(1, \sin 1) = (c, c\sin 1)$ is not in W. Hence, W is not closed under scalar multiplication, and it cannot be a subspace.

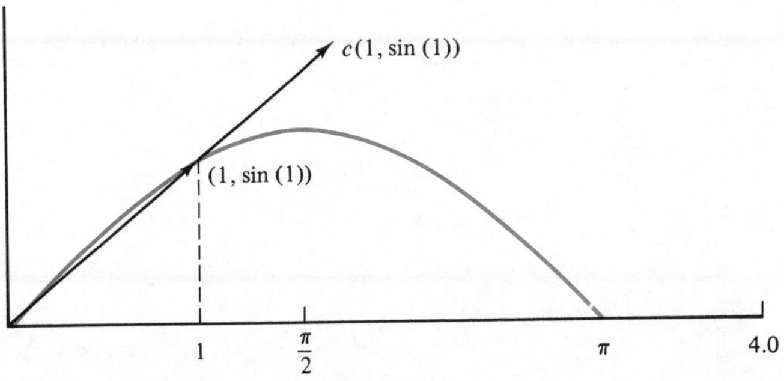

Figure 2.12

■

EXAMPLE 9 Let A be an $m \times n$ matrix. Let $K = \{\mathbf{x}\colon \mathbf{x} \text{ is in } \mathbb{R}^n \text{ and } A\mathbf{x} = \mathbf{0}\}$. Thus K is the solution set of the system of linear homogeneous equations whose coefficient matrix is A. Show that K is a subspace of $\mathbb{R}^n$. Notice that $\mathbb{R}^n$ is being thought of as the set of $n \times 1$ matrices.

Solution. We first note that $\mathbf{0}$ is in K, since $A\mathbf{0} = \mathbf{0}$. Thus K is nonempty. Suppose now that $\mathbf{x}_1$ and $\mathbf{x}_2$ are in K. Then

$$A(\mathbf{x}_1 + \mathbf{x}_2) = A\mathbf{x}_1 + A\mathbf{x}_2 = \mathbf{0} + \mathbf{0} = \mathbf{0}$$

and K is closed under addition. To see that K is closed under scalar multiplication we have, if $\mathbf{x}$ is in K and c is any number:

$$A(c\mathbf{x}) = cA(\mathbf{x}) = c\mathbf{0} = \mathbf{0}$$

Hence, by Theorem 2.6, K is a subspace of $\mathbb{R}^n$. ■

EXAMPLE 10 Let $V = \mathbb{R}^4$ and let $W = \{(x_1, x_2, x_3, x_4)\colon 2x_1 - x_2 + x_3 = 0\}$. Show W is a subspace of $\mathbb{R}^4$.

Solution. Note that W is the solution set of a system of homogeneous equations. Thus, by the previous example W is a subspace. ■

EXAMPLE 11 Show that $F = \left\{ \begin{bmatrix} 1 & -1 \\ 0 & 0 \end{bmatrix} \begin{bmatrix} 1 & 0 \\ 0 & 0 \end{bmatrix} \begin{bmatrix} 0 & 0 \\ 1 & 1 \end{bmatrix} \begin{bmatrix} 0 & 0 \\ 1 & 0 \end{bmatrix} \right\}$ spans M_{22}.

Solution. Let $A = \begin{bmatrix} a_1 & a_2 \\ a_3 & a_4 \end{bmatrix}$ be an arbitrary vector (matrix) in M_{22}. Then we have to find constants c_1, c_2, c_3, and c_4 such that

$$\begin{bmatrix} a_1 & a_2 \\ a_3 & a_4 \end{bmatrix} = c_1 \begin{bmatrix} 1 & -1 \\ 0 & 0 \end{bmatrix} + c_2 \begin{bmatrix} 1 & 0 \\ 0 & 0 \end{bmatrix} + c_3 \begin{bmatrix} 0 & 0 \\ 1 & 1 \end{bmatrix} + c_4 \begin{bmatrix} 0 & 0 \\ 1 & 0 \end{bmatrix}$$

Thus, we have the equations

$$a_1 = c_1 + c_2 \qquad a_2 = -c_1 \qquad a_3 = c_3 + c_4 \qquad a_4 = c_3$$

Hence,

$$\begin{bmatrix} a_1 & a_2 \\ a_3 & a_4 \end{bmatrix} = (-a_2) \begin{bmatrix} 1 & -1 \\ 0 & 0 \end{bmatrix} + (a_1 + a_2) \begin{bmatrix} 1 & 0 \\ 0 & 0 \end{bmatrix} + a_4 \begin{bmatrix} 0 & 0 \\ 1 & 1 \end{bmatrix} + (a_3 - a_4) \begin{bmatrix} 0 & 0 \\ 1 & 0 \end{bmatrix}$$

Thus, $M_{22} = S[F]$. ■

PROBLEM SET 2.3

1. Show that $S[(1,0),(0,1)] = \mathbb{R}^2$.
2. Show that $S[(1,0,0),(0,1,0)]$ equals the x_1, x_2 plane in $\mathbb{R}^3$.
3. Show that $S[(1,0,1),(1,2,-1)] = \{(x_1,x_2,x_3)\colon x_1 - x_2 - x_3 = 0\}$.
4. Show that $S[(1,1)] = \{(x_1,x_2)\colon x_1 - x_2 = 0\}$.
5. Let $A = \{(1,-1,0),(-1,0,1)\}$. Determine which if any of the following vectors is in $S[A]$: $(1,-1,1)$, $(0,-1,1)$, $(2,3,-1)$.
6. Let $A = \{(1,2),(6,3)\}$. Show that $S[A] = \mathbb{R}^2$.
7. Let $\mathbf{x} = (x_1,x_2)$ be any nonzero vector in $\mathbb{R}^2$ Show that $S[\mathbf{x}]$ is a straight line passing through the origin and the point with coordinates (x_1,x_2).
8. What is $S[(0,0)]$?
9. Let V be a vector space. Let W be that subset of V which consists of just the zero vector. Prove the following:
   a. W is a subspace of V.
   b. Show that V is a subspace of itself.
10. Let A be any set of vectors in some vector space V. Show that A is contained in $S[A]$. Show that if H is any subspace of V containing A, then $S[A]$ is also in H. Thus, $S[A]$ is the smallest subspace of V which contains A.
11. Let $V = \mathbb{R}^3$. Let $A_1 = \{(1,1,0)\}$, $A_2 = \{(1,1,0),(0,0,1)\}$. Show that $S[A_1]$ is contained in $S[A_2]$ and describe these two subspaces geometrically.
12. Let $\mathbf{f}(x) = ax$ for some constant a. Show that the graph of $\mathbf{f} = \{(x, \mathbf{f}(x))\colon x$ is any real number$\}$ is a subspace of $\mathbb{R}^2$. Conversely suppose we have a function $\mathbf{f}(x)$ whose graph is a subspace of $\mathbb{R}^2$. Show that $\mathbf{f}(x) = ax$ for some constant a. (Hint: What must a equal?)
13. Let $A = \begin{bmatrix} 1 & 0 & 3 \\ 4 & 4 & 6 \end{bmatrix}$. Let K be the solution set of $A\mathbf{x} = 0$, for $\mathbf{x}$ in $\mathbb{R}^3$. From Example 9 we know that K is a subspace of $\mathbb{R}^3$. Find a vector $\mathbf{x}_0$ such that $S[\mathbf{x}_0] = K$.
14. Let $W = \{(x_1,x_2,x_3)\colon x_1 - x_2 + x_3 = c\}$. For which values of c, if any, is W a subspace of $\mathbb{R}^3$?
15. Example 11 exhibited a spanning set of M_{22} with four vectors. Can you find a spanning set of M_{22} with five vectors? With three vectors?
16. Find spanning sets for each of the following vector spaces:
   a. $V = \{(x_1,x_2,x_3)\colon x_1 + x_2 - 6x_3 = 0\}$
   b. $V = \{\mathbf{p}\colon \mathbf{p}$ is in P_2 and $\mathbf{p}(1) = 0\}$
   c. $V = \{A = [a_{ij}]\colon A$ is in M_{23} and $\Sigma_{j=1}^{3} a_{ij} = 0$ for $i = 1, 2\}$
17. Let V be the vector space of all $n \times n$ matrices. Show that the following subsets of V are subspaces:
   a. $W =$ all $n \times n$ scalar matrices $= \{cI_n\colon c$ is any number$\}$
   b. $W =$ all $n \times n$ diagonal matrices $= \{[d_j\delta_{jk}]\colon d_j$ any number$\}$
   c. $W =$ all $n \times n$ upper triangular matrices
   d. $W =$ all $n \times n$ lower triangular matrices
18. Let W be the set of all $n \times n$ invertible matrices. Show that W is not a subspace of the vector space of all $n \times n$ matrices.

19. Let W by any subspace of the vector space M_{mn} of all $m \times n$ matrices. Show that W^T, that subset of M_{nm} consisting of the transposes of all the matrices in W, is a subspace of M_{nm}.

20. Let P_2 be all polynomials of degree 2 or less. Let $W = \{a + bt$: a and b arbitrary real numbers$\}$. Is W a subspace of P_2?

21. Let V be a vector space. Let W_1 and W_2 be any two subspaces of V. Let $W = W_1 \cap W_2 = \{\mathbf{x}$: $\mathbf{x}$ belongs to both W_1 and $W_2\}$. Show that W is a subspace of V.

22. Let V, W_1, and W_2 be as in problem 21. Let $W = W_1 \cup W_2 = \{\mathbf{x}$: $\mathbf{x}$ belongs to W_1 or $W_2\}$. Show that W need not be a subspace of V.

23. Let V, W_1, and W_2 be as in problem 21. Define $W_1 + W_2$ as that subset of V which consists of all possible sums of the vectors in W_1 and W_2, that is

$$W_1 + W_2 = \{\mathbf{u} + \mathbf{v}: \mathbf{u} \text{ and } \mathbf{v} \text{ any vectors in } W_1 \text{ and } W_2\}$$

Show $W_1 + W_2$ is a subspace of V and that any other subspace of V which contains W_1 and W_2 must also contain $W_1 + W_2$.

24. Let $V = C[a,b]$. Let W be that subset of V which consists of all polynomials. Is W a subspace of V?

25. Let $W = \{\ldots, -2, -1, 0, 1, 2, \ldots\}$. Is W a subspace of $\mathbb{R}^1$?

26. Let $W = \{(x_1,x_2)$: x_1 or x_2 equals zero$\}$. Let $W_1 = \{(x_1,0)\}$ and $W_2 = \{(0,x_2)\}$. Thus, $W = W_1 \cup W_2$. Show that both W_1 and W_2 are subspaces of $\mathbb{R}^2$, and that W is not a subspace.

27. Let $W = \{(x_1,x_2): x_1^2 + x_2^2 \le c\}$. For which values of c is W a subspace of $\mathbb{R}^2$?

28. Let $W_1 = \{\mathbf{p}$: $\mathbf{p}$ is in P_2 and $\mathbf{p}(1) = 0\}$. Let $W_2 = \{\mathbf{p}$: $\mathbf{p}$ is in P_2 and $\mathbf{p}(2) = 0\}$.
a. Find a spanning set for $W_1 \cap W_2$.
b. Find spanning sets for W_1 and W_2 each of which contains the spanning set you found in part a.

29. Example 9 shows that the solution sets of homogeneous systems of equations are subspaces. Find spanning sets for each of the solution spaces of the following systems of equations:
a. $2x_1 + 6x_2 - x_3 = 0$
b. $2x_1 + 6x_2 - x_3 = 0,\ x_2 + 4x_3 = 0$
c. $2x_1 + 6x_2 - x_3 = 0,\ x_2 + 4x_3 = 0,\ x_1 + x_2 = 0$

30. Let $A = \begin{bmatrix} 2 & 6 & 1 \\ -3 & 1 & 0 \end{bmatrix}$. Let $V = \{B: B \in M_{34} \text{ and } AB = 0_{24}\}$.
a. Show V is a subspace of M_{34}.
b. Find a spanning set for V.

2.4 LINEAR INDEPENDENCE

In this section we discuss what it means to say that a set of vectors is linearly independent. We encourage the reader to go over this material several times. The ideas discussed here are important, but hard to digest, thus needing rumination.

Consider the two vectors $\mathbf{x} = (1,0,-1)$ and $\mathbf{y} = (0,1,1)$ in $\mathbb{R}^3$. They have the following property: Any vector $\mathbf{z}$ in their span can be written in *only one way* as a linear combination of these two vectors. Let's check the details of this. Suppose there are constants c_1, c_2, c_3, and c_4 such that

$$\mathbf{z} = c_1\mathbf{x} + c_2\mathbf{y} = c_3\mathbf{x} + c_4\mathbf{y} \tag{2.2}$$

What we want to show is that $c_1 = c_3$ and $c_2 = c_4$; that is, the coefficients of $\mathbf{x}$ must be equal and the coefficients of $\mathbf{y}$ must also be equal. The two expressions for $\mathbf{z}$ lead to the equation

$$\mathbf{0} = \mathbf{z} - \mathbf{z} = (c_1 - c_3)\mathbf{x} + (c_2 - c_4)\mathbf{y} \tag{2.3}$$

Thus, we need to show that anytime $\mathbf{0} = a\mathbf{x} + b\mathbf{y}$ we must have $a = 0 = b$. Suppose now that

$$\mathbf{0} = (0,0,0) = a(1,0,-1) + b(0,1,1) = (a,b,-a+b)$$

Clearly this can occur only if $a = 0 = b$. This property of the two vectors $\mathbf{x}$ and $\mathbf{y}$ is given a special name.

Definition 2.7. A set of vectors $\{\mathbf{x}_k\colon k = 1, \ldots, n\}$ in a vector space V is said to be linearly independent if whenever

$$\mathbf{0} = c_1\mathbf{x}_1 + c_2\mathbf{x}_2 + \cdots + c_n\mathbf{x}_n = \sum_{k=1}^{n} c_k x_k$$

then

$$c_1 = c_2 = \cdots = c_n = 0$$

That is, the zero vector can be written in only one way as a linear combination of these vectors.

A set of vectors is said to be *linearly dependent* if it is not linearly independent. It is clear that a set of vectors $\{\mathbf{x}_k\colon k = 1, \ldots, n\}$ is linearly dependent only if there are constants $c_1, c_2, \ldots, c_n$ *not all zero* such that $\mathbf{0} = c_1\mathbf{x}_1 + c_2\mathbf{x}_2 + \cdots + c_n\mathbf{x}_n$.

EXAMPLE 1 Show that the vectors $\{(1,0,1), (2,0,3), (0,0,1)\}$ are linearly dependent.

Solution. We need to find three constants c_1, c_2, c_3 not all zero such that

$$c_1(1,0,1) + c_2(2,0,3) + c_3(0,0,1) = (0,0,0)$$

Thus we have to find a nontrivial solution to the following system of equations:

$$\begin{aligned} c_1 + 2c_2 \phantom{{}+c_3} &= 0 \\ c_1 + 3c_2 + c_3 &= 0 \end{aligned}$$

This system has many nontrivial solutions. A particular one is $c_1 = 2$, $c_2 = -1$, and $c_3 = 1$. Since we have found a nontrivial linear combination of these vectors, namely, $2(1,0,1) - (2,0,3) + (0,0,1)$, which equals zero, they are linearly dependent. ■

EXAMPLE 2 Determine whether the set $\{(1,0,-1,1,2), (0,1,1,2,3), (1,1,0,1,0)\}$ is linearly dependent or independent.

Solution. We have to decide whether or not there are constants c_1, c_2, c_3 not all zero such that

$$(0,0,0,0,0) = c_1(1,0,-1,1,2) + c_2(0,1,1,2,3) + c_3(1,1,0,1,0)$$

The linear system that arises from this is

$$\begin{aligned} c_1 + c_3 &= 0 \\ c_2 + c_3 &= 0 \\ -c_1 + c_2 &= 0 \\ c_1 + 2c_2 + c_3 &= 0 \\ 2c_1 + 3c_2 &= 0 \end{aligned}$$

The only solution is the trivial one, $c_1 = c_2 = c_3 = 0$. Thus the vectors are linearly independent. ■

To say that a set of vectors is linearly independent is to say that the zero vector can be written in only one way as a linear combination of these vectors; that is, every coefficient c_k must equal zero. The following theorem shows that even more is true for a linearly independent set of vectors.

Theorem 2.7. If $A = \{\mathbf{x}_k \colon k = 1, \ldots, p\}$ is a linearly independent set of vectors, then every vector in $S[A]$ can be written in only one way as a linear combination of the vectors in A.

Proof. Let $\mathbf{y}$ be any vector in $S[A]$. Suppose that we have

$$\mathbf{y} = \sum_{k=1}^{p} a_k \mathbf{x}_k = \sum_{k=1}^{p} b_k \mathbf{x}_k$$

We wish to show that $a_k = b_k$ for each k. Subtracting $\mathbf{y}$ from itself we have

$$\mathbf{0} = \mathbf{y} - \mathbf{y} = \sum_{k=1}^{p} a_k \mathbf{x}_k - \sum_{k=1}^{p} b_k \mathbf{x}_k = \sum_{k=1}^{p} (a_k - b_k)\mathbf{x}_k$$

Since the set A is linearly independent we must have

$$a_1 - b_1 = 0 \qquad a_2 - b_2 = 0 \qquad \cdots \qquad a_p - b_p = 0$$

Thus we can write any vector in $S[A]$ as a linear combination of the linearly independent vectors $\mathbf{x}_k$ in one and only one way.

The unique constants a_k that are associated with any vector $\mathbf{x}$ in $S[A]$ are given a special name.

Definition 2.8. Given a linearly independent set of vectors $A = \{\mathbf{x}_k\colon k = 1, \ldots, n\}$. If $\mathbf{x} = \Sigma_{k=1}^{n} a_k \mathbf{x}_k$, the coefficients a_k, $1 \le k \le n$, are called the coordinates of $\mathbf{x}$ with respect to A. Note that there is an explicit ordering of the vectors in the set A.

EXAMPLE 3 Show that the set $A = \{(1,1,-1), (0,1,0), (-1,0,2)\}$ is linearly independent, and then find the coordinates of $(2,-4,6)$ with respect to A.

Solution. To see that A is linearly independent suppose

$$c_1(1,1,-1) + c_2(0,1,0) + c_3(-1,0,2) = (0,0,0)$$

The system of equations that is equivalent to this vector equation, namely,

$$\begin{aligned} c_1 \quad\quad - c_3 &= 0 \\ c_1 + c_2 \quad\quad &= 0 \\ -c_1 \quad\quad + 2c_3 &= 0 \end{aligned}$$

has only the trivial solution. Thus A is a linearly independent set of vectors. We next want to write, if possible, $(2,-4,6)$ as a linear combination of the vectors in A. Thus we want constants c_k, $k = 1, 2, 3$, such that

$$(2,-4,6) = c_1(1,1,-1) + c_2(0,1,0) + c_3(-1,0,2)$$

The associated system is

$$\begin{aligned} c_1 \quad\quad - c_3 &= 2 \\ c_1 + c_2 \quad\quad &= -4 \\ -c_1 \quad\quad + 2c_3 &= 6 \end{aligned}$$

and its unique solution is $c_1 = 10$, $c_2 = -14$, $c_3 = 8$. Thus $(2,-4,6) = 10(1,1,-1) - 14(0,1,0) + 8(-1,0,2)$ and the coordinates of $\mathbf{x} = (2,-4,6)$ with respect to A are $[10,-14,8]$. ■

EXAMPLE 4 Show that the set F is a linearly independent subset of M_{22}.

$$F = \left\{\begin{bmatrix} 1 & -1 \\ 0 & 0 \end{bmatrix}\begin{bmatrix} 1 & 0 \\ 0 & 0 \end{bmatrix}\begin{bmatrix} 0 & 0 \\ 1 & 1 \end{bmatrix}\begin{bmatrix} 0 & 0 \\ 1 & 0 \end{bmatrix}\right\}$$

Solution. Suppose there are constants $c_1, \ldots, c_4$ such that

$$\begin{bmatrix} 0 & 0 \\ 0 & 0 \end{bmatrix} = c_1\begin{bmatrix} 1 & -1 \\ 0 & 0 \end{bmatrix} + c_2\begin{bmatrix} 1 & 0 \\ 0 & 0 \end{bmatrix} + c_3\begin{bmatrix} 0 & 0 \\ 1 & 1 \end{bmatrix} + c_4\begin{bmatrix} 0 & 0 \\ 1 & 0 \end{bmatrix}$$

Thus, since the equations

$$c_1 + c_2 = 0 \qquad -c_1 = 0 \qquad c_3 + c_4 = 0 \qquad c_3 = 0$$

have only the trivial solution, F is linearly independent. From Example 11 in Section 2.3 we know that $S[F] = M_{22}$, and that

$$\begin{bmatrix} 1 & 0 \\ 0 & 0 \end{bmatrix} = 0\begin{bmatrix} 1 & -1 \\ 0 & 0 \end{bmatrix} + 1\begin{bmatrix} 1 & 0 \\ 0 & 0 \end{bmatrix} + 0\begin{bmatrix} 0 & 0 \\ 1 & 1 \end{bmatrix} + 0\begin{bmatrix} 0 & 0 \\ 1 & 0 \end{bmatrix}$$

Thus, the coordinates of $\begin{bmatrix} 1 & 0 \\ 0 & 0 \end{bmatrix}$ with respect to F are $[0,1,0,0]$. ■

EXAMPLE 5 Show that the set $F = \{1 + t, t^2, t - 2\}$ is a linearly independent subset of P_2.

Solution. Suppose there are constants c_1, c_2, and c_3 such that

$$0 = c_1(1 + t) + c_2(t^2) + c_3(t - 2)$$

Then these constants satisfy the equations

$$c_1 - 2c_3 = 0 \qquad c_1 + c_3 = 0 \qquad c_2 = 0$$

Since the only solution to this system is the trivial one, F is a linearly independent subset of P_2. ■

Theorem 2.8. A set of vectors $\{\mathbf{x}_k : k = 1, \ldots, n\}$ is linearly dependent if and only if one of its vectors can be written as a linear combination of the remaining $n - 1$ vectors.

Proof. Suppose one of the vectors, say $\mathbf{x}_1$, can be written as a linear combination of the others. Then we have

$$\mathbf{x}_1 = c_2\mathbf{x}_2 + c_3\mathbf{x}_3 + \cdots + c_n\mathbf{x}_n$$

for some constants c_k. But then

$$\mathbf{0} = \mathbf{x}_1 - c_2\mathbf{x}_2 - \cdots - c_n\mathbf{x}_n$$

and we have found a nontrivial (the coefficient of x_1 is 1) linear combination of these n vectors that equals the zero vector. Thus, this set of vectors is linearly dependent. Suppose now that there is a nontrivial linear combination of these vectors that equals $\mathbf{0}$; that is

$$\mathbf{0} = c_1\mathbf{x}_1 + \cdots + c_n\mathbf{x}_n$$

and not all the constants c_k are zero. We may suppose without loss of generality that c_1 is not zero (merely rename the vectors). Solving the above equation for $\mathbf{x}_1$, we have

$$\mathbf{x}_1 = \frac{1}{c_1}(-c_2\mathbf{x}_2 - \cdots - c_n\mathbf{x}_n)$$

Hence, we have written one of the vectors in our linearly dependent set as a linear combination of the remaining vectors.

If a linearly dependent set contains exactly two vectors, then one of them must be a multiple of the other. In $\mathbb{R}^3$, three nonzero vectors are linearly dependent if and only if they are coplanar, i.e., one of them lies in the plane spanned by the other two. For example, the three vectors (1,0,1), (2,0,3), and (0,0,1), as we saw in Example 1, are linearly dependent and all three of them lie in the plane generated by (1,0,1) and (2,0,3); that is, the plane $x_2 = 0$.

PROBLEM SET 2.4

1. Show that (1,0,1) is in the span of {(2,0,0), (0,0,−1)}.
2. Show that $S[(2,0,0), (0,0,-1)] = S[(2,0,0), (0,0,-1), (1,0,1)]$.
3. Let A be a collection of vectors in a vector space V. Suppose that $\mathbf{x}$ is any vector in $S[A]$. Show $S[A,\mathbf{x}] = S[A]$.
4. Determine whether the following sets of vectors are linearly independent.
   a. {(1,3), (1,−4), (8,12)}
   b. {(1,1,4), (8,−3,2), (0,2,1)}
   c. {(1,1,2,−3), (−2,3,0,4), (8,−7,4,−18)}
5. Which of the following sets of vectors is linearly dependent?
   a. {(1,2,3,0), (−6,1,0,0), (1,2,0,3)}
   b. {(1,−1,6), (4,2,0), (1,1,1), (−1,0,−1)}
   c. {(2,1,4), (1,1,1), (−1,1,1)}
6. Show that any set of vectors containing the zero vector must be linearly dependent.
7. Suppose that A is a linearly dependent set of vectors and B is any set containing A. Show that B must also be linearly dependent.
8. Show that any nonempty subset of a linearly independent set is linearly independent.
9. Show that any set of four or more vectors in $\mathbb{R}^3$ must be linearly dependent.
10. Let $A = \{(0,1,1), (1,0,1), (1,1,0)\}$. Is A linearly dependent? $S[A] = ?$
11. Let $A = \{1 + t, 1 - t^2, t^2\}$. Is A a linearly independent subset of P_2? $S[A] = ?$
12. Let $A = \{\sin t, \cos t, \sin(t + \pi)\}$. Is A a linearly independent subset of $C[0,1]$? Does A span $C[0,1]$?
13. Let $A = \{(x_1,x_2)\colon x_1^2 + x_2^2 = 1\}$. Is A a linearly independent subset of $\mathbb{R}^2$? Is A a spanning subset of $\mathbb{R}^2$?
14. Show that $F = \{1 + t, t^2 - t, 2\}$ is a linearly independent subset of P_2. Show $S[F] = P_2$.
15. Show that $\{\sin t, \sin 2t, \cos t\}$ is a linearly independent subset of $C[0,1]$. Does it span $C[0,1]$?
16. Let $V = \{\mathbf{p}(t)\colon \mathbf{p}$ is in P_2 and $\int_0^1 \mathbf{p}(t)\,dt = 0\}$. Find a spanning set of V that has exactly two vectors in it. Show that your set is also linearly independent.

17. Let $V = \{\mathbf{p}(t)$: $\mathbf{p}$ is in P_3 and $\mathbf{p}'(0) = 0$, $\mathbf{p}(1) - \mathbf{p}(0) = 0\}$. Find a spanning set of V that has exactly two vectors. Is the set also linearly independent?

18. Let $V = \{A$: A is in M_{23} and $AB = AC = 0_{22}\}$, where

$$B = \begin{bmatrix} 1 & 1 \\ 1 & -1 \\ 1 & 1 \end{bmatrix} \qquad C = \begin{bmatrix} 1 & -1 \\ -1 & 1 \\ 1 & -1 \end{bmatrix}$$

Find a linearly independent spanning set of V. How many vectors are in your set?

19. Let A be a subset of a vector space V. We say that A is linearly independent if, whenever $\{\mathbf{x}_1, \ldots, \mathbf{x}_p\}$ is a finite subset of A and $c_1\mathbf{x}_1 + \cdots + c_p\mathbf{x}_p = 0$, then each $c_k = 0$.

a. Show that if A is a finite set of vectors to begin with, then Definition 2.7 and the above definition are equivalent.

b. Show Theorem 2.7 is also true using this definition of linear independence. Note: $S[A]$ is the collection of all possible finite linear combinations of the vectors in A.

20. Let $V = C[0,1]$, the space of real-valued continuous functions defined on $[0,1]$. Let $A = \{1, t, \ldots, t^n, \ldots\}$.

a. Show that A is a linear independent set as defined in problem 19.

b. What is $S[A]$?

21. Let V be the set of all polynomials with only even powers of t (constants have even degree). Thus $1 + t$ is not in V, $1 + t + t^2$ is not in V, but $1 + t^2$ and $t^4 + 5t^8$ are in V. Let $A = \{1, t^2, t^4, \ldots, t^{2n}, \ldots\}$.

a. Show that V is a vector space and that A is a linearly independent subset of V.

b. $S[A] = ?$

2.5 BASES

In the preceding two sections we discussed two different types of subsets of a vector space V. In Section 2.3 we covered spanning sets, i.e., subsets A with the property that any vector in V can be written as a linear combination of vectors in A. Section 2.4 was concerned with sets that are linearly independent. This section discusses properties of sets that are both linearly independent and spanning. Such subsets are called bases of V. Besides their defining properties all bases of a vector space share one other common feature: They have exactly the same number of vectors in them. That is, if $A = \{\mathbf{x}_1, \ldots, \mathbf{x}_p\}$ and $B = \{\mathbf{y}_1, \ldots, \mathbf{y}_n\}$ are linearly independent and they both span V, then $n = p$. Thus the number of vectors in a basis of V depends only on V and not on which linearly independent spanning set we happen to have chosen.

Before proving this useful and interesting fact, we prove two lemmas. These lemmas aid us in the construction of bases.

LEMMA 2.1 Suppose $A = \{\mathbf{x}_1, \ldots, \mathbf{x}_n\}$ is linearly dependent. Then it is possible to remove one of the vectors $\mathbf{x}_k$ from A without changing $S[A]$; that is, $S[\mathbf{x}_1, \ldots, \mathbf{x}_{k-1}, \mathbf{x}_{k+1}, \ldots, \mathbf{x}_n] = S[A]$.

Proof. Since A is linearly dependent, there are constants c_k, $1 \le k \le n$, not all zero, such that

$$\mathbf{0} = c_1\mathbf{x}_1 + \cdots + c_n\mathbf{x}_n \tag{2.4}$$

Suppose, for convenience, that $c_1 \neq 0$. Then we may solve (2.4) for x_1 and obtain

$$\mathbf{x}_1 = -\frac{1}{c_1}[c_2\mathbf{x}_2 + \cdots + c_n\mathbf{x}_n] \tag{2.5}$$

We now claim that $S[A] = S[\mathbf{x}_2, \ldots, \mathbf{x}_n]$. Let $B = \{\mathbf{x}_2, \ldots, \mathbf{x}_n\}$. Since B is a subset of A, we have $S[B] \subseteq S[A]$. Thus, to show these two subspaces of V are equal, we only need show that any vector in $S[A]$ is also in $S[B]$. To this end, let $\mathbf{y}$ be in $S[A]$. Then there are constants a_k, $1 \le k \le n$, such that

$$\begin{aligned}
\mathbf{y} &= a_1\mathbf{x}_1 + a_2\mathbf{x}_2 + \cdots + a_n\mathbf{x}_n \\
&= a_1\left(-\frac{1}{c_1}\right)[c_2\mathbf{x}_2 + \cdots + c_n\mathbf{x}_n] + a_2\mathbf{x}_2 + \cdots + a_n\mathbf{x}_n \\
&= \left(a_2 - \frac{a_1c_2}{c_1}\right)\mathbf{x}_2 + \cdots + \left(a_n - \frac{a_1c_n}{c_1}\right)\mathbf{x}_n
\end{aligned}$$

Since $\mathbf{y}$ can be written as a linear combination of the vectors in B, $\mathbf{y}$ is in $S[B]$.

EXAMPLE 1 Let $A = \{(1,1), (3,2), (1,-1)\}$. It's easy to see that the span of A is $\mathbb{R}^2$. Moreover since

$$(0,0) = 5(1,1) - 2(3,2) + (1,-1)$$

the set A is linearly dependent. Thus we may remove one of the three vectors from A and still have a set with the same span, i.e., $\mathbb{R}^2$. Which vector can we delete? Since each of the vectors in the linear combination above has a nonzero coefficient, we may delete any one of the three. Let's remove the second one. Then we know that

$$S[(1,1), (1,-1)] = \mathbb{R}^2$$

We note that these last two vectors are linearly independent. In fact this is one way in which a linearly independent spanning set of a vector space can be constructed. That is, from a linearly dependent spanning set remove, one by one, those vectors that depend linearly on the remaining vectors, until a linearly independent set is obtained. ■

Lemma 2.2. Let $A = \{\mathbf{x}_1, \ldots, \mathbf{x}_n\}$ be a linearly independent set. Suppose that the vector $\mathbf{y}$ is not in $S[A]$. Then the set obtained by adding $\mathbf{y}$ to A is linearly independent.

Proof. Suppose there are constants $c, c_1, \ldots, c_n$ such that

$$\mathbf{0} = c\mathbf{y} + c_1\mathbf{x}_1 + \cdots + c_n\mathbf{x}_n \qquad (2.6)$$

We wish to show that each of these $n + 1$ constants is zero. Suppose that c is not zero. Then we may solve (2.6) for $\mathbf{y}$ as a linear combination of the vectors in A, contradicting our assumption that $\mathbf{y}$ is not in the span of A. Thus $c = 0$. But then (2.6) reduces to a linear combination, of the vectors in A, which equals the zero vector. Since A is linearly independent, each of the constants $c_k = 0$. Thus, the set $\{\mathbf{y}, \mathbf{x}_1, \ldots, \mathbf{x}_n\}$ is linearly independent.

This last lemma is also used to construct linearly independent spanning subsets of a vector space. We start out with a linearly independent subset. If its span is not V, we add to it a vector $\mathbf{y}$ not in its span. If the new set spans V, we're done. If not, another vector is added. This process continues until we reach a spanning set.

For most of the vector spaces we've seen either of the above procedures is successful in constructing bases. There are, however, some vector spaces for which this does not work. Those which are not finite dimensional. See Definition 2.10.

EXAMPLE 2 Let $A = \{(1,1,-1), (0,1,1)\}$. Find a vector $\mathbf{y}$ such that when $\mathbf{y}$ is added to A, we have a linearly independent spanning subset of $\mathbb{R}^3$.

$$\begin{aligned} S[A] &= \{c_1(1,1,-1) + c_2(0,1,1)\} \\ &= \{(c_1, c_1 + c_2, c_2 - c_1)\} \\ &= \{(x_1, x_2, x_3)\colon x_2 - 2x_1 = x_3\} \end{aligned}$$

From this last description of $S[A]$, it is easy to find vectors that are not in the span of A. For example, $(1,0,0)$ is one such vector. Since A is linearly independent, Lemma 2.2 now tells us that the set

$$\{(1,0,0), (1,1,-1), (0,1,1)\}$$

is also linearly independent. The reader may easily verify, and should do so, that this set does indeed span $\mathbb{R}^3$. ■

Definition 2.9. Let V be a vector space. A subset B of V is called a basis of V if it satisfies:

1. B is linearly independent.
2. B is a spanning set for V; that is, every vector in V can be written as a linear combination of the vectors in B.

It can be shown that every vector space containing more than the zero vector has a basis. We shall not prove this theorem for arbitrary spaces but instead will exhibit a basis for most of the vector spaces discussed in this text.

EXAMPLE 3

a. For $\mathbb{R}^1$, the vector space of real numbers, any set consisting of just one nonzero number is a basis.

b. In $\mathbb{R}^2$, let $\mathbf{e}_1 = (1,0)$, $\mathbf{e}_2 = (0,1)$. The set $\{\mathbf{e}_1, \mathbf{e}_2\}$ is a basis of $\mathbb{R}^2$ and is the standard basis of $\mathbb{R}^2$.

c. If $V = \mathbb{R}^n$, define $\mathbf{e}_j$, $1 \le j \le n$ by

$$\begin{aligned} \mathbf{e}_1 &= (1,0, \ldots ,0) \\ \mathbf{e}_2 &= (0,1,0, \ldots ,0) \\ &\vdots \\ \mathbf{e}_j &= (0, \ldots , 0,1,0, \ldots ,0), \text{ a 1 is in the } j\text{th slot.} \end{aligned}$$

The set $\{\mathbf{e}_j : 1 \le j \le n\}$ is a basis for $\mathbb{R}^n$, as the reader can easily show. This particular set is the standard basis of $\mathbb{R}^n$. ■

Theorem 2.9. Let V be a vector space that has a basis consisting of n vectors. Then:

a. Any set with more than n vectors is linearly dependent.
b. Any spanning set has at least n vectors.
c. Any linearly independent set has at most n vectors.
d. Every basis of V contains exactly n vectors.

Proof. Part a. Suppose that $A = \{\mathbf{f}_1, \ldots , \mathbf{f}_n\}$ is the given basis of V with n vectors. Let $B = \{\mathbf{x}_1, \ldots , \mathbf{x}_n, \mathbf{x}_{n+1}\}$ be any subset of V with $n + 1$ vectors. We want to show that B is linearly dependent. Since A is a basis of V, it is a spanning set. Hence, there are constants a_{jk} such that

$$\mathbf{x}_k = a_{1k}\mathbf{f}_1 + \cdots + a_{nk}\mathbf{f}_n$$

We want to find constants c_k, $1 \le k \le n + 1$, not all zero such that

$$\begin{aligned} \mathbf{0} &= c_1\mathbf{x}_1 + \cdots + c_{n+1}\mathbf{x}_{n+1} = \sum_{k=1}^{n+1} c_k x_k \\ &= \sum_{k=1}^{n+1} c_k \sum_{j=1}^{n} a_{jk}\mathbf{f}_j = \sum_{j=1}^{n} \left(\sum_{k=1}^{n+1} c_k a_{jk} \right) \mathbf{f}_j \end{aligned}$$

If we pick the constants c_k in such a manner that each of the coefficients multiplying $\mathbf{f}_j$ is zero, this linear combination of the vectors in B will equal the zero vector. In other words, we wish to find a nontrivial solution of the following system of equations:

$$a_{j1}c_1 + a_{j2}c_2 + \cdots + a_{j(n+1)}c_{n+1} = 0$$

for $j = 1, 2, \ldots, n$. This is a homogeneous system of equations with more unknowns ($n + 1$) than equations (n). Therefore, there is a nontrivial solution, and the set B is linearly dependent. Notice that we did not use the fact that A is also linearly independent. The only hypothesis needed to show the linear dependence of any set with $n + 1$ vectors was the existence of a spanning set with fewer than $n + 1$ vectors.

Part b. Let $B = \{\mathbf{x}_1, \ldots, \mathbf{x}_m\}$ be any spanning set of V. We wish to show that B must contain at least n elements. Suppose it doesn't, i.e., m is less than n. But then by part a, since B is a spanning set, any set with more than m vectors must be linearly dependent. In particular the set A would have to be linearly dependent. This contradicts the assumption that A is linearly independent.

Part c. Suppose $B = \{\mathbf{x}_1, \ldots, \mathbf{x}_p\}$ is linearly independent; then clearly part a implies that $p \leq n$. That is, any linearly independent set of vectors cannot have more than n vectors in it.

Part d. Let $A_1 = \{\mathbf{g}_1, \ldots, \mathbf{g}_p\}$ be any other basis of V. By part b, since A_1 is spanning, we must have $p \geq n$. But reversing the roles of A and A_1, we have by the same reasoning that $p \leq n$. Thus every basis of V must have the same number of vectors.

Definition 2.10. Let V be a vector space. We say that the dimension of V is n, $\dim(V) = n$, if V has a basis of n vectors. If V is the vector space that consists of the zero vector only, we define the dimension of V to be 0. If for any n, V has a set of n linearly independent vectors, we say that the dimension of V is infinite.

Most of the vector spaces we discuss in this book will have a basis consisting of a finite number of vectors. There are, however, a large number of vector spaces that do not have such a basis. Two examples of such spaces are the set of all polynomials, and the set of all real-valued continuous functions defined on some interval of real numbers, i.e., $C[a,b]$. Both of these spaces are infinite-dimensional.

Example 3 exhibited a basis of $\mathbb{R}^n$ that contained n vectors. Theorem 2.9 now tells us that any subset of $\mathbb{R}^2$ with more than two vectors must be linearly dependent. We also have that $\dim(\mathbb{R}^n) = n$, $n = 1, 2, \ldots$.

EXAMPLE 4

a. Show that $\dim(P_n) = n + 1$. Since P_n consists of all polynomials with real coefficients of degree less than or equal to n, we have

$$P_n = \{a_0 + a_1 t + \cdots + a_n t^n\colon a_k \text{ arbitrary real numbers}\}$$

We claim that the set $A = \{1, t, t^2, \ldots, t^n\}$ is a basis for P_n. It is clear that any polynomial of degree less than or equal to n can be written as a linear combination of the vectors in A. We leave it to the reader to show that A is also linearly independent. Since A has $n+1$ vectors, dim $P_n = n+1$.

b. The vector space M_{mn}, consisting of all $m \times n$ matrices, has dimension equal to mn. The reader is asked in the problems to show that the matrices E_{jk}, $1 \le j \le m$, $1 \le k \le n$, which contain a 1 in the j, k position and zeros everywhere else, form a basis for M_{mn}. ■

EXAMPLE 5 Consider the following system of homogeneous equations:

$$\begin{aligned} 2x_1 - x_3 + x_4 &= 0 \\ x_2 + x_4 &= 0 \end{aligned} \tag{2.7}$$

Find a basis for the solution space.

Solution. The coefficient matrix $\begin{bmatrix} 2 & 0 & -1 & 1 \\ 0 & 1 & 0 & 1 \end{bmatrix}$ is row equivalent to $\begin{bmatrix} 1 & 0 & -\frac{1}{2} & \frac{1}{2} \\ 0 & 1 & 0 & 1 \end{bmatrix}$. Thus every solution to (2.7) must satisfy

$$x_1 = \frac{x_3 - x_4}{2}$$

$$x_2 = -x_4$$

We have two free parameters x_3 and x_4 which we set equal to 1 and 0 to get one solution. To get a second solution, which is linearly independent of the first, we set x_3 and x_4 equal to 0 and 1, respectively. Thus we have the set $S = \{(\frac{1}{2},0,1,0),(-\frac{1}{2},-1,0,1)\}$ which is contained in the solution space of (2.7). Clearly S is linearly independent. [The last two slots look like (1,0) and (0,1).] Moreover, every solution of (2.7) can be written as a linear combination of these two vectors; for suppose (x_1,x_2,x_3,x_4) satisfies (2.7). Then we have

$$\begin{aligned} (x_1,x_2,x_3,x_4) &= \left(\frac{x_3 - x_4}{2}, -x_4, x_3, x_4\right) \\ &= x_3(\tfrac{1}{2},0,1,0) + x_4(-\tfrac{1}{2},-1,0,1) \end{aligned}$$

Thus, S is a basis for the solution space of (2.7), and the dimension of this subspace of $\mathbb{R}^4$ is 2. ■

EXAMPLE 6 Consider the system of linear equations

$$\begin{aligned} x_1 - 2x_2 + x_3 &= 0 \\ x_2 - x_3 &= 0 \\ 2x_1 + x_2 + x_3 &= 0 \end{aligned}$$

Find the dimension of the solution space of this system.

Solution. The coefficient matrix

$$\begin{bmatrix} 1 & -2 & 1 \\ 0 & 1 & -1 \\ 2 & 1 & 1 \end{bmatrix}$$

of this system is row equivalent to

$$\begin{bmatrix} 1 & -2 & 1 \\ 0 & 1 & -1 \\ 0 & 0 & 4 \end{bmatrix}$$

This second system has only the trivial solution. Therefore, the solution space K consists of just the zero vector and we have $\dim(K) = 0$. ■

EXAMPLE 7 Let $V = \mathbb{R}^3$. Let $W = \{(x_1,x_2,x_3): x_3 = 2x_1 + x_2\}$. Find a basis for W and then extend it to a basis for V; i.e., add enough vectors to the basis of W to get a basis for V.

Solution. Since $W = \{(x_1,x_2,2x_1 + x_2)$: x_1 and x_2 arbitrary$\}$, there are two free parameters x_1 and x_2. Let's alternately set them equal to 0 and 1 to get a basis for W. This gives us

$$\mathbf{f}_1 = (1,0,2): x_1 = 1, x_2 = 0$$
$$\mathbf{f}_2 = (0,1,1): x_1 = 0, x_2 = 1$$

Both $\mathbf{f}_1$ and $\mathbf{f}_2$ belong to W, and they are linearly independent. To see that they span W, let $\mathbf{x}$ be in W. Then for some x_1 and x_2

$$\begin{aligned} \mathbf{x} &= (x_1,x_2,2x_1 + x_2) = (x_1,0,2x_1) + (0,x_2,x_2) \\ &= x_1(1,0,2) + x_2(0,1,1) = x_1\mathbf{f}_1 + x_2\mathbf{f}_2 \end{aligned}$$

Since the set $\{\mathbf{f}_1,\mathbf{f}_2\}$ is linearly independent and spans W, it is a basis. We now face the problem of extending this basis to one for $\mathbb{R}^3$. Since $\dim(\mathbb{R}^3) = 3$, we need one additional vector. Thus, we want to find a vector $\mathbf{y}$ such that the set $\{\mathbf{f}_1,\mathbf{f}_2,\mathbf{y}\}$ is linearly independent. Let $\mathbf{y} = (1,1,0)$. One easily sees that $\mathbf{y}$ is not in W. Hence, by Lemma 2.2, we may conclude that $\{\mathbf{f}_1,\mathbf{f}_2,\mathbf{y}\}$ is linearly independent. We leave it for the reader to show that this set spans $\mathbb{R}^3$. (Hint: Suppose not. Then we should be able to find a linearly independent subset of $\mathbb{R}^3$ with four vectors in it! See c of Theorem 2.9.) ■

Figure 2.13 shows $\mathbf{f}_1,\mathbf{f}_2$, and $\mathbf{y}$. It is clear that $W = S[\mathbf{f}_1,\mathbf{f}_2]$ is the plane in $\mathbb{R}^3$ passing through the three points (1,0,2), (0,1,1), and (0,0,0). The vector $\mathbf{y} = (1,1,0)$ clearly does not lie in this plane.

The preceding example used Theorem 2.10 to infer that a certain set is a basis.

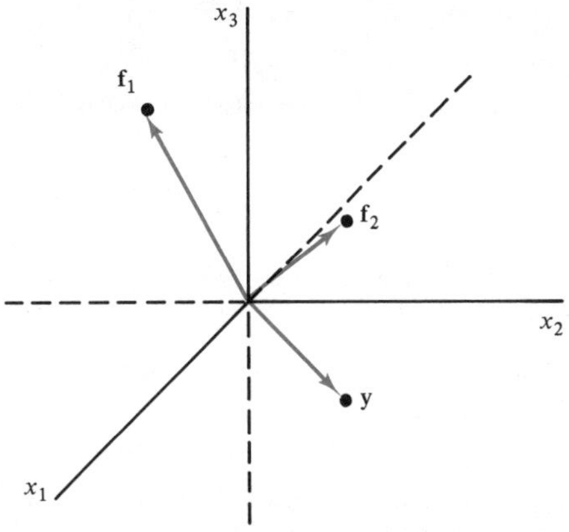

Figure 2.13

Theorem 2.10. Let V be a vector space with $\dim(V) = n$. Then

a. Any linearly independent set with n vectors is a basis of V.
b. Any spanning set with n vectors is a basis of V.

Proof. Suppose first that A is a linearly independent set of n vectors such that $S[A]$ is not all of V. Then there must be a vector $\mathbf{y}$ not in $S[A]$. But then Lemma 2.2 implies that the set consisting of A and $\mathbf{y}$ must be linearly independent. This contradicts a of Theorem 2.9. Therefore, any linearly independent set with n vectors must also span the vector space. Hence it is a basis. Suppose next that A is a spanning set of n vectors. Then, by Lemma 2.1, if A is not linearly independent, we may discard a vector from A, obtaining a set with fewer than n vectors, and which still spans V. This contradicts b of Theorem 2.9.

EXAMPLE 8 Let $A = \begin{bmatrix} 3 & 2 & 16 & 23 \\ 2 & 1 & 10 & 13 \end{bmatrix}$. Find a basis for the solution space of $A\mathbf{X} = 0$. If this is not a basis for $\mathbb{R}^4$, find a basis of $\mathbb{R}^4$ that contains the basis of the solution space.

Solution. The matrix A is row equivalent to the matrix

$$\begin{bmatrix} 1 & 0 & 4 & 3 \\ 0 & 1 & 2 & 7 \end{bmatrix}$$

Thus, if $\mathbf{X} = (x_1, \ x_2, \ x_3, \ x_4)^T$ solves $A\mathbf{X} = \mathbf{0}$, we must have $x_2 = -2x_3 - 7x_4$ and $x_1 = -4x_3 - 3x_4$. Setting $x_3 = 1$, $x_4 = 0$, and then $x_3 = 0$, $x_4 = 1$, we have two linearly independent solutions to our system:

$$\mathbf{f}_1 = (-4,-2,1,0) \quad \text{and} \quad \mathbf{f}_2 = (-3,-7,0,1)$$

Clearly the set $\{\mathbf{f}_1, \mathbf{f}_2\}$ is a basis for the solution space. To extend this set to a basis of $\mathbb{R}^4$ we have to find two more vectors $\mathbf{g}_1$ and $\mathbf{g}_2$ such that $\{\mathbf{f}_1, \mathbf{f}_2, \mathbf{g}_1, \mathbf{g}_2\}$ is linearly independent. $\mathbb{R}^4$ has as a basis $\{\mathbf{e}_1, \mathbf{e}_2, \mathbf{e}_3, \mathbf{e}_4\}$ where $\mathbf{e}_1 = (1,0,0,0)$, etc. Two of these $\mathbf{e}_k$'s will not depend linearly on $\mathbf{f}_1$ and $\mathbf{f}_2$. We need to find such a pair. A quick inspection of the form of $\mathbf{f}_1$ and $\mathbf{f}_2$ shows us that $\mathbf{e}_1$ and $\mathbf{e}_2$ will work. We leave the details of showing that $\{\mathbf{f}_1, \mathbf{f}_2, \mathbf{e}_1, \mathbf{e}_2\}$ is linearly independent to the reader. ■

We next outline a systematic procedure that can be used to extend any linearly independent set to a basis. Suppose $A = \{\mathbf{x}_1, \ldots, \mathbf{x}_r\}$ is a linearly independent subset of a vector space V, where $\dim(V) = n$ and $r < n$. We need to find $n - r$ vectors to adjoin to A to form a basis. Suppose $\{\mathbf{f}_1, \ldots, \mathbf{f}_n\}$ is any basis of V. Consider the set $\{\mathbf{x}_1, \ldots, \mathbf{x}_r, \mathbf{f}_1, \ldots, \mathbf{f}_n\}$. It is linearly dependent and its span is all of V. Now just discard those $\mathbf{f}_k$ which depend on the $\mathbf{x}_j$ and the remaining $\mathbf{f}_p$, until r vectors have been discarded. The vectors that are left form a basis containing A.

EXAMPLE 9 Extend the linearly independent set of vectors $B = \{(2,0,-1,0,0), (1,1,0,1,1)\}$ to a basis of $\mathbb{R}^5$.

Solution. Let A be the 7×5 matrix whose first two rows are the vectors in B and whose last five rows are the standard basis vectors of $\mathbb{R}^5$.

$$A = \begin{bmatrix} 2 & 0 & -1 & 0 & 0 \\ 1 & 1 & 0 & 1 & 1 \\ 1 & 0 & 0 & 0 & 0 \\ 0 & 1 & 0 & 0 & 0 \\ 0 & 0 & 1 & 0 & 0 \\ 0 & 0 & 0 & 1 & 0 \\ 0 & 0 & 0 & 0 & 1 \end{bmatrix}$$

Now use elementary row operations but not row interchanges to find a matrix that is row equivalent to A and that has two rows of zeros. Zeroing out rows corresponds to discarding those vectors in the standard basis which depend linearly on the remaining vectors. After a few row operations, we have A row equivalent to the matrix

$$\begin{bmatrix} 2 & 0 & -1 & 0 & 0 \\ 1 & 1 & 0 & 1 & 1 \\ 1 & 0 & 0 & 0 & 0 \\ 0 & 1 & 0 & 0 & 0 \\ 0 & 0 & 0 & 0 & 0 \\ 0 & 0 & 0 & 1 & 0 \\ 0 & 0 & 0 & 0 & 0 \end{bmatrix}$$

Thus the set $\{(2,0,-1,0,0), (1,1,0,1,1), \mathbf{e}_1, \mathbf{e}_2, \mathbf{e}_4\}$ is a basis of $\mathbb{R}^5$ that contains our original set A. ■

EXAMPLE 10 Find a basis for $V = \{\mathbf{p}: \mathbf{p} \text{ is in } P_2, \mathbf{p}(1) = 0\}$. If $\mathbf{p}$ is in P_2, then $\mathbf{p}(t) = a_0 + a_1 t + a_2 t^2$, and $\mathbf{p}(1) = 0$ implies

$$\mathbf{p}(1) = 0 = a_0 + a_1 + a_2$$

Thus, $F = \{1 - t^2, t - t^2\}$ is a subset of V. F is easily shown to be linearly independent. To see that F spans V, let $\mathbf{p}$ be in V. Then

$$\begin{aligned}\mathbf{p}(t) &= a_0 + a_1 t + a_2 t^2 \\ &= a_0 + a_1 t + (-a_0 - a_1)t^2 \\ &= a_0(1 - t^2) + a_1(t - t^2)\end{aligned}$$

Thus, $\mathbf{p}(t)$ is in $S[F]$; F is a basis of V and $\dim(V)$ equals 2. ■

PROBLEM SET 2.5

1. Find a basis for $S[(2,6,0,1,3), (1,1,0,-1,4), (0,0,0,1,1)]$.
2. Let $W = \{(x_1,x_2,x_3,x_4): x_1 - x_2 = x_3\}$. Show that W is a subspace of $\mathbb{R}^4$ and find a basis of W.
3. Let K be the solution space of the following homogeneous system of equations:

$$\begin{aligned}2x_1 - 6x_2 + 3x_3 \quad\quad &= 0 \\ x_2 + \ x_3 + x_4 &= 0\end{aligned}$$

   Find any two linearly independent vectors in K. Show that every other vector in K can be written as a linear combination of your two vectors, and conclude that these two vectors are a basis of K. Extend your basis to a basis of $\mathbb{R}^4$.
4. Show that any subset of $\mathbb{R}^3$ containing at most two vectors will not span all of $\mathbb{R}^3$.
5. Find a basis of $\mathbb{R}^3$ that is not the standard one given in the text.
6. Find a basis of P_2 that is different from the standard basis $\{1,t,t^2\}$.
7. Let $K = \{(x_1,x_2,x_3): 2x_1 + x_2 = 0, x_1 + x_2 - x_3 = 0\}$.
   a. Show that every vector in K can be written as a scalar multiple of $(-1,2,1)$. Thus, $\dim(K) = 1$.
   b. Find two vectors $\mathbf{y}_1$ and $\mathbf{y}_2$ such that $\{(-1,2,1),\mathbf{y}_1,\mathbf{y}_2\}$ is a basis of $\mathbb{R}^3$.
8. Show that the matrices $\begin{bmatrix} 1 & 0 \\ 0 & 0 \end{bmatrix}, \begin{bmatrix} 0 & 1 \\ 0 & 0 \end{bmatrix}, \begin{bmatrix} 0 & 0 \\ 1 & 0 \end{bmatrix}, \begin{bmatrix} 0 & 0 \\ 0 & 1 \end{bmatrix}$ form a basis of M_{22}, the set of 2×2 matrices.
9. Show that the $m \times n$ matrices E_{ij} described in Example 4b form a basis of M_{mn}, the set of $m \times n$ matrices. Since there are mn such matrices, the dimension of the set of $m \times n$ matrices is mn.
10. Let $K = \{(x_1,x_2,x_3): x_1 - x_2 + 3x_3 = 0\}$.
    a. Find a basis for K.
    b. $\dim(K) = ?$
    c. Extend your basis of K to a basis of $\mathbb{R}^3$.
11. Consider the complex numbers as a real vector space. Remember complex numbers are things of the form $a + bi$, where $i^2 = -1$ and a and b are arbitrary real numbers. Find a basis for the complex numbers.

12. Let $K = \{(x_1, x_2, x_3): x_1 - x_2 = x_3,\ x_2 + 2x_3 = 0\}$.
a. Find a basis B for K.
b. $\dim(K) = ?$
c. How many vectors have to be added to B to get a basis for $\mathbb{R}^3$?

13. Let V be the vector space of all $n \times n$ matrices. Then we know that $\dim(V) = n^2$. Find a basis and determine the dimension of each of the following subspaces of V.
a. Scalar matrices
b. Diagonal matrices
c. Upper triangular matrices
d. Lower triangular matrices

14. Verify that the subset $A = \{\mathbf{e}_k: k = 1, \ldots, n\}$ is a basis of $\mathbb{R}^n$; cf. Example 3.

15. Show that the set $A = \{(1,0,0), (1,1,-1), (0,1,1)\}$ spans $\mathbb{R}^3$; cf. Example 2. Is A a basis of $\mathbb{R}^3$?

16. Show that the subset $\{1, t, \ldots, t^n\}$ of P_n is a basis for that vector space.

17. In problem 19 of Section 2.4 we defined what it means to say that an infinite set of vectors is linearly independent. Using that definition, we say that B is a basis of a vector space V, if B is linearly independent and if $S[B]$ equals V. Again, the span of a set is all possible finite linear combinations of vectors from that set. Let $V = C[0,1]$. Let $B = \{1, e^t, e^{2t}, \ldots, e^{nt}, \ldots\}$. Is B a basis for V?

18. Let V be the set of all polynomials with real coefficients. Let $B = \{1, t, \ldots, t^n, \ldots\}$. Is B a basis for V?

19. Let V be a three-dimensional space. Let $\mathbf{x}$ be any nonzero vector in V.
a. Show that there are vectors $\mathbf{y}_1$ and $\mathbf{y}_2$ such that $\{\mathbf{x}, \mathbf{y}_1, \mathbf{y}_2\}$ is a basis of V.
b. Show that any linearly independent subset A of V can be extended to a basis of V.

20. Let $V = \left\{\begin{bmatrix} a & b \\ c & d \end{bmatrix}: a + b + c = 0\right\}$. Find a basis for V and thus determine $\dim(V)$.

21. Let $V = \{\mathbf{p}: \mathbf{p} \text{ is in } P_4,\ \int_0^1 \mathbf{p}(t)\,dt = 0,\ \mathbf{p}'(2) = 0\}$. Find a basis for V.

22. Show that $\left\{\begin{bmatrix} 0 & 1 \\ 1 & 1 \end{bmatrix}\begin{bmatrix} 1 & 0 \\ 1 & 1 \end{bmatrix}\begin{bmatrix} 1 & 1 \\ 0 & 1 \end{bmatrix}\begin{bmatrix} 1 & 1 \\ 1 & 0 \end{bmatrix}\right\}$ is a basis for M_{22}.

23. Is $\left\{\begin{bmatrix} 1 & 1 \\ 0 & 0 \end{bmatrix}\begin{bmatrix} 0 & 1 \\ 1 & 0 \end{bmatrix}\begin{bmatrix} 0 & 0 \\ 1 & 1 \end{bmatrix}\begin{bmatrix} 1 & 0 \\ 0 & 1 \end{bmatrix}\right\}$ a basis of M_{22}?

24. Let $V = \{\mathbf{p}: \mathbf{p} \text{ is in } P_4,\ \int_0^1 \mathbf{p}(t)\,dt = 0 = \int_1^2 \mathbf{p}(t)\,dt\}$. Find a basis for V.

25. Let $A = \{(1,0,0), (0,1,1), (0,2,3), (0,3,4), (1,-1,3)\}$.
a. Show that A is linearly dependent.
b. What is the largest number of vectors in A that can form a linearly independent set?
c. For which vectors $\mathbf{x}$ in A is it true that $S[A] = S[A \backslash \{\mathbf{x}\}]$?
d. Find a subset B of A such that B is linearly independent and for which $S[B] = S[A]$.

2.6 COORDINATES OF A VECTOR

We have seen in an earlier section (Theorem 2.7) that given a basis for a vector space, every vector can be uniquely written as a linear combination of the basis vectors. The constants that are used in this sum are called the coordinates of the vector with respect to the basis. In this section we show how these coordinates change when we change our bases. For convenience, we again state the definition of the coordinates of a vector.

Definition 2.11. Let $B = \{\mathbf{f}_1, \mathbf{f}_2, \ldots, \mathbf{f}_n\}$ be a basis of a vector space V, $\dim(V) = n$. Then the coordinates of a vector $\mathbf{x}$ in V with respect to the basis B are the constants needed to write $\mathbf{x}$ as a linear combination of the basis vectors. Thus if

$$\mathbf{x} = c_1\mathbf{f}_1 + c_2\mathbf{f}_2 + \cdots + c_n\mathbf{f}_n$$

then the coordinates of $\mathbf{x}$ with respect to B are the scalars $c_1, c_2, \ldots, c_n$. We write this as

$$[\mathbf{x}]_B = [c_1, c_2, \ldots, c_n]$$

Note that the ordering of the vectors in B is important. If we change this order, we change the order in which we list the coordinates.

EXAMPLE 1 Let $V = \mathbb{R}^3$.

a. Let $S = \{\mathbf{e}_1, \mathbf{e}_2, \mathbf{e}_3\}$. Find the coordinates of $\mathbf{x} = (1,-1,2)$ with respect to S. Clearly,

$$\mathbf{x} = (1,-1,2) = (1,0,0) - (0,1,0) + 2(0,0,1)$$

Thus,

$$[\mathbf{x}]_S = [1,-1,2]$$

Note that since S is the standard basis, the coordinates of $\mathbf{x}$ with respect to S are the same three numbers used to define $\mathbf{x}$.

b. Let $G = \{(0,0,1), (0,1,0), (1,0,0)\}$. Thus, G is just a reordering of the standard basis of $\mathbb{R}^3$. Then

$$[\mathbf{x}]_G = [(1,-1,2)]_G = [2,-1,1]$$

c. Let $G = \{(2,-1,3), (0,1,1), (1,-1,0)\}$. Find the coordinates of $\mathbf{x} = (1,-1,2)$ with respect to G. We need to find constants c_1, c_2, and c_3 such that

$$(1,-1,2) = c_1(2,-1,3) + c_2(0,1,1) + c_3(1,-1,0)$$

The system of equations derived from this vector equation is

$$\begin{aligned} 1 &= 2c_1 \phantom{{}+c_2} + c_3 \\ -1 &= -c_1 + c_2 - c_3 \\ 2 &= 3c_1 + c_2 \end{aligned}$$

The solution is $c_1 = 1$, $c_2 = -1$, $c_3 = -1$. Hence

$$[\mathbf{x}]_G = [1,-1,-1]$$

■

We now discuss the general situation. Let $F = \{\mathbf{f}_1, \ldots, \mathbf{f}_n\}$ and $G = \{\mathbf{g}_1, \ldots, \mathbf{g}_n\}$ be any two bases of a vector space V. Then each vector in F can be written uniquely in terms of the vectors in G and each vector in G can be written uniquely in terms of the basis F. Let $[p_{jk}]$, $1 \le j, k \le n$, be such that

$$\mathbf{f}_j = p_{1j}\mathbf{g}_1 + p_{2j}\mathbf{g}_2 + \cdots + p_{nj}\mathbf{g}_n \tag{2.8}$$

Thus $[\mathbf{f}_j]_G = [p_{1j}, p_{2j}, \ldots, p_{nj}]$ for $1 \le j \le n$. Let $[q_{jk}]$, $1 \le j, k \le n$, be such that

$$\mathbf{g}_j = q_{1j}\mathbf{f}_1 + q_{2j}\mathbf{f}_2 + \cdots + q_{nj}\mathbf{f}_n \tag{2.9}$$

That is, $[\mathbf{g}_j]_F = [q_{1j}, q_{2j}, \ldots, q_{nj}]$.

Thus, the first column of the $n \times n$ matrix $P = [p_{jk}]$ consists of the coordinates of $\mathbf{f}_1$ and the jth column of P consists of the coordinates of $\mathbf{f}_j$, both with respect to the basis G. Similarly, the kth column of $Q = [q_{jk}]$ is formed using the coordinates of $\mathbf{g}_k$ with respect to the basis F.

Since P gives the coordinates of the vectors in F with respect to G and Q gives the coordinates of the vectors in G with respect to F, these two matrices should be related to each other. In fact, we have

Theorem 2.11. Let F and G be two bases of a vector space V. Let the matrices P and Q be as defined above. Then

$$Q = P^{-1}$$

Proof. From (2.8) and (2.9) we have

$$\mathbf{f}_j = \sum_{k=1}^{n} p_{kj}\mathbf{g}_k = \sum_{k=1}^{n} p_{kj}\left(\sum_{i=1}^{n} q_{ik}\mathbf{f}_i\right)$$
$$= \sum_{i=1}^{n}\left(\sum_{k=1}^{n} q_{ik}p_{kj}\right)\mathbf{f}_i$$

Thus the terms $\Sigma_{k=1}^{n} q_{ik}p_{kj}$ are the coordinates of $\mathbf{f}_j$ with respect to F. But clearly

$$\mathbf{f}_j = 0\mathbf{f}_1 + 0\mathbf{f}_2 + \cdots + (1)\mathbf{f}_j + \cdots + 0\mathbf{f}_n \qquad \text{for } j = 1, 2, \ldots, n$$

Thus

$$\sum_{k=1}^{n} q_{ik}p_{kj} = \delta_{ij} \tag{2.10}$$

But this summand is the i, j entry of the matrix QP. Hence we have $QP = I_n$. In a similar manner we can show that $PQ = I_n$. Therefore, $Q = P^{-1}$.

In the following we refer to P or P^{-1} as the change of basis matrix. Which one is meant will be clear from the context.

EXAMPLE 2 Let $F = \{(1,0,0), (0,1,0), (0,0,1)\}$. Let $G = \{(2,-1,3), (0,1,1), (1,-1,0)\}$. Thus $\mathbf{f}_1$, $\mathbf{f}_2$, and $\mathbf{f}_3$ are our standard basis vectors, and $\mathbf{g}_1 = (2,-1,3)$, $\mathbf{g}_2 = (0,1,1)$, and $\mathbf{g}_3 = (1,-1,0)$.

We clearly have

$$[\mathbf{g}_1]_F = [2,-1,3] \qquad [\mathbf{g}_2]_F = [0,1,1] \qquad [\mathbf{g}_3]_F = [1,-1,0]$$

Thus, the matrix $Q = P^{-1}$ must equal

$$\begin{bmatrix} 2 & 0 & 1 \\ -1 & 1 & -1 \\ 3 & 1 & 0 \end{bmatrix}$$

We next want to find the coordinates of $\mathbf{f}_j$ with respect to the basis G. There are two ways to go; one is to actually find the p_{ij} such that $\mathbf{f}_1 = p_{11}\mathbf{g}_1 + p_{21}\mathbf{g}_2 + p_{31}\mathbf{g}_3$, etc., or we can just compute Q^{-1}. The first way is instructive; so we will find the first column of P by computing the coordinates of $\mathbf{f}_1$ with respect to G. We want to find constants c_k, $k = 1, 2, 3$, such that

$$(1,0,0) = c_1(2,-1,3) + c_2(0,1,1) + c_3(1,-1,0)$$

The solutions are $c_1 = -\frac{1}{2}$, $c_2 = \frac{3}{2}$, $c_3 = 2$. Thus $(1,0,0) = (-\frac{1}{2})\mathbf{g}_1 + (\frac{3}{2})\mathbf{g}_2 + 2\mathbf{g}_3$. If we write out the above equations for the c_j as a matrix equation we have

$$\begin{bmatrix} 2 & 0 & 1 \\ -1 & 1 & -1 \\ 3 & 1 & 0 \end{bmatrix}\begin{bmatrix} c_1 \\ c_2 \\ c_3 \end{bmatrix} = \begin{bmatrix} 1 \\ 0 \\ 0 \end{bmatrix}$$

or

$$Q\begin{bmatrix} c_1 \\ c_2 \\ c_3 \end{bmatrix} = \begin{bmatrix} 1 \\ 0 \\ 0 \end{bmatrix}$$

Thus, $(c_1 \quad c_2 \quad c_3)^T = Q^{-1}(1 \quad 0 \quad 0)^T$, and clearly the first column of P must be the first column of Q^{-1}, and

$$Q^{-1} = \begin{bmatrix} -\frac{1}{2} & -\frac{1}{2} & \frac{1}{2} \\ \frac{3}{2} & \frac{3}{2} & -\frac{1}{2} \\ 2 & 1 & -1 \end{bmatrix}$$

Reading down the columns of $Q^{-1} = P$ we have

$$\mathbf{f}_1 = (-\tfrac{1}{2})\mathbf{g}_1 + (\tfrac{3}{2})\mathbf{g}_2 + 2\mathbf{g}_3$$
$$\mathbf{f}_2 = (-\tfrac{1}{2})\mathbf{g}_1 + (\tfrac{3}{2})\mathbf{g}_2 + (1)\mathbf{g}_3$$
$$\mathbf{f}_3 = (\tfrac{1}{2})\mathbf{g}_1 + (-\tfrac{1}{2})\mathbf{g}_2 + (-1)\mathbf{g}_3$$ ■

We next discuss how to use the change of basis matrix P to express the coordinates of $\mathbf{x}$ with respect to G in terms of the coordinates of $\mathbf{x}$ with respect to

F. Thus let $F = \{\mathbf{f}_1, \ldots, \mathbf{f}_n\}$ and $G = \{\mathbf{g}_1, \ldots, \mathbf{g}_n\}$. Suppose $[\mathbf{x}]_F = [x_1, \ldots, x_n]$. Then

$$\begin{aligned}\mathbf{x} &= x_1\mathbf{f}_1 + x_2\mathbf{f}_2 + \cdots + x_n\mathbf{f}_n \\ &= x_1\left(\sum_{j=1}^{n} p_{j1}\mathbf{g}_j\right) + x_2\left(\sum_{j=1}^{n} p_{j2}\mathbf{g}_j\right) + \cdots + x_n\left(\sum_{j=1}^{n} p_{jn}\mathbf{g}_j\right) \\ &= \sum_{k=1}^{n}\left(x_k \sum_{j=1}^{n} p_{jk}\mathbf{g}_j\right) = \sum_{j=1}^{n}\left(\sum_{k=1}^{n} p_{jk}x_k\right)\mathbf{g}_j\end{aligned}$$

Thus,

$$[\mathbf{x}]_G = \left[\sum_{k=1}^{n} p_{1k}x_k, \sum_{k=1}^{n} p_{2k}x_k, \ldots, \sum_{k=1}^{n} p_{nk}x_k\right] \tag{2.11}$$

Note that the jth coordinate of $\mathbf{x}$ with respect to G is the jth column of the matrix product $P([x_1, \ldots, x_n])^T$. Thus, (2.11) implies the following equation:

$$[\mathbf{x}]_G^T = P[\mathbf{x}]_F^T \tag{2.12}$$

In the sequel we use equation (2.12) to specify P; that is, we ask for the change of basis matrix P such that $[\mathbf{x}]_G^T = P[\mathbf{x}]_F^T$. The following summarizes these relationships:

$$\begin{array}{ll} F = \{\mathbf{f}_1, \ldots, \mathbf{f}_n\} & G = \{\mathbf{g}_1, \ldots, \mathbf{g}_n\} \\ \mathbf{f}_j = \sum_{k=1}^{n} p_{kj}\mathbf{g}_k & \mathbf{g}_j = \sum_{k=1}^{n} q_{kj}\mathbf{f}_k \\ [\mathbf{x}]_F = [x_1, \ldots, x_n] & [\mathbf{x}]_G = [y_1, \ldots, y_n] \end{array}$$

$$y_j = \sum_{k=1}^{n} p_{jk}x_k$$

$$[\mathbf{x}]_G^T = P[\mathbf{x}]_F^T \qquad [\mathbf{x}]_F^T = P^{-1}[\mathbf{x}]_G^T$$

One way of remembering how to compute a change of basis matrix is to realize that we really want a matrix P such that, given two bases F and G, the coordinates of any vector $\mathbf{x}$ with respect to F and G are related by equation (2.12):

$$[\mathbf{x}]_G^T = P[\mathbf{x}]_F^T \tag{2.12}$$

Let us start from this equation and see what P must equal. Let $F = \{\mathbf{f}_1, \ldots, \mathbf{f}_n\}$. If (2.12) holds for all $\mathbf{x}$ in V, it must also hold for $\mathbf{x} = \mathbf{f}_1$. What are the coordinates of $\mathbf{f}_1$ with respect to F? Clearly we have

$$[\mathbf{f}_1]_F = [1,0, \ldots, 0]$$

But then $P[\mathbf{f}_1]_F^T$ is just the first column of P; that is, the first column of P equals

$[\mathbf{f}_1]_G^T$. This is, as we would expect, equation (2.8) for $j = 1$. Similarly we see that the kth column of P must equal $[\mathbf{f}_k]_G^T$.

EXAMPLE 3 Let V be a vector space. Let F_1, F_2, and F_3 each denote a basis of V. Suppose P and Q denote change of basis matrices between F_1 and F_2, and between F_3 and F_2, respectively, such that

$$[\mathbf{x}]_{F_1}^T = P[\mathbf{x}]_{F_2}^T \tag{2.13}$$

and

$$[\mathbf{x}]_{F_3}^T = Q[\mathbf{x}]_{F_2}^T \tag{2.14}$$

Find the change of basis matrix R, relating F_1 and F_3, such that

$$[\mathbf{x}]_{F_3}^T = R[\mathbf{x}]_{F_1}^T$$

From (2.13) and (2.14) we have $[\mathbf{x}]_{F_3}^T = Q[\mathbf{x}]_{F_2}^T = QP^{-1}[\mathbf{x}]_{F_1}^T$. By problem 11 at the end of this section we must have

$$R = QP^{-1}$$ ■

The reader might find this fact useful when computing change of basis matrices, when neither basis is the standard basis. We illustrate this below.

EXAMPLE 4 Let $V = \mathbb{R}^2$. Let $F = \{(7,8), (-9,20)\}$. Let $G = \{(6,-5), (1,1)\}$. Find the change of basis matrix R such that

$$[\mathbf{x}]_G^T = R[\mathbf{x}]_F^T$$

Let S be the standard basis of $\mathbb{R}^2$. Let P be such that $[\mathbf{x}]_S^T = P[\mathbf{x}]_F^T$. Let Q be such that $[\mathbf{x}]_S^T = Q[\mathbf{x}]_G^T$. Thus, the columns of P and Q are the coordinates of the vectors in F and G, respectively, with respect to the standard basis. Hence,

$$P = \begin{bmatrix} 7 & -9 \\ 8 & 20 \end{bmatrix} \quad \text{and} \quad Q = \begin{bmatrix} 6 & 1 \\ -5 & 1 \end{bmatrix}$$

and we have

$$\begin{aligned} [\mathbf{x}]_G^T &= Q^{-1}[\mathbf{x}]_S^T \\ &= Q^{-1}(P[\mathbf{x}]_F^T) \\ &= (Q^{-1}P)[\mathbf{x}]_F^T \end{aligned}$$

Thus,

$$\begin{aligned} R = Q^{-1}P &= (\tfrac{1}{11})\begin{bmatrix} 1 & -1 \\ 5 & 6 \end{bmatrix}\begin{bmatrix} 7 & -9 \\ 8 & 20 \end{bmatrix} \\ &= (\tfrac{1}{11})\begin{bmatrix} -1 & 11 \\ 83 & 75 \end{bmatrix} \end{aligned}$$

To check our computations, we verify that the first column of R consists of the coordinates of $\mathbf{f}_1$ with respect to G.

$$-(\tfrac{1}{11})(6,-5) + (\tfrac{83}{11})(1,1) = (\tfrac{77}{11},\tfrac{88}{11}) = (7,8) = \mathbf{f}_1$$ ■

PROBLEM SET 2.6

1. Find the coordinates of $\mathbf{x} = (1,-1,2)$ with respect to each of the following sets:
   a. $\{(1,0,0), (0,1,0), (0,0,1)\}$
   b. $\{(1,1,1), (2,1,4), (-1,1,1)\}$
   c. $\{(2,0,6), (4,2,0), (0,3,2)\}$
2. Find the coordinates of $(1,2,0,-4)$ with respect to each of the following bases:
   a. The standard basis of $\mathbb{R}^4$
   b. $\{(0,1,1,1), (1,0,1,1), (1,1,0,1), (1,1,1,0)\}$
3. Find the coordinates of $\mathbf{x} = (6,-4)$ with respect to each of the following bases of $\mathbb{R}^2$:
   a. $\{(1,0), (0,1)\}$ b. $\{(6,-4), (9,17)\}$
   c. $\{(1,1), (-1,0)\}$ d. $\{(9,17), (6,-4)\}$
4. Let $F = \{(1,-2), (8,3)\}$.
   a. If $[\mathbf{x}]_F = [1,0]$, then $\mathbf{x} = ?$
   b. If $[\mathbf{x}]_F = [1,1]$, then $\mathbf{x} = ?$
   c. If $[\mathbf{x}]_F = [0,0]$, then $\mathbf{x} = ?$
   d. If $\mathbf{x} = (1,1)$, then $[\mathbf{x}]_F = ?$
5. Let $V = M_{22}$. Let F be the set:

$$\left\{\begin{bmatrix} 0 & 1 \\ 1 & 1 \end{bmatrix}\begin{bmatrix} 1 & 0 \\ 1 & 1 \end{bmatrix}\begin{bmatrix} 1 & 1 \\ 0 & 1 \end{bmatrix}\begin{bmatrix} 1 & 1 \\ 1 & 0 \end{bmatrix}\right\}$$

   a. Show that F is a basis of V.
   b. Let $\mathbf{x} = \begin{bmatrix} 6 & -4 \\ 3 & 2 \end{bmatrix}$; $[\mathbf{x}]_F = ?$
   c. If $[\mathbf{x}]_F = [0,1,0,0]$, then $\mathbf{x}$ equals?
6. Let $V = P_2$. Let $F = \{1, 1 + t, 1 + t + t^2\}$.
   a. Show that F is a basis of V.
   b. If $[\mathbf{x}]_F = [-2,3,7]$, then $\mathbf{x}$ equals?
   c. $[6 - t^2]_F = ?$
7. Let S be the standard basis of $\mathbb{R}^2$, and let $G = \{(1,6), (2,3)\}$.
   a. Show that G is a basis.
   b. Find the change of basis matrix P such that $[\mathbf{x}]_S^T = P[\mathbf{x}]_G^T$.
   c. Find the change of basis matrix Q such that $[\mathbf{x}]_G^T = Q[\mathbf{x}]_S^T$.
   d. Compute PQ and QP.
8. Let $\mathbf{x} = (-8,4)$. Let S and G be the bases in problem 7.
   a. Find $[\mathbf{x}]_S$ and $[\mathbf{x}]_G$.
   b. Show that $[\mathbf{x}]_G^T = P[\mathbf{x}]_S^T$.
9. Let $F = \{(-1,7), (2,-3)\}$ and $G = \{(1,2), (1,3)\}$.
   a. Show that both F and G are bases of $\mathbb{R}^2$.
   b. Find the change of basis matrix P such that $[\mathbf{x}]_G^T = P[\mathbf{x}]_F^T$.

10. Let $\mathbf{x} = (3,7)$. Let F and G be the bases in problem 9.
  a. Find $[\mathbf{x}]_F$ and $[\mathbf{x}]_G$.
  b. Show that $[\mathbf{x}]_G^T = P[\mathbf{x}]_F^T$.

11. Let A and B be the two $m \times n$ matrices. Suppose that $A\mathbf{x} = B\mathbf{x}$ for every $\mathbf{x}$ in $\mathbb{R}^n$. Show $A = B$. Note that it will be sufficient to show that if $A\mathbf{x} = 0$ for every $\mathbf{x}$ in $\mathbb{R}^n$, then A must be the $m \times n$ zero matrix. Why?

12. Let $F = \{(1,1), (-1,2)\}$, $G = \{(-1,2), (1,1)\}$.
  a. Find the change of basis matrix P such that $[\mathbf{x}]_G^T = P[\mathbf{x}]_F^T$.
  b. If $\mathbf{x} = (1,1)$, find $[\mathbf{x}]_F$ and $[\mathbf{x}]_G$.
  c. Show $[\mathbf{x}]_F^T = P^{-1}[\mathbf{x}]_G^T$.

13. Let S be the standard basis of $\mathbb{R}^3$ and let $G = \{(6,0,1), (1,-1,-1), (0,3,1)\}$.
  a. Show that G is a basis of $\mathbb{R}^3$.
  b. Find the change of basis matrix P such that $[\mathbf{x}]_G^T = P[\mathbf{x}]_S^T$.
  c. Write $\mathbf{e}_k$, $k = 1, 2, 3$, as linear combinations of the vectors in G.

14. Let $\mathbf{x} = (5,-3,4)$. Let S and G be the bases in problem 13. Find $[\mathbf{x}]_G$ in two different ways.

15. Let $F = \{\mathbf{f}_1, \mathbf{f}_2\}$ be a basis of $\mathbb{R}^2$. Let $G = \{\mathbf{f}_2, \mathbf{f}_1\}$. Find the change of basis matrix P, such that $[\mathbf{x}]_G^T = P[\mathbf{x}]_S^T$.

16. Let F be any basis of some n-dimensional vector space. Let G consist of the same vectors that belong to F, but in perhaps a different order. Show that the matrix P relating these two bases is a permutation matrix; cf. problem 13 in section 1.5.

17. Let $V = M_{22}$. Let $F = \left\{\begin{bmatrix}1 & 0\\0 & 0\end{bmatrix}\begin{bmatrix}0 & 1\\0 & 0\end{bmatrix}\begin{bmatrix}0 & 0\\1 & 0\end{bmatrix}\begin{bmatrix}0 & 0\\0 & 1\end{bmatrix}\right\}$.

  Let $G = \left\{\begin{bmatrix}0 & 1\\1 & 1\end{bmatrix}\begin{bmatrix}1 & 0\\1 & 1\end{bmatrix}\begin{bmatrix}1 & 1\\0 & 1\end{bmatrix}\begin{bmatrix}1 & 1\\1 & 0\end{bmatrix}\right\}$.

  a. Let $\mathbf{x} = \begin{bmatrix}2 & -6\\7 & 5\end{bmatrix}$. $[\mathbf{x}]_F = ?$ $[\mathbf{x}]_G = ?$
  b. Find the change of basis matrix P such that $[\mathbf{x}]_G^T = P[\mathbf{x}]_F^T$.
  c. Check that $[\mathbf{x}]_G^T = P[\mathbf{x}]_F^T$ for the vector $\mathbf{x}$ of part a.
  d. If $[\mathbf{x}]_G^T = [-1,3,2,4]$, then $\mathbf{x}$ equals?

18. Let $V = P_3$. Let $F = \{-2, t + t^2, t^2 - 1, t^3 + 1\}$. Let S denote the standard basis of V.
  a. Verify that F is a basis of V.
  b. Find the change of basis matrix P such that $[\mathbf{x}]_S^T = P[\mathbf{x}]_F^T$.
  c. Compute the coordinates of t^3, t^2, t, and 1 with respect to the basis F.

19. Let $S = \{(1,0), (0,1)\}$ and let $B = \{(2,-3), (1,6)\}$.
  a. Show that both S and B are bases of $\mathbb{R}^2$.
  b. Find the coordinates of $(-6,3)$ with respect to S.
  c. Find the coordinates of $(-6,3)$ with respect to B.

20. Let S and B be the same sets as in problem 19. The coordinates of $(2,-3)$ and $(1,6)$ with respect to S are $[2,-3]$ and $[1,6]$, respectively. Show that the coordinates of $(1,0)$ and $(0,1)$ with respect to B are $[\frac{2}{5},\frac{1}{5}]$ and $[-\frac{1}{15},\frac{2}{15}]$, respectively. Let

$$P = \begin{bmatrix}\frac{2}{5} & -\frac{1}{15}\\ \frac{1}{5} & \frac{2}{15}\end{bmatrix} \quad \text{and} \quad Q = \begin{bmatrix}2 & 1\\-3 & 6\end{bmatrix}$$

  a. Identify the columns of P and Q with the above coordinates.
  b. Show that $PQ = QP = I_2$.

21. Let S and B be the sets defined in problem 19. In that problem, we saw that the coordinates of $(-6,3)$ with respect to B are $[-\frac{13}{5},-\frac{4}{5}]$. Show that

$$\begin{bmatrix} -\frac{13}{5} \\ -\frac{4}{5} \end{bmatrix} = P\begin{bmatrix} -6 \\ 3 \end{bmatrix}$$

22. Find the coordinates of $(2,6)$ with respect to each of the following bases of $\mathbb{R}^2$:
a. $\{(1,0), (0,1)\}$ **b.** $\{(0,1), (1,0)\}$
c. $\{(2,6), (3,0)\}$ **d.** $\{(3,0), (2,6)\}$

SUPPLEMENTARY PROBLEMS

1. Define each of the following terms and then give at least two examples of each one:
a. A vector space of dimension 5
b. Linearly independent set
c. Basis
d. A vector space of dimension $100 = 10^2$
e. Spanning set

2. Let V be the vector space of all polynomials in t. Show that there is no finite set of vectors in V that spans V.

3. Let $F = \{(1,-1,2), (0,1,1), (1,1,1)\}$.
a. Show that F is a basis of $\mathbb{R}^3$.
b. Let $\mathbf{x} = (1,0,0)$. Find $[\mathbf{x}]_F$.
c. Find the change of basis matrix P, relating F to the standard basis, for which $[\mathbf{x}]_F^T = P[\mathbf{x}]_S^T$.
d. Explain why the first column of P is your answer to part b.

4. Show that any nontrivial subspace W of a finite-dimensional vector space V has a basis and $\dim(W) \leq \dim(V)$.

5. Let V be the vector space consisting of all real-valued functions defined on $[0,1]$. Which of the following subsets are subspaces, and which of these are finite-dimensional?
a. All polynomials of degree ≤ 4
b. All $f(t)$ for which $f(\frac{1}{2}) = 0$
c. All $f(t)$ such that $f(0) = f(1)$
d. All $f(t)$ such that $f(0) + f(1) = 1$
e. All functions of the form ce^{rt} for some constants c and r

6. Let A be an $m \times m$ matrix. Show that A is invertible if and only if its rows form a linearly independent set of vectors in $\mathbb{R}^m$. Hint: A is invertible if and only if A is row equivalent to I_m.

7. Suppose $\mathbf{X}_1, \ldots, \mathbf{X}_m$ are linearly independent vectors. Let $A = [a_{jk}]$ be an $m \times m$ matrix. Define

$$\mathbf{Y}_j = \sum_{k=1}^{m} a_{jk}\mathbf{X}_k \qquad j = 1, \ldots, m$$

Show that the set of vectors $\{\mathbf{Y}_1, \ldots, \mathbf{Y}_m\}$ is linearly independent if and only if the matrix A is invertible.

8. If a, b, and c are any three distinct constants, show that $\sin(a+t)$, $\sin(b+t)$, and $\sin(c+t)$ are linearly dependent on $[0,1]$.

9. Let $V = \left\{ \begin{bmatrix} a & b \\ c & d \end{bmatrix} : ad - bc = 1 \right\}$. Is V a vector space?

10. Let W be that subspace of M_{22} spanned by

$$\begin{bmatrix} 1 & 1 \\ 0 & 1 \end{bmatrix} \quad \text{and} \quad \begin{bmatrix} 1 & -1 \\ 0 & 1 \end{bmatrix}$$

a. Show that $\dim(W) = 2$ and that $W = \left\{ \begin{bmatrix} a & b \\ c & d \end{bmatrix} : a = d \text{ and } c = 0 \right\}$.

b. Show that there is no real matrix A such that

$$A \begin{bmatrix} 1 & 1 \\ 0 & 1 \end{bmatrix} A^{-1} = \begin{bmatrix} 1 & -1 \\ 0 & 1 \end{bmatrix}$$

11. Let $\mathbf{x}_1$ and $\mathbf{x}_2$ be any two vectors. Let $\mathbf{y}_1$, $\mathbf{y}_2$, and $\mathbf{y}_3$ be any three vectors in $S[\mathbf{x}_1, \mathbf{x}_2]$. Show that these three vectors form a linearly dependent set.

12. Let $\{\mathbf{x}_1, \ldots, \mathbf{x}_r\}$ be any set of vectors. Let F be any set of m vectors contained in the span of the given set of r vectors. Show that if $m > r$, then F must be a linearly dependent set of vectors.

13. Let $V = M_{22}$. Let $C = \{A: AB = BA \text{ for every } B \text{ in } V\}$. That is, C consists of all 2×2 matrices that commute with every other 2×2 matrix.

a. Show that C is a subspace of V.

b. Find a basis of C and hence determine $\dim(C)$.

c. Let $V = M_{nn}$, define C as above, and repeat parts a and b.

14. For which numbers x are the vectors $(2,3)$ and $(1,x)$ linearly independent?

3 LINEAR TRANSFORMATIONS

In your previous mathematics courses you undoubtedly studied real-valued functions of one or more variables. For example, when you discussed parabolas the function $f(x) = x^2$ appeared, or when you talked about straight lines the function $f(x) = 2x$ arose. In this chapter we study functions of several variables, that is, functions of vectors. Moreover, their values will be vectors rather than scalars. The particular transformations that we study also satisfy a "linearity" condition that will be made precise later.

3.1 DEFINITION AND EXAMPLES

Before defining a linear transformation we look at two examples. The first is not a linear transformation and the second one is.

EXAMPLE 1 Let $V = \mathbb{R}^2$ and let $W = \mathbb{R}$. Define $f: V \to W$ by $f(x_1, x_2) = x_1 x_2$. Thus, f is a function defined on a vector space of dimension 2, with values in a one-dimensional space. The notation is highly suggestive; that is, $f: V \to W$ indicates that f does something to a vector in V to get a vector in W. For example, $f(1,-1) = -1$, $f(1,2) = 2$, etc. We will see later that this function does not satisfy the "linearity" condition and hence is not a linear transformation. ■

EXAMPLE 2 Let $V = \mathbb{R}^2$ and $W = \mathbb{R}^3$. Define $L: V \to W$ by $L(x_1, x_2) = (x_1, x_2 - x_1, x_2)$. Here the function L takes a vector in $\mathbb{R}^2$ and transforms it into a vector in $\mathbb{R}^3$. For example, $L(1,-1) = (1,-2,-1)$ and $L(2,6) = (2,4,6,)$. This particular function satisfies the linearity condition below, and so would be called a linear transformation from $\mathbb{R}^2$ to $\mathbb{R}^3$. ■

Definition 3.1. Let V and W be two vector spaces. Let L be a function defined on V with values in W. L will be called a linear transformation if it satisfies the following two conditions:

1. $L(\mathbf{x} + \mathbf{y}) = L(\mathbf{x}) + L(\mathbf{y})$, for any two vectors $\mathbf{x}$ and $\mathbf{y}$ in V.
2. $L(c\mathbf{x}) = cL(\mathbf{x})$, for any scalar c and vector $\mathbf{x}$ in V.

Let's go back to Example 2 and verify that we did indeed have a linear transformation. For any $\mathbf{x} = (x_1, x_2)$ and $\mathbf{y} = (y_1, y_2)$, we have

$$\begin{aligned} L(\mathbf{x} + \mathbf{y}) &= L[(x_1 + y_1, x_2 + y_2)] = [x_1 + y_1, (x_2 + y_2) - (x_1 + y_1), x_2 + y_2] \\ &= (x_1, x_2 - x_1, x_2) + (y_1, y_2 - y_1, y_2) = L(\mathbf{x}) + L(\mathbf{y}). \end{aligned}$$

Thus, condition 1 holds. Moreover

$$L(c\mathbf{x}) = L[(cx_1, cx_2)] = (cx_1, cx_2 - cx_1, cx_2) = c(x_1, x_2 - x_1, x_2) = cL(\mathbf{x})$$

Since 1 and 2 hold, L is a linear transformation from $\mathbb{R}^2$ to $\mathbb{R}^3$. The reader should now check that the function in Example 1 does not satisfy either of these two conditions.

EXAMPLE 3 Define $L: \mathbb{R}^3 \to \mathbb{R}^2$ by $L(x_1, x_2, x_3) = (x_3 - x_1, x_1 + x_2)$.

a. Compute $L(\mathbf{e}_1)$, $L(\mathbf{e}_2)$, and $L(\mathbf{e}_3)$.
b. Show L is a linear transformation.
c. Show $L(x_1, x_2, x_3) = x_1 L(\mathbf{e}_1) + x_2 L(\mathbf{e}_2) + x_3 L(\mathbf{e}_3)$.

a. $L[(1,0,0)] = (-1,1)$, $L[(0,1,0)] = (0,1)$, $L[(0,0,1)] = (1,0)$.
b.
$$\begin{aligned} L(\mathbf{x} + \mathbf{y}) &= L[(x_1 + y_1, x_2 + y_2, x_3 + y_3)] \\ &= ((x_3 + y_3) - (x_1 + y_1), (x_1 + y_1) + (x_2 + y_2)) \\ &= (x_3 - x_1, x_1 + x_2) + (y_3 - y_1, y_1 + y_2) \\ &= L(\mathbf{x}) + L(\mathbf{y}) \end{aligned}$$

$$\begin{aligned} L(c\mathbf{x}) &= L[(cx_1, cx_2, cx_3)] = (cx_3 - cx_1, cx_1 + cx_2) \\ &= c(x_3 - x_1, x_1 + x_2) = cL(\mathbf{x}) \end{aligned}$$

Thus L satisfies conditions 1 and 2 of Definition 3.1, and it is a linear transformation.

c.
$$\begin{aligned} L(x_1,x_2,x_3) &= L(x_1\mathbf{e}_1 + x_2\mathbf{e}_2 + x_3\mathbf{e}_3) \\ &= L(x_1\mathbf{e}_1) + L(x_2\mathbf{e}_2) + L(x_3\mathbf{e}_3) \\ &= x_1 L(\mathbf{e}_1) + x_2 L(\mathbf{e}_2) + x_3 L(\mathbf{e}_3) \end{aligned}$$

Notice that c implies that once $L(\mathbf{e}_k)$, $k = 1, 2, 3$, are known, the fact that L is a linear transformation completely determines $L(\mathbf{x})$ for any vector $\mathbf{x}$ in $\mathbb{R}^3$. ■

We collect a few facts about linear transformations in the next theorem.

Theorem 3.1. Let L be a linear transformation from a vector space V into a vector space W. Then

1. $L(\mathbf{0}) = \mathbf{0}$
2. $L(-\mathbf{x}) = -L(\mathbf{x})$
3. $L\left(\sum_{k=1}^{n} a_k\mathbf{x}_k\right) = \sum_{k=1}^{n} a_k L(\mathbf{x}_k)$

Proof

1. Let $\mathbf{x}$ be any vector in V. Then $L(\mathbf{x}) = L(\mathbf{x} + \mathbf{0}) = L(\mathbf{x}) + L(\mathbf{0})$. Adding $-L(\mathbf{x})$ to both sides, we have $L(\mathbf{0}) = \mathbf{0}$, where the zero vector on the left-hand side is in V while the zero vector on the right-hand side is in W.
2. $\mathbf{0} = L(\mathbf{0}) = L(\mathbf{x} - \mathbf{x}) = L(\mathbf{x}) + L(-\mathbf{x})$. Thus $L(-\mathbf{x}) = -L(\mathbf{x})$.
3. We show that this formula is true for $n = 3$ and leave the details of an induction argument to the reader.

$$\begin{aligned} L(a_1\mathbf{x}_1 + a_2\mathbf{x}_2 + a_3\mathbf{x}_3) &= L(a_1\mathbf{x}_1 + a_2\mathbf{x}_2) + L(a_3\mathbf{x}_3) \\ &= L(a_1\mathbf{x}_1) + L(a_2\mathbf{x}_2) + L(a_3\mathbf{x}_3) \\ &= a_1 L(\mathbf{x}_1) + a_2 L(\mathbf{x}_2) + a_3 L(\mathbf{x}_3) \end{aligned}$$

EXAMPLE 4 Let $L: \mathbb{R}^3 \to \mathbb{R}^4$ be a linear transformation. Suppose we know that $L(1,0,1) = (-1,1,0,2)$, $L(0,1,1) = (0,6,-2,0)$, and $L(-1,1,1) = (4,-2,1,0)$. Determine $L(1,2,-1)$.

Solution. The trick is to realize that the three vectors for which we know L form a basis F of $\mathbb{R}^3$. Thus, all we need to do is find the coordinates of $(1,2,-1)$ with respect to F, and then use 3 of Theorem 3.1. The change of basis matrix P below is such that $[\mathbf{x}]_F^T = P[\mathbf{x}]_S^T$.

$$P = \begin{bmatrix} 1 & 0 & -1 \\ 0 & 1 & 1 \\ 1 & 1 & 1 \end{bmatrix}^{-1} = \begin{bmatrix} 0 & -1 & 1 \\ 1 & 2 & -1 \\ -1 & -1 & 1 \end{bmatrix}$$

Using this matrix to find the coordinates of $(1,2,-1)$ with respect to F, we have

$$[1,2,-1]_F^T = P[1,2,-1]_S^T = \begin{bmatrix} 0 & -1 & 1 \\ 1 & 2 & -1 \\ -1 & -1 & 1 \end{bmatrix} \begin{bmatrix} 1 \\ 2 \\ -1 \end{bmatrix} = \begin{bmatrix} -3 \\ 6 \\ -4 \end{bmatrix}$$

Thus

$$(1,2,-1) = -3(1,0,1) + 6(0,1,1) + (-4)(-1,1,1)$$

and

$$\begin{aligned} L(1,2,-1) &= -3L(1,0,1) + 6L(0,1,1) + (-4)L(-1,1,1) \\ &= -3(-1,1,0,2) + 6(0,6,-2,0) - 4(4,-2,1,0) \\ &= (-13,41,-16,-6) \end{aligned}$$

■

A standard method of defining a linear transformation from $\mathbb{R}^n$ to $\mathbb{R}^m$ is by matrix multiplication. thus, if $\mathbf{x} = (x_1, \ldots, x_n)$ is any vector in $\mathbb{R}^n$ and $A = [a_{jk}]$ is an $m \times n$ matrix, define $L(\mathbf{x}) = A\mathbf{x}^T$. Then $L(\mathbf{x})$ is an $m \times 1$ matrix that we think of as a vector in $\mathbb{R}^m$. The various properties of matrix multiplication that were proved in Theorem 1.3 are just the statements that L is a linear transformation from $\mathbb{R}^n$ to $\mathbb{R}^m$.

EXAMPLE 5 Let $A = \begin{bmatrix} 1 & -1 & 2 \\ 4 & 1 & 3 \end{bmatrix}$. If L is the linear transformation defined

by A, compute the following:

a. $L(x_1,x_2,x_3)$ **b.** $L(1,0,0)$, $L(0,1,0)$, $L(0,0,1)$

$$L(x_1,x_2,x_3) = \begin{bmatrix} 1 & -1 & 2 \\ 4 & 1 & 3 \end{bmatrix} \begin{bmatrix} x_1 \\ x_2 \\ x_3 \end{bmatrix} = \begin{bmatrix} x_1 - x_2 + 2x_3 \\ 4x_1 + x_2 + 3x_3 \end{bmatrix}$$

$$L(1,0,0) = (1,4) \qquad L(0,1,0) = (-1,1) \qquad L(0,0,1) = (2,3)$$

The reader should note that $L(\mathbf{e}_1)$ is the first column of A, $L(\mathbf{e}_2)$ is the second column of A, and $L(\mathbf{e}_3)$ is the third column. ■

In general, if A is an $m \times n$ matrix and $L(\mathbf{x}) = A\mathbf{x}$, then $L(\mathbf{e}_k)$ will be the kth column of the matrix A.

Until now we've thought of a linear transformation as an expression combining n variables to produce a vector in R^m. If we limit ourselves to this algebraic viewpoint we miss a fuller appreciation of linear transformations. For

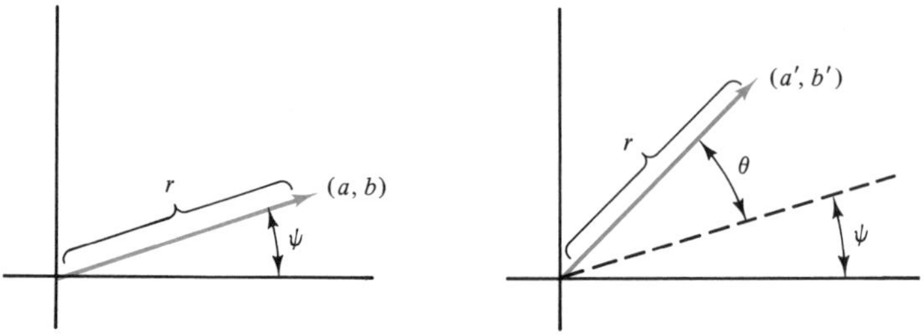

Figure 3.1

example, consider the mapping that rotates the points in the plane through an angle θ about the origin. Thus, if the point (a,b) is rotated through an angle θ to the position (a',b'), it turns out that the formulas relating (a',b') to (a,b) imply that this is a linear transformation. In fact (cf. Figure 3.1), setting $r = (a^2+b^2)^{1/2} = [(a')^2+(b')^2]^{1/2}$, we have that

$$\begin{aligned} a' &= r\cos(\theta+\psi) = r(\cos\theta\cos\psi - \sin\theta\sin\psi) \\ &= a\cos\theta - b\sin\theta \\ b' &= r\sin(\theta+\psi) = r(\sin\theta\cos\psi + \sin\psi\cos\theta) \\ &= a\sin\theta + b\cos\theta \end{aligned}$$

Thus we have

$$\begin{bmatrix} a' \\ b' \end{bmatrix} = \begin{bmatrix} \cos\theta & -\sin\theta \\ \sin\theta & \cos\theta \end{bmatrix} \begin{bmatrix} a \\ b \end{bmatrix}$$

Now whenever we see a matrix A of the form $\begin{bmatrix} a & -b \\ b & a \end{bmatrix}$, where $a^2+b^2=1$, we can think of A as defining a linear transformation from $\mathbb{R}^2$ to $\mathbb{R}^2$ that rotates the plane about the origin through an angle θ, where $\cos\theta = a$, $\sin\theta = b$. Note that $A^T = A^{-1}$ corresponds to a rotation of $-\theta$.

In the succeeding pages we sometimes describe a linear transformation in a geometrical manner as well as algebraically, and the reader should try to visualize what the particular transformation is doing.

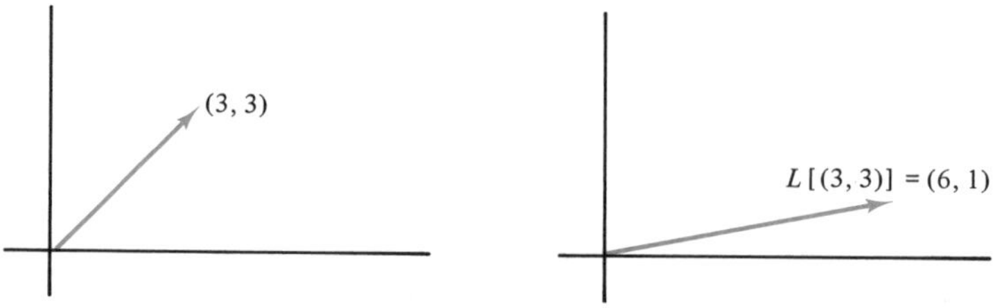

Figure 3.2 ■

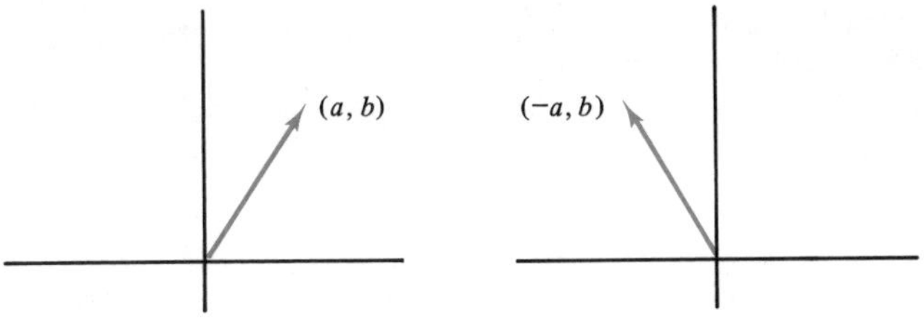

Figure 3.3

EXAMPLE 6 Describe in geometrical terms the linear transformation defined by the following matrices:

a. $A = \begin{bmatrix} 0 & 1 \\ -1 & 0 \end{bmatrix}$. This is a clockwise rotation of the plane about the origin through 90 degrees.

b. $A = \begin{bmatrix} 2 & 0 \\ 0 & \frac{1}{3} \end{bmatrix}$

$$A[x_1, x_2]^T = (2x_1, \tfrac{1}{3}x_2)^T$$

This linear transformation stretches the vectors in the subspace $S[\mathbf{e}_1]$ by a factor of 2 and at the same time compresses the vectors in the subspace $S[\mathbf{e}_2]$ by a factor of $\frac{1}{3}$. See Figure 3.2.

c. $A = \begin{bmatrix} -1 & 0 \\ 0 & 1 \end{bmatrix}$. For this A, the pair (a,b) gets sent to the pair $(-a,b)$. Hence this linear transformation reflects $\mathbb{R}^2$ through the x_2 axis. See Figure 3.3. ■

PROBLEM SET 3.1

1. Let $L(x_1, x_2, x_3) = x_1 - x_2 + x_3$.
  a. Show that L is a linear transformation from $\mathbb{R}^3$ to $\mathbb{R}$.
  b. Find a 1×3 matrix A such that $L(\mathbf{x}) = A\mathbf{x}$ for every $\mathbf{x}$ in $\mathbb{R}^3$.
  c. Compute $L(\mathbf{e}_k)$ for $k = 1, 2, 3$.
  d. Find a basis for the subspace $K = \{\mathbf{x}: A\mathbf{x} = \mathbf{0}\}$.

2. Let L be a linear transformation from $\mathbb{R}^3$ to $\mathbb{R}^2$ such that $L(\mathbf{e}_1) = (-1,6)$, $L(\mathbf{e}_2) = (0,2)$, $L(\mathbf{e}_3) = (8,1)$.
  a. $L(1,2,-6) = ?$
  b. $L(x_1, x_2, x_3) = ?$
  c. Find a matrix A such that $L(\mathbf{x}) = A\mathbf{x}$.

3. Let L be a linear transformation from $\mathbb{R}^3$ to $\mathbb{R}^5$. Suppose that $L(1,0,1) = (0,1,0,2,0)$, $L(0,-1,2) = (-1,6,2,0,1)$, and $L(1,1,2) = (2,-3,1,4,0)$. Notice that the three vectors for which we know L form a basis of $\mathbb{R}^3$.
  a. Compute $L(\mathbf{e}_k)$ for $k = 1, 2, 3$.
  b. $L(x_1, x_2, x_3) = ?$
  c. Find a matrix A such that $L(\mathbf{x}) = A\mathbf{x}$.

4. For each of the following functions f determine an appropriate V and W. Then decide if f is a linear transformation from V to W.
a. $f(x_1,x_2) = (x_1,0,1)$
b. $f(x_1,x_2) = (x_1 - x_2,x_1)$
c. $f(x) = (x,x)$
d. $f(x_1,x_2,x_3) = (x_1x_2,x_2x_3,x_3x_1)$

5. Let L: $V \to W$ be a linear transformation. Let K be any subspace of V. Define $L(K) = \{L(\mathbf{x})$: $\mathbf{x}$ is any vector in $K\}$. Show that $L(K)$ is a subspace of W.

6. Let L: $V \to W$ be a linear transformation. Let H be any subspace of W. Define $L^{-1}(H) = \{\mathbf{x}$: $L(\mathbf{x})$ is in $H\}$. Show that $L^{-1}(H)$ is a subspace of V.

7. Show that the function defined in Example 1 is not a linear transformation.

8. Let L_1 and L_2 both be linear transformations from V into W. Let $B = \{\mathbf{f}_k, k = 1, \ldots, n\}$ be any basis of V. Suppose that $L_1(\mathbf{f}_k) = L_2(\mathbf{f}_k)$ for each k. Show that $L_1(\mathbf{x}) = L_2(\mathbf{x})$ for every vector $\mathbf{x}$ in V.

9. Let $A = [a_{jk}]$ be an $m \times n$ matrix. If $L(\mathbf{x}) = A\mathbf{x}$, show that $L(\mathbf{e}_k)$ is the kth column of A.

10. Let $S = cI_2$, be an arbitrary 2×2 scalar matrix. Describe the geometrical effect that the linear transformation $S\mathbf{x}$ has on $\mathbb{R}^2$.

11. Suppose $D = \begin{bmatrix} d_1 & 0 \\ 0 & d_2 \end{bmatrix}$ is an arbitrary 2×2 diagonal matrix. Describe what happens to $\mathbf{x}$ under the linear transformation $D\mathbf{x}$ for various values of d_j.

12. If D is any invertible 2×2 diagonal matrix, describe geometrically the effects of the linear transformations defined by the two matrices D^{-1} and $D^{-1}D$.

13. Let P_n be the vector space of all polynomials of degree at most n. Define $L(\mathbf{p}) = t^2\mathbf{p}$ for each $\mathbf{p}$ in P_n. Then L can be thought of as a mapping from P_n to some vector space W. List some of these vector spaces, and then show that L is a linear transformation for each of your W's.

14. Let $V = C[0,1]$, the vector space of continuous functions defined on $[0,1]$.
a. Define $L[\mathbf{f}] = \int_0^1 \mathbf{f}(t)\, dt$. Show that L is a linear transformation from V to $\mathbb{R}^1$.
b. Define $T[\mathbf{f}](t) = \int_0^t \mathbf{f}(s)\, ds$, for each t in $[0,1]$. Show that T is a linear transformation from V to V.

15. Show that the operation of differentiation can be viewed as a linear transformation from P_n to P_{n-1}.

16. Let $V = C[0,1]$.
a. Let L: $V \to V$ be defined by $L[\mathbf{f}](x) = \mathbf{f}(x) \sin x$. Is L a linear transformation?
b. Let L: $V \to V$ be defined by $L[\mathbf{f}](x) = \sin x + \mathbf{f}(x)$. Is L a linear transformation?

17. Let $A = \begin{bmatrix} 2 & -3 \\ 1 & 5 \end{bmatrix}$. Let $V = M_{22}$.
a. Define L: $V \to V$ by $L(\mathbf{x}) = \mathbf{x}A$ (matrix multiplication). Compute $L(\mathbf{e}_j)$ for $j = 1, 2, 3, 4$, where the $\mathbf{e}_j$'s denote the standard basis vectors of M_{22}.
b. Show that L is a linear transformation.
c. Repeat parts a and b for L: $V \to V$ defined by $L(\mathbf{x}) = A\mathbf{x}$.

3.2 MATRIX REPRESENTATIONS

In the preceding section, matrices were used to define linear transformations. In this section we show that every linear transformation between two finite-dimensional vector spaces can be represented by a matrix. Suppose first that L is a linear transformation from $\mathbb{R}^n$ to $\mathbb{R}^m$. To find a matrix that can be used to represent L we do the following: let $\{\mathbf{e}_k\}$, $k = 1, 2, \ldots, n$ be the standard basis of $\mathbb{R}^n$. Then

$$L(\mathbf{e}_k) = (a_{1k}, a_{2k}, \ldots, a_{mk}) \tag{3.1}$$

for some constants a_{1k}, a_{2k}, etc. The subscript convention is important to remember when forming the matrix A, that will represent L. Thus, $A = [a_{jk}]$ is an $m \times n$ matrix, and the entries in the kth column of A are the coordinates of $L(\mathbf{e}_k)$ with respect to the standard basis in R^m.

EXAMPLE 1 Let L: $\mathbb{R}^3 \to \mathbb{R}^4$ be defined by

$$L(x_1, x_2, x_3) = (-6x_2 + 2x_3, x_1 - x_2 + x_3, \\ -x_1 + x_2 - 6x_3, 3x_1 - x_2 + 4x_3)$$

Then

$$L(\mathbf{e}_1) = L(1,0,0) = (0,1,-1,3) = (a_{11}, a_{21}, a_{31}, a_{41})$$
$$L(\mathbf{e}_2) = L(0,1,0) = (-6,-1,1,-1) = (a_{12}, a_{22}, a_{32}, a_{42})$$
$$L(\mathbf{e}_3) = L(0,0,1) = (2,1,-6,4) = (a_{13}, a_{23}, a_{33}, a_{43})$$

Thus,

$$A = \begin{bmatrix} 0 & -6 & 2 \\ 1 & -1 & 1 \\ -1 & 1 & -6 \\ 3 & -1 & 4 \end{bmatrix}$$

The next task is to show how to use this matrix in computing $L(\mathbf{x})$. Let $\mathbf{x} = (x_1, \ldots, x_n)$. Then

$$L(\mathbf{x}) = L\left(\sum_{k=1}^{n} x_k \mathbf{e}_k\right) = \sum_{k=1}^{n} x_k L(\mathbf{e}_k)$$
$$= \sum_{k=1}^{n} x_k \left[\sum_{j=1}^{n} a_{jk} \mathbf{e}_j\right] = \sum_{j=1}^{n} \left[\sum_{k=1}^{n} a_{jk} x_k\right] \mathbf{e}_j$$

Note: The coordinates of $L(\mathbf{x})$ with respect to the standard basis in $\mathbb{R}^m$ can be found by computing the matrix product $A\mathbf{x}^T$, where $[x_1, \ldots, x_n] = [\mathbf{x}]_S$. ■

EXAMPLE 2 In Example 1 we had

$$L(x_1, x_2, x_3) = (-6x_2 + 2x_3, x_1 - x_2 + x_3, -x_1 + x_2 - 6x_3, 3x_1 - x_2 + 4x_3)$$

with matrix representation:

$$A = \begin{bmatrix} 0 & -6 & 2 \\ 1 & -1 & 1 \\ -1 & 1 & -6 \\ 3 & -1 & 4 \end{bmatrix}$$

Computing $A\mathbf{x}$, we have

$$\begin{bmatrix} 0 & -6 & 2 \\ 1 & -1 & 1 \\ -1 & 1 & -6 \\ 3 & -1 & 4 \end{bmatrix} \begin{bmatrix} x_1 \\ x_2 \\ x_3 \end{bmatrix} = \begin{bmatrix} -6x_2 + 2x_3 \\ x_1 - x_2 + x_3 \\ -x_1 + x_2 - 6x_3 \\ 3x_1 - x_2 + 4x_3 \end{bmatrix}$$

Thus, $A\mathbf{x}$ gives us the coordinates of $L(\mathbf{x})$ in $\mathbb{R}^4$. ■

The preceding computations were based upon the vector spaces being $\mathbb{R}^n$ and $\mathbb{R}^m$ and using the standard bases. None of this is necessary in order for us to interpret L as matrix multiplication.

Let $L: V \rightarrow W$ be a linear transformation from V into W. Let $F = \{\mathbf{f}_k: k = 1, \ldots, n\}$, and $G = \{\mathbf{g}_j: j = 1, 2, \ldots, m\}$ be bases of V and W, respectively. Proceeding as before, we write $L(\mathbf{f}_k)$ as a linear combination of the basis vectors in G.

$$L(\mathbf{f}_k) = a_{1k}\mathbf{g}_1 + a_{2k}\mathbf{g}_2 + \cdots + a_{mk}\mathbf{g}_m = \sum_{j=1}^{m} a_{jk}\mathbf{g}_j \tag{3.2}$$

Let $A = [a_{jk}]$ be the $m \times n$ matrix whose kth column is $[L(\mathbf{f}_k)]_G^T$, the *coordinates* of $L[\mathbf{f}_k]$ with respect to the basis G. This matrix A depends upon three things:

1. The linear transformation L
2. The basis F in V
3. The basis G in W

If we change either of the bases picked, the matrix representation A will also change.

The next calculation illustrates how to use the representation A to calculate the coordinates of $L[\mathbf{x}]$. Let $\mathbf{x}$ be any vector in V. Let $[\mathbf{x}]_F = [x_1, \ldots, x_n]$.

We want to show that the coordinates of $L[\mathbf{x}]$ with respect to G are given by the matrix product $A[\mathbf{x}]_F^T$. Computing $L[\mathbf{x}]$ we have

$$\begin{aligned} L[\mathbf{x}] &= L\left[\sum_{k=1}^{n} x_k \mathbf{f}_k\right] = \sum_{k=1}^{n} x_k L[\mathbf{f}_k] \\ &= \sum_{k=1}^{n} x_k \left(\sum_{j=1}^{m} a_{jk}\mathbf{g}_j\right) \\ &= \sum_{j=1}^{m} \left(\sum_{k=1}^{n} (a_{jk}x_k)\right) \mathbf{g}_j \end{aligned}$$

This equation says that the jth coordinate of $L[\mathbf{x}]$ with respect to the basis G is $\Sigma_{k=1}^{n} a_{jk}x_k$, but this is just the jth row in the $m \times 1$ matrix $A[\mathbf{x}]_F^T$. In other words, when using A we do not compute $L[\mathbf{x}]$ directly, but rather the coordinates of $L[\mathbf{x}]$ with respect to the basis G, that is,

$$[L(\mathbf{x})]_G^T = A[\mathbf{x}]_F^T \tag{3.3}$$

EXAMPLE 3 Let $V = \mathbb{R}^2$ and $W = \mathbb{R}^3$. Define $L: V \to W$ by $L(x_1,x_2) = (x_1 - x_2, x_1, x_2)$. Let $F = \{(1,1),(-1,1)\}$, and let $G = \{(1,0,1),(0,1,1),(1,1,0)\}$.

a. Find the matrix representation of L using the standard bases in both V and W.

b. Find the matrix representation of L using the standard basis in V and the basis G in W.

c. Find the matrix representation of L using the basis F in $\mathbb{R}^2$ and the standard basis in $\mathbb{R}^3$.

d. Find the matrix representation of L using the bases F and G.

Solution

a. $L(\mathbf{e}_1) = L(1,0) = (1,1,0) = \mathbf{e}_1 + \mathbf{e}_2$
$L(\mathbf{e}_2) = L(0,1) = (-1,0,1) = -\mathbf{e}_1 + \mathbf{e}_3$

$$A = \begin{bmatrix} 1 & -1 \\ 1 & 0 \\ 0 & 1 \end{bmatrix}$$

b. $L(\mathbf{e}_1) = (1,1,0) = 0(1,0,1) + 0(0,1,1) + (1,1,0) = 0\mathbf{g}_1 + 0\mathbf{g}_2 + \mathbf{g}_3$
$L(\mathbf{e}_2) = (-1,0,1) = 0(1,0,1) + (0,1,1) - (1,1,0) = 0\mathbf{g}_1 + \mathbf{g}_2 - \mathbf{g}_3$

$$A = \begin{bmatrix} 0 & 0 \\ 0 & 1 \\ 1 & -1 \end{bmatrix}$$

c. $L(\mathbf{f}_1) = L(1,1) = (0,1,1) = \mathbf{e}_2 + \mathbf{e}_3$
$L(\mathbf{f}_2) = L(-1,1) = (-2,-1,1) = -2\mathbf{e}_1 - \mathbf{e}_2 + \mathbf{e}_3$

$$A = \begin{bmatrix} 0 & -2 \\ 1 & -1 \\ 1 & 1 \end{bmatrix}$$

d. $L(\mathbf{f}_1) = 0\mathbf{g}_1 + \mathbf{g}_2 + 0\mathbf{g}_3$
$L(\mathbf{f}_2) = 0\mathbf{g}_1 + \mathbf{g}_2 - 2\mathbf{g}_3$

$$A = \begin{bmatrix} 0 & 0 \\ 1 & 1 \\ 0 & -2 \end{bmatrix}$$

■

It is clear from this example that the matrix representation of a linear transformation depends upon which bases are used in V and in W. If V and W are the same vector spaces, then normally (in this text always) only one basis is used, rather than two different bases for the same vector space.

EXAMPLE 4 Let L: $\mathbb{R}^2 \to \mathbb{R}^2$ be a linear transformation. Let $F = \{(1,6), (-2,3)\}$. Suppose the matrix representation of L with respect to F is

$$A = \begin{bmatrix} 2 & 8 \\ -1 & -4 \end{bmatrix}$$

Compute $L(\mathbf{x})$ for any vector $\mathbf{x}$ in $\mathbb{R}^2$.

Solution. Let $\mathbf{x} = (x_1, x_2)$ be any vector in $\mathbb{R}^2$. To compute $L(\mathbf{x})$ using the matrix A we need to find $[\mathbf{x}]_F$, the coordinates of $\mathbf{x}$ with respect to the basis F. The change of basis matrix P below gives the basis F in terms of the standard basis

$$P = \begin{bmatrix} 1 & -2 \\ 6 & 3 \end{bmatrix}$$

Using P^{-1} we calculate $[\mathbf{x}]_F^T$

$$[\mathbf{x}]_F = P^{-1}\begin{bmatrix} x_1 \\ x_2 \end{bmatrix} = \frac{1}{15}\begin{bmatrix} 3 & 2 \\ -6 & 1 \end{bmatrix}\begin{bmatrix} x_1 \\ x_2 \end{bmatrix}$$
$$= \frac{1}{15}\begin{bmatrix} 3x_1 + 2x_2 \\ -6x_1 + x_2 \end{bmatrix}$$

Thus, the coordinates of $L(\mathbf{x})$ with respect to F are

$$A[\mathbf{x}]_F = \begin{bmatrix} 2 & 8 \\ -1 & -4 \end{bmatrix}\begin{bmatrix} (3x_1 + 2x_2)/15 \\ (-6x_1 + x_2)/15 \end{bmatrix}$$
$$= \frac{1}{15}\begin{bmatrix} -42x_1 + 12x_2 \\ 21x_1 - 6x_2 \end{bmatrix}$$

Thus

$$
\begin{aligned}
L(\mathbf{x}) &= (\tfrac{1}{15})(-42x_1 + 12x_2)\mathbf{f}_1 + (\tfrac{1}{15})(21x_1 - 6x_2)\mathbf{f}_2 \\
&= (\tfrac{1}{15})(-42x_1 + 12x_2)(1,6) + (\tfrac{1}{15})(21x_1 - 6x_2)(-2,3) \\
&= \frac{6x_2 - 21x_1}{15}\{2(1,6) - (-2,3)\} \\
&= \frac{6x_2 - 21x_1}{15}(4,9)
\end{aligned}
$$

■

EXAMPLE 5 Let $F = \left\{\begin{bmatrix} 1 & 0 \\ 0 & 0 \end{bmatrix}\begin{bmatrix} 0 & 1 \\ 0 & 0 \end{bmatrix}\begin{bmatrix} 0 & 0 \\ 1 & 0 \end{bmatrix}\begin{bmatrix} 0 & 0 \\ 0 & 1 \end{bmatrix}\right\}$. Thus F is the standard basis of M_{22}. Let $B = \begin{bmatrix} -2 & 1 \\ 3 & 4 \end{bmatrix}$. Define $L\colon M_{22} \to M_{22}$ by $L(\mathbf{x}) = B\mathbf{x}$. Find the matrix representation of L with respect to the standard basis F of M_{22}.

$$
\begin{aligned}
L(\mathbf{f}_1) = B\mathbf{f}_1 &= \begin{bmatrix} -2 & 1 \\ 3 & 4 \end{bmatrix}\begin{bmatrix} 1 & 0 \\ 0 & 0 \end{bmatrix} = \begin{bmatrix} -2 & 0 \\ 3 & 0 \end{bmatrix} \\
&= -2\mathbf{f}_1 + 3\mathbf{f}_3 \\
L(\mathbf{f}_2) = B\mathbf{f}_2 &= \begin{bmatrix} -2 & 1 \\ 3 & 4 \end{bmatrix}\begin{bmatrix} 0 & 1 \\ 0 & 0 \end{bmatrix} = \begin{bmatrix} 0 & -2 \\ 0 & 3 \end{bmatrix} \\
&= -2\mathbf{f}_2 + 3\mathbf{f}_4 \\
L(\mathbf{f}_3) = B\mathbf{f}_3 &= \begin{bmatrix} -2 & 1 \\ 3 & 4 \end{bmatrix}\begin{bmatrix} 0 & 0 \\ 1 & 0 \end{bmatrix} = \begin{bmatrix} 1 & 0 \\ 4 & 0 \end{bmatrix} \\
&= \mathbf{f}_1 + 4\mathbf{f}_3 \\
L(\mathbf{f}_4) = B\mathbf{f}_4 &= \begin{bmatrix} -2 & 1 \\ 3 & 4 \end{bmatrix}\begin{bmatrix} 0 & 0 \\ 0 & 1 \end{bmatrix} = \begin{bmatrix} 0 & 1 \\ 0 & 4 \end{bmatrix} \\
&= \mathbf{f}_2 + 4\mathbf{f}_4.
\end{aligned}
$$

Thus, the matrix representation of L is

$$
\begin{bmatrix} -2 & 0 & 1 & 0 \\ 0 & -2 & 0 & 1 \\ 3 & 0 & 4 & 0 \\ 0 & 3 & 0 & 4 \end{bmatrix}
$$

■

In the following pages, when we say A, an $m \times n$ matrix, is a linear transformation or represents a linear transformation without specifically mentioning a basis or vector spaces, it is to be understood that $V = \mathbb{R}^n$, $W = \mathbb{R}^m$, and that the standard bases in both V and W are being used.

PROBLEM SET 3.2

1. Let $L(x_1,x_2) = (3x_1 + 6x_2, -2x_1 + x_2)$
   a. Find the matrix representation of L using the standard bases.

b. Find the matrix representation of L using the basis $F = \{(-4,1), (2,3)\}$.

2. Let $L: \mathbb{R}^2 \to \mathbb{R}^4$ have matrix representation $A = \begin{bmatrix} 6 & 1 \\ -4 & 0 \\ 2 & 9 \\ 8 & -3 \end{bmatrix}$ with respect to the standard bases.

a. $L(\mathbf{e}_1) = ?$, $L(\mathbf{e}_2) = ?$

b. $L(-3,7) = ?$ **c.** $L(x_1,x_2) = ?$

3. Let L be a linear transformation from $\mathbb{R}^2$ into $\mathbb{R}^2$. Define $L^2(\mathbf{x}) = L(L(\mathbf{x}))$, $L^3(\mathbf{x}) = L(L^2(\mathbf{x}))$, and $L^{n+1}(\mathbf{x}) = L(L^n(\mathbf{x}))$. Suppose $L(x_1,x_2) = (ax_1 + bx_2, cx_1 + dx_2)$.

a. Find the matrix representation A of L with respect to the standard bases.

b. Show that the matrix representation of L^2 with respect to the standard bases is A^2, and in general the matrix representation of L^n with respect to the standard bases is A^n.

c. What can you say if some basis other than the standard basis is used?

4. Let $V = \mathbb{R}^3$ and let $F = \{(1,2,-3), (1,0,0), (0,1,0)\}$. Suppose that the matrix A represents a linear transformation $L: \mathbb{R}^3 \to \mathbb{R}^3$ with respect to the basis F, where

$$A = \begin{bmatrix} 0 & 1 & -2 \\ 2 & 1 & 0 \\ 5 & 0 & 1 \end{bmatrix}$$

a. $L(1,2,-3) = ?$ **b.** $L(1,0,0) = ?$ **c.** $L(x_1,x_2,x_3) = ?$

5. Let V be a vector space of dimension n. Define $L: V \to V$ by $L(\mathbf{x}) = c\mathbf{x}$, where c is any constant. Let F be any basis of V. What is the matrix representation of L with respect to this basis?

6. Let $L_1(x_1,x_2) = (x_1 - x_2, 2x_1 + 3x_2)$ and let $L_2(x_1,x_2) = (2x_1 - 5x_2, 3x_1 - x_2)$. Define $(L_1 + L_2)(\mathbf{x}) = L_1(\mathbf{x}) + L_2(\mathbf{x})$.

a. Find the matrix representations A_1 and A_2 of L_1 and L_2, respectively, with respect to the standard basis of R^2.

b. Show that the matrix representation of $L_1 + L_2$ is $A_1 + A_2$.

7. Let L_1 and L_2 be two linear transformations mapping $\mathbb{R}^2$ into $\mathbb{R}^2$. Let F be any basis of $\mathbb{R}^2$. Let A_1 and A_2 be the matrix representations of L_1 and L_2, with respect to the basis F, respectively. Show that the matrix representation of $L_1 + L_2$ with respect to F is $A_1 + A_2$.

8. Let $L(x_1,x_2,x_3) = (x_2 + x_3, 6x_1 - x_2 + 3x_3, 2x_1 + 3x_2 - 7x_3, 2x_1 + 6x_3)$.

a. Compute $L[\mathbf{e}_k]$ for $k = 1,2,3$.

b. Find the matrix representation A, of L, with respect to the standard bases in $\mathbb{R}^3$ and $\mathbb{R}^4$.

c. Compute $A[\mathbf{x}]$ for any vector $\mathbf{x}$ in $\mathbb{R}^3$.

9. Define $L: \mathbb{R}^4 \to \mathbb{R}^2$ by $L(x_1,x_2,x_3,x_4) = (x_2 + 2x_3 + 3x_4, 2x_1 - 6x_4)$.

a. Compute $L(\mathbf{e}_k)$ for $k = 1, 2, 3, 4$.

b. Find the matrix representation A, of L, with respect to the standard bases in $\mathbb{R}^4$ and $\mathbb{R}^2$.

c. Compute $A[\mathbf{x}]$ for any vector $\mathbf{x}$ in $\mathbb{R}^4$.

10. Let L be a linear transformation from $\mathbb{R}^3$ to $\mathbb{R}^2$. Let $F = \{(1,1,1), (0,1,1), (1,1,0)\} = \{\mathbf{f}_1,\mathbf{f}_2,\mathbf{f}_3\}$. Let $G = \{(1,2), (2,3)\} = \{\mathbf{g}_1,\mathbf{g}_2\}$. Suppose that the matrix representation of L with respect to these bases is $\begin{bmatrix} 2 & 0 & 4 \\ 1 & -2 & 1 \end{bmatrix}$.

a. For $k = 1, 2, 3$, $[L(\mathbf{f}_k)]_G = ?$
b. Compute $L(\mathbf{f}_k)$ for $k = 1, 2, 3$.
c. Find the matrix representation of L using the standard basis S in $\mathbb{R}^3$ and the basis G in $\mathbb{R}^2$.
d. Find the matrix representation of L using the standard basis S in both $\mathbb{R}^3$ and $\mathbb{R}^2$.

11. Let $L: \mathbb{R}^2 \to \mathbb{R}^2$. Let $F = \{\mathbf{f}_1, \mathbf{f}_2\}$ be a basis of $\mathbb{R}^2$. Suppose that

$$A = \begin{bmatrix} -2 & 0 \\ 0 & 3 \end{bmatrix}$$

is the matrix representation of L with respect to the basis F. What is $L(\mathbf{f}_k)$ for $k = 1, 2$?

12. Let L be the linear transformation that rotates the plane through an angle of θ degrees. Let A be the matrix representation of L. Then as we saw in Section 3.1

$$A = \begin{bmatrix} \cos\theta & -\sin\theta \\ \sin\theta & \cos\theta \end{bmatrix}$$

Find the matrix representations of $L^2, L^3, \ldots, L^n$. (Hint: What is L^2 geometrically?)

13. Let L_1 and L_2 be two linear transformations from $\mathbb{R}^2$ to $\mathbb{R}^2$. Define the composition of L_1 with L_2 by $L_1 \circ L_2(\mathbf{x}) = L_1(L_2(\mathbf{x}))$.
a. Show that the composition of two linear transformations is also a linear transformation.
b. If A_1 and A_2 are the matrix representations of L_1 and L_2 with respect to the same basis, respectively, show that the matrix representation of the composition $L_1 \circ L_2$ is given by the matrix product $A_1 A_2$.

14. Find the matrix representations for each of the following linear transformations with respect to the standard basis of the vector space in question:
a. $L: P_n \to P_{n-1}$ by $L(\mathbf{p}) = \mathbf{p}'$, i.e., $L(\mathbf{p})$ is the derivative of the polynomial $\mathbf{p}$.
b. $L: P_n \to P_n$ by $L(\mathbf{p}) = \mathbf{p}'$.
c. $L: P_n \to P_{n+2}$ by $L(\mathbf{p}) = t^2\mathbf{p}$.

15. Define $L[\mathbf{p}](t) = \int_0^t \mathbf{p}(s)ds$, for each t in $[0,1]$. Then $L: P_n \to P_{n+1}$. Find a matrix representation for L using the standard bases.

16. If A is an $m \times n$ matrix, we can think of A as a linear transformation from $\mathbb{R}^n$ to $\mathbb{R}^m$. What spaces are appropriate for each of the following matrices to be thought of as a linear transformation?
a. A^T b. A^TA c. AA^T

17. Let L be a linear transformation from a vector space V to a vector space W. Suppose that $L(\mathbf{x}) = \mathbf{0}$ for every vector $\mathbf{x}$ in V. What must any matrix representation of L equal?

18. Let $V = \{\sum_{j=1}^{2} a_j \cos jx + b_j \sin jx : a_j \text{ and } b_j \text{ arbitrary}\}$. Define $L: V \to V$ by

$$L\left(\sum_{j=1}^{2} a_j \cos jx + b_j \sin jx\right) = \sum_{j=1}^{2} (-ja_j \sin jx + jb_j \cos jx)$$

a. Find a basis F for the vector space V.
b. Find the matrix representation A of L with respect to your basis.

19. Using the same notation as in Example 5, define $L: M_{22} \to M_{22}$ by $L(\mathbf{x}) = \mathbf{x}B$. Find the matrix representation of L with respect to the standard basis of M_{22}.

20. Let $G = \left\{ \begin{bmatrix} 0 & 1 \\ 1 & 1 \end{bmatrix} \begin{bmatrix} 1 & 0 \\ 1 & 1 \end{bmatrix} \begin{bmatrix} 1 & 1 \\ 0 & 1 \end{bmatrix} \begin{bmatrix} 1 & 1 \\ 1 & 0 \end{bmatrix} \right\}$. Define L: $M_{22} \to M_{23}$ by $L[\mathbf{x}] = \mathbf{x}B = \mathbf{x}\begin{bmatrix} -1 & 3 & 4 \\ 2 & 0 & 1 \end{bmatrix}$. Using the standard basis in M_{23} and the basis G in M_{22}, find the matrix representation of L.

21. Let B and G be as in problem 20. Define L: $M_{32} \to M_{22}$ by $L[\mathbf{x}] = B\mathbf{x}$. Using the standard basis in M_{32} and the basis G in M_{22}, find the matrix representation of L.

22. Let L: $M_{22} \to M_{22}$ be defined by

$$L\left(\begin{bmatrix} a & b \\ c & d \end{bmatrix}\right) = \begin{bmatrix} a & 0 \\ 0 & 0 \end{bmatrix}$$

a. Show that L is a linear transformation.
b. Find the matrix representation of L with respect to the standard basis of M_{22}.
c. Show that there is no 2×2 matrix B such that $L[\mathbf{x}] = B\mathbf{x}$ ($L[\mathbf{x}] = \mathbf{x}B$) for all $\mathbf{x}$ in M_{22}.

23. Let V be the vector space in problem 18. For each $\mathbf{f}$ in V define $L[\mathbf{f}](t) = \int_0^t f(s)ds$. Show that L is a linear transformation from V to V. Find its matrix representation with respect to the basis found in problem 18. Show that the product of the matrix found in problem 18 with the matrix found in this problem equals I_4.

3.3 KERNEL AND RANGE OF A LINEAR TRANSFORMATION

For any linear transformation L, mapping V into W, there are two important subspaces associated with L. The first is a subspace of V called the kernel of L; the second is a subspace of W called the range of L. In this section we define these two subspaces and describe their relation to the solution set of a system of equations.

Definition 3.2. Let L: $V \to W$. The kernel of L is the set of vectors $\mathbf{x}$ in V for which $L(\mathbf{x}) = \mathbf{0}$. Letting ker($L$) represent the kernel of L, we have ker(L) = $\{\mathbf{x}: L(\mathbf{x}) = \mathbf{0}\}$.

EXAMPLE 1 Let $A = \begin{bmatrix} 2 & -6 & 4 \\ 1 & -1 & 2 \end{bmatrix}$ be the matrix representation of L. Find the kernel K of this linear transformation.

Solution. Since A is a 2×3 matrix, A: $\mathbb{R}^3 \to \mathbb{R}^2$. We are asked to find those $\mathbf{x} = (x_1, x_2, x_3)$ such that

$$A\mathbf{x} = \begin{bmatrix} 2 & -6 & 4 \\ 1 & -1 & 2 \end{bmatrix}\begin{bmatrix} x_1 \\ x_2 \\ x_3 \end{bmatrix} = \begin{bmatrix} 2x_1 - 6x_2 + 4x_3 \\ x_1 - x_2 + 2x_3 \end{bmatrix} = \begin{bmatrix} 0 \\ 0 \end{bmatrix}$$

Thus, $\mathbf{x}$ is in the kernel of A if and only if $2x_1 - 6x_2 + 4x_3 = 0 = x_1 - x_2 + 2x_3$. Hence $K = \{(x_1, x_2, x_3)\colon x_1 + 2x_3 = 0 = x_2\}$. ■

This example demonstrates that the kernel is just the solution set of a homogeneous system of linear equations. We note that K has dimension equal to 1 and that $(-2,0,1)$ is a basis of K.

Definition 3.3. Let L be a linear transformation mapping V into W. The range of L is the set of vectors $\mathbf{w}$ in W such that $L(\mathbf{x}) = \mathbf{w}$, for some vector $\mathbf{x}$ in V. Thus, $\text{Rg}(L) = \text{range}(L) = \{\mathbf{w}\colon \mathbf{w} = L(\mathbf{x}) \text{ for some } \mathbf{x} \text{ in } V\}$.

EXAMPLE 2 Find the range of the linear transformation in Example 1.

Solution. Since $A\colon \mathbb{R}^3 \to \mathbb{R}^2$, the range of A consists of those $\mathbf{w}$ in $\mathbb{R}^2$ such that $A\mathbf{x} = \mathbf{w}$ has a solution. The augmented matrix of the associated system is

$$\begin{bmatrix} 2 & -6 & 4 & w_1 \\ 1 & -1 & 2 & w_2 \end{bmatrix}$$

It is clear that this system has a solution regardless of the values of w_1 and w_2, e.g., $x_1 = (6w_2 - w_1)/4$, $x_2 = (2w_2 - w_1)/4$, and $x_3 = 0$. Thus, $\text{Rg}(L) = R^2$. ■

Theorem 3.2. Let $L\colon V \to W$ be a linear transformation. Then

a. $\ker(L)$ is a subspace of V.
b. $\text{Rg}(L)$ is a subspace of W.

Proof. Since $L(\mathbf{0}) = \mathbf{0}$, we know that both the kernel and the range are nonempty. Thus, to show that these two sets are subspaces we may use Theorem 2.6. Hence, suppose that $\mathbf{x}$ and $\mathbf{y}$ are in $K = \ker(L)$. Then $L(\mathbf{x} + \mathbf{y}) = L(\mathbf{x}) + L(\mathbf{y}) = \mathbf{0} + \mathbf{0} = \mathbf{0}$. Thus, K is closed under vector addition. Now let a be any scalar; then $L(a\mathbf{x}) = aL(\mathbf{x}) = a\mathbf{0} = \mathbf{0}$, and we see that K is also closed under scalar multiplication. This shows that K is a subspace. To see that $\text{Rg}(L)$ is a subspace, suppose that $\mathbf{u}$ and $\mathbf{v}$ are any two vectors in $\text{Rg}(L)$. Then there are two vectors $\mathbf{x}$ and $\mathbf{y}$ in V such that $L(\mathbf{x}) = \mathbf{u}$ and $L(\mathbf{y}) = \mathbf{v}$. Then $L(\mathbf{x} + \mathbf{y}) = L(\mathbf{x}) + L(\mathbf{y}) = \mathbf{u} + \mathbf{v}$ and $\text{Rg}(L)$ is closed under addition. Similarly if a is any constant, then $a\mathbf{u} = aL(\mathbf{x}) = L(a\mathbf{x})$. Since $\text{Rg}(L)$ is closed under vector addition and scalar multiplication, it is a subspace of W.

Consider a system of m linear equations in n unknowns

$$\begin{aligned} a_{11}x_1 + \cdots + a_{1n}x_n &= b_1 \\ \cdots\cdots\cdots&\cdots\cdots \\ a_{m1}x_1 + \cdots + a_{mn}x_n &= b_m \end{aligned} \tag{3.4}$$

Let L be the linear transformation from $\mathbb{R}^n$ to $\mathbb{R}^m$ defined by $L[\mathbf{x}] = A\mathbf{x}$, where A is the coefficient matrix $[a_{jk}]$ of (3.4). Then the kernel of L is just the solution set of the homogeneous system associated with (3.4). For $\mathbf{x}$ is in $\ker(L)$ if and only if $L(\mathbf{x}) = \mathbf{0}$, but $L(\mathbf{x}) = \mathbf{0}$ if and only if $A\mathbf{x} = \mathbf{0}$. That is, $\mathbf{x}$ is in $\ker(L)$ if and only if x is a solution of (3.4) with $b_j = 0$ for $j = 1, 2, \ldots, m$. The range of L consists of those vectors $\mathbf{b}$ in $\mathbb{R}^m$ such that there is an $\mathbf{x}$ in $\mathbb{R}^n$ for which $L(\mathbf{x}) = \mathbf{b}$, i.e., $A\mathbf{x} = \mathbf{b}$. That is, $\mathbf{b}$ is in the range of L if and only if (3.4) has a solution.

EXAMPLE 3 Consider the following system of equations:

$$\begin{aligned} -x_1 + 2x_2 + 3x_4 &= b_1 \\ 2x_1 + 3x_2 + 7x_3 + 8x_4 &= b_2 \\ 4x_1 - 2x_2 + 6x_3 &= b_3 \end{aligned} \tag{3.5}$$

Find the kernel and range of the coefficient matrix of the above system of equations. Then determine the dimensions of these two subspaces.

Solution. The coefficient matrix A equals

$$\begin{bmatrix} -1 & 2 & 0 & 3 \\ 2 & 3 & 7 & 8 \\ 4 & -2 & 6 & 0 \end{bmatrix}$$

and is row equivalent to the matrix

$$\begin{bmatrix} 1 & 0 & 2 & 1 \\ 0 & 1 & 1 & 2 \\ 0 & 0 & 0 & 0 \end{bmatrix}$$

Thus, $\mathbf{x}$ is a solution to the homogeneous system, i.e., $\mathbf{x}$ is in $\ker(A)$ if and only if $x_2 = -x_3 - 2x_4$ and $x_1 = -2x_3 - x_4$. Thus, $\ker(A) = \{(x_1, x_2, x_3, x_4){:}x_1 = -2x_3 - x_4, x_2 = -x_3 - 2x_4\}$. A basis for $\ker(A)$ is $\{(-2,-1,1,0), (-1,-2,0,1)\}$. Thus, $\dim(\ker(A)) = 2$.

The augmented matrix of (3.5)

$$\begin{bmatrix} -1 & 2 & 0 & 3 & b_1 \\ 2 & 3 & 7 & 8 & b_2 \\ 4 & -2 & 6 & 0 & b_3 \end{bmatrix}$$

is row equivalent to

$$\begin{bmatrix} -1 & 2 & 0 & 3 & b_1 \\ 0 & 1 & 1 & 2 & (b_2 + 2b_1)/7 \\ 0 & 0 & 0 & 0 & (26b_1 - 2b_2 + 7b_3)/14 \end{bmatrix}$$

(3.5) has a solution if and only if

$$26b_1 - 2b_2 + 7b_3 = 0$$

Thus, $\text{Rg}(A) = \{(b_1, b_2, b_3){:}\ 26b_1 - 2b_2 + 7b_3 = 0\}$. A basis for $\text{Rg}(A)$ is $\{(1,13,0), (7,0,-26)\}$ and $\dim(\text{Rg}(A)) = 2$. ■

In the previous example $A: \mathbb{R}^4 \to \mathbb{R}^3$, $\dim(\ker(A)) = 2$, and $\dim(\mathrm{Rg}(A)) = 2$. It is not a coincidence that we have the following relationship: $\dim(\ker(A)) + \dim(\mathrm{Rg}(A)) = \dim(\mathbb{R}^4)$.

Theorem 3.3. Let L be a linear transformation from V to W, where V is a finite dimensional vector space. Then

$$\dim(\ker(L)) + \dim(\mathrm{Rg}(L)) = \dim(V) \tag{3.6}$$

Proof. Let $\dim(V) = n$. Suppose that $\dim(\ker(L)) = k$, where we assume first that $0 < k < n$. Let $\mathbf{x}_j$, $j = 1, \ldots, k$ be a basis for $\ker(L)$ and let $\mathbf{y}_j$, $j = 1, \ldots, n-k$ be a set of $n-k$ linearly independent vectors in V such that $S = \{\mathbf{x}_1, \ldots, \mathbf{x}_k, \mathbf{y}_1, \ldots, \mathbf{y}_{n-k}\}$ is a basis of V. Then $\{L(\mathbf{x}_1), \ldots, L(\mathbf{y}_1), \ldots, L(\mathbf{y}_{n-k})\}$ is a spanning set of $\mathrm{Rg}(L)$. Since the $\mathbf{x}_j$ are in $\ker(L)$, we have $L(\mathbf{x}_j) = \mathbf{0}$ for $j = 1, \ldots, k$. Thus, $\{L(\mathbf{y}_1), \ldots, L(\mathbf{y}_{n-k})\}$ must span $\mathrm{Rg}(L)$. We wish to show that this set is linearly independent, and hence forms a basis for $\mathrm{Rg}(L)$. To this end suppose that

$$c_1 L(\mathbf{y}_1) + \cdots + c_{n-k} L(\mathbf{y}_{n-k}) = \mathbf{0}$$

Setting $\mathbf{z} = c_1 \mathbf{y}_1 + \cdots + c_{n-k}\mathbf{y}_{n-k}$, we have $L(\mathbf{z}) = \mathbf{0}$. Thus, $\mathbf{z}$ is in $\ker(L)$ and there are constants a_j such that

$$a_1 \mathbf{x}_1 + \cdots + a_k \mathbf{x}_k = \mathbf{z} = c_1 \mathbf{y}_1 + \cdots + c_{n-k}\mathbf{y}_{n-k}$$

Since the set S is linearly independent, every one of the constants must be zero. In particular $c_1 = c_2 = \cdots = c_{n-k} = 0$, and we conclude that the set $\{L(\mathbf{y}_1), \ldots, L(\mathbf{y}_{n-k})\}$ is a basis for $\mathrm{Rg}(L)$. Hence we have

$$\dim(\ker(L)) + \dim(\mathrm{Rg}(L)) = k + (n - k) = n = \dim(V)$$

This equation remains to be verified in the two cases $k = 0$ and $k = n$. We leave the details as an exercise for the reader.

In the next section we show how one can easily determine the dimension of $\mathrm{Rg}(L)$. This technique coupled with the above theorem gives us an effective means of determining how large the solution space of a set of homogeneous linear equations is, and hence the size of the solution set for any system of linear equations, homogeneous or not; cf. Theorem 1.10.

Before starting the next section, we define several terms.

Definition 3.4. Let $L: V \to W$ be a linear transformation. L is said to be one-to-one if $L(\mathbf{x}) = L(\mathbf{y})$ implies that $\mathbf{x} = \mathbf{y}$.

Definition 3.5. Let $L: V \to W$ be a linear transformation. L is said to be onto if $\mathrm{Rg}(L) = W$.

EXAMPLE 4 Let $L: \mathbb{R}^3 \to \mathbb{R}^2$ have matrix representation A, where A is given below. Show L is onto but not one-to-one.

$$A = \begin{bmatrix} 1 & 2 & 3 \\ 0 & 1 & 2 \end{bmatrix}$$

Solution. To see that L is not one-to-one we observe that the vector $(1,-2,1)$ is in $\ker(L)$; that is, $L(1,-2,1) = (0,0) = L(0,0,0)$, but $(1,-2,1) \neq (0,0,0)$. Hence, L is not one-to-one. To see that L is onto we have to show that $\text{Rg}(L) = \mathbb{R}^2$. Thus, let (w_1,w_2) be any vector in $\mathbb{R}^2$. Our task is to find $\mathbf{x} = (x_1,x_2,x_3)$ such that $L(\mathbf{x}) = (w_1,w_2)$. Equivalently we need to solve the following system of equations:

$$\begin{aligned} x_1 + 2x_2 + 3x_3 &= w_1 \\ x_2 + 2x_3 &= w_2 \end{aligned}$$

A solution to this system is $x_1 = w_1 - 2w_2$, $x_2 = w_2$, and $x_3 = 0$. ■

Theorem 3.4. Let $L: V \to W$ be a linear transformation. Assume, for parts **b** and **c**, that V and W are finite dimensional. Then

a. L is one-to-one if and only if $\ker(L) = \{\mathbf{0}\}$.
b. L is one-to-one if and only if $\dim(V) = \dim(\text{Rg}(L))$.
c. L is onto if and only if $\dim(\text{Rg}(L)) = \dim(W)$.

Proof. Suppose L is one-to-one. We want to show that $K = \ker(L) = \{\mathbf{0}\}$. Thus, suppose that $\mathbf{x}$ is in K. Then $L(\mathbf{x}) = \mathbf{0} = L(\mathbf{0})$ and we must have $\mathbf{x} = \mathbf{0}$. Conversely, suppose $K = \{\mathbf{0}\}$. Then if $L(\mathbf{x}) = L(\mathbf{y})$, we must have $L(\mathbf{x} - \mathbf{y}) = 0$. Hence, $\mathbf{x} - \mathbf{y}$ is in K, and we conclude that $\mathbf{x} = \mathbf{y}$. Part b of this theorem is an easy consequence of part a and Theorem 3.3. Suppose that L is one-to-one. Then we have $\dim(V) = \dim(\text{Rg}(L)) + \dim(\ker(L)) = \dim(\text{Rg}(L)) + 0 = \dim(\text{Rg}(L))$. Conversely, if $\dim(\text{Rg}(L)) = \dim(V)$, then $\dim(\ker(L)) = 0$, and we have $\ker(L) = \{0\}$. The verification of part c is left to the reader as an exercise.

PROBLEM SET 3.3

1. Each of the matrices below represents a linear transformation from $\mathbb{R}^n$ to $\mathbb{R}^m$. Determine the values of n and m for each matrix. Then determine their kernels and ranges and find a basis for each of these subspaces.

a. $\begin{bmatrix} 1 & 0 & 1 \end{bmatrix}$ **b.** $\begin{bmatrix} 1 \\ 0 \\ 1 \end{bmatrix}$ **c.** $\begin{bmatrix} 1 & 1 \\ 0 & 1 \end{bmatrix}$ **d.** $\begin{bmatrix} 1 & 1 & 1 \\ 1 & 0 & 1 \\ 1 & 1 & 1 \\ 1 & 0 & -1 \end{bmatrix}$

2. Each of the matrices below represents a linear transformation from $\mathbb{R}^n$ to $\mathbb{R}^m$. Determine the values of n and m for each matrix, and their respective kernels and ranges.

a. $[1 \quad 2]$ **b.** $\begin{bmatrix} 1 & 2 \\ -1 & 0 \\ 0 & 1 \end{bmatrix}$ **c.** $\begin{bmatrix} 0 \\ 1 \end{bmatrix}$

3. Let A be the coefficient matrix of the system of equations below. If L is the linear transformation defined by A, what is the range of L and what is its kernel? Does this particular equation have a solution; i.e., is $(-2,1)$ in the range of L?

$$\begin{aligned} 2x_1 - 5x_2 + 3x_3 &= -2 \\ x_1 + 3x_2 \qquad &= 1 \end{aligned}$$

4. For each of the matrices below determine the dimension of its range and the dimension of its kernel. Then decide if the linear transformations represented by these matrices are onto and/or one-to-one.

a. $[1 \quad 2]$ **b.** $\begin{bmatrix} 1 & 2 \\ -1 & 0 \\ 0 & 1 \end{bmatrix}$ **c.** $\begin{bmatrix} 1 \\ 2 \end{bmatrix}$

5. For each of the matrices below determine the dimensions of their range and kernel. Then determine if the linear transformations represented by these matrices are onto and/or one-to-one.

a. $\begin{bmatrix} 1 & 0 & -1 \\ 0 & 0 & 4 \\ 1 & 0 & 0 \\ 0 & 0 & 1 \end{bmatrix}$ **b.** $\begin{bmatrix} 1 & 2 & -1 & 3 \\ 1 & -1 & 1 & -1 \\ 0 & 1 & 0 & 1 \\ 1 & 0 & 1 & 0 \\ 1 & 1 & 0 & 0 \end{bmatrix}$

6. Verify part c of Theorem 3.4. Remember, the range of L is always a subspace of W.

7. Let $L: V \to W$ be a linear transformation. Let $\{\mathbf{x}_j: j = 1, \ldots, n\}$ be a basis of V. Show that the set $\{L(\mathbf{x}_j): j = 1, \ldots, n\}$ is a spanning set for Rg(L).

8. Let $L: \mathbb{R}^n \to \mathbb{R}^m$ be a linear transformation.
   a. Show that if $n > m$, then L must have a nontrivial kernel, i.e., $\dim(\ker(L)) > 0$.
   b. If $n \leq m$ does L have to be one-to-one?

9. Let $L: \mathbb{R}^n \to \mathbb{R}^m$ be a linear transformation.
   a. If $n < m$, show that L cannot be onto.
   b. If $n \geq m$, must L be onto?

10. Let $L: V \to W$ be a linear transformation. Let Q be that subset of W that contains all vectors in W not in the range of L, i.e., $Q = W \backslash \text{Rg}(L)$. Is Q a subspace of W?

11. Let S be any $n \times n$ scalar matrix, i.e., $S = cI_n$ for some constant c. Determine the kernel and range of S for various values of c.

12. Let D be any $n \times n$ diagonal matrix. Determine the kernel and range of D for various values of the diagonal entries. For example, what happens if the entry in the 1,1 position of D is zero? Is nonzero?

13. Characterize the kernel and range for the linear transformations in problems 13, 14, and 15 in Problem Set 3.1.

14. For each of the matrices A in problem 1, compute A^T. Then determine the kernel and range of A^T.

15. For each of the matrices A in problem 1, compute A^TA. Determine if these product matrices are one-to-one and/or onto. Compare the kernels of A^TA and A.

16. Let $L: \mathbb{R}^n \to \mathbb{R}^m$ be a linear transformation. Show that if L is both onto and one-to-one, then $m = n$.

17. Verify Theorem 3.3 for the two cases $\ker(L) = \{\mathbf{0}\}$ and $\ker(L) = V$.

18. Let $L: P_2 \to P_3$ be defined be $L[\mathbf{p}](t) = \int_0^t \mathbf{p}(s)ds$. Find a matrix representation for L using the standard basis in each of the vector spaces. Find a basis for the range and kernel of L.

19. Let $L: P_3 \to P_3$ be defined by $L[\mathbf{p}](t) = \mathbf{p}'(t)$. Find a matrix representation for L using the standard basis in each of the vector spaces. Find a basis for the range and kernel of L.

20. Let $B = \begin{bmatrix} -1 & 2 & 5 \\ 2 & 3 & 1 \end{bmatrix}$.

a. Let $L[\mathbf{x}] = \mathbf{x}B$ for any $\mathbf{x}$ in M_{22}. Then L is a linear transformation from M_{22} to M_{23}. Find a basis for the kernel of L and also a basis for the range of L.
b. Let $L[\mathbf{x}] = B\mathbf{x}$ for any $\mathbf{x}$ in M_{32}. Then L is a linear transformation from M_{32} to M_{22}. Find a basis for the kernel of L and also a basis for the range of L.
c. Find a matrix representation for each of the above two linear transformations. Use the standard bases.

3.4 RANK OF A MATRIX

In the last section we proved a theorem that said the dimensions of the kernel, range, and domain of a linear transformation are related by the equation $\dim(V) = \dim(\ker) + \dim(\mathrm{Rg})$. We have also seen that the kernel is just the solution set for a system of homogeneous equations. In this section we show that the dimension of the range of L is the same as the maximum number of linearly independent rows or columns in a matrix representation of L. Since this number is easy to calculate, we have an effective method for computing the dimension of $\mathrm{Rg}(L)$ and hence an efficient method of determining the size of the solution set of a system of linear equations.

Definition 3.6. Let $A = [a_{jk}]$ be an $m \times n$ matrix. Each row of A can be thought of as a vector in $\mathbb{R}^n$ and each column of A can be considered a vector in $\mathbb{R}^m$. The row space of A is that subspace of $\mathbb{R}^n$ spanned by the row vectors of A, and the column space of A is that subspace of $\mathbb{R}^m$ spanned by the column vectors of A.

Definition 3.7. Let $A = [a_{jk}]$ be an $m \times n$ matrix. The row rank of A is the dimension of the row space of A and the column rank of A is the dimension of the column space of A.

EXAMPLE 1 Let $A = \begin{bmatrix} -2 & -1 \\ 1 & 1 \\ 0 & 4 \end{bmatrix}$. The row space of A is that subspace of $\mathbb{R}^2$ spanned by the set $\{(-2,-1),(1,1),(0,4)\}$. Clearly this set spans a subspace of $\mathbb{R}^2$ of dimension 2. Hence the row space of A is $\mathbb{R}^2$, and the row rank is 2. The column space of A is that subspace of $\mathbb{R}^3$ spanned by the set $\{(-2,1,0), (-1,1,4)\}$. Since this set is linearly independent the column space has dimension 2. Thus the column rank is 2. ■

The fact that the row rank of A was equal to its column rank was no accident, as the following theorem shows.

Theorem 3.5. Let $A = [a_{ik}]$ be an $m \times n$ matrix. Then the column rank and the row rank of A are equal.

Proof. Suppose the column rank of A is p. Then $0 \le p \le n$. If $p = 0$, every column of A is the zero vector in $\mathbb{R}^m$, and hence every row of A is the zero vector in $\mathbb{R}^n$. Thus the row space is the zero vector and we have row rank equal to 0 also. Now assume that $p > 0$. Let $\{\mathbf{z}_j : j = 1, \ldots, p\}$ be a basis for the column space of A, where

$$\mathbf{z}_j = (z_{1j}, z_{2j}, \ldots, z_{mj})$$

Then if $\mathbf{C}_k$ is the kth column of A, i.e., $\mathbf{C}_k = (a_{1k}, a_{2k}, \ldots, a_{mk})^T$, there are constants b_{jk}, $1 \le j \le p$, such that

$$\mathbf{C}_k = \sum_{j=1}^{p} b_{jk}\mathbf{z}_j^T$$

Equating components, we have the following:

$$a_{ik} = \sum_{j=1}^{p} b_{jk} z_{ij} \qquad 1 \le k \le n,\ 1 \le i \le m$$

Thus if r_i is the ith row of A, we have

$$\begin{aligned} r_i &= (a_{i1}, a_{i2}, a_{i3}, \ldots, a_{in}) \\ &= \left(\sum_{j=1}^{p} b_{j1} z_{ij}, \sum_{j=1}^{p} b_{j2} z_{ij}, \ldots, \sum_{j=1}^{p} b_{jn} z_{ij} \right) \\ &= \sum_{j=1}^{p} z_{ij}(b_{j1}, b_{j2}, \ldots, b_{jn}) \end{aligned}$$

Thus, the p row vectors $(b_{j1}, \ldots, b_{jn})$, $1 \le j \le p$, form a spanning set for the row space of A. Hence, row rank $\le p$. This shows that the row rank of any matrix must be less than or equal to its column rank. Since the row rank of A^T is the column rank of A, and the column rank of A^T is the row rank of A, we also have that the column rank is less than or equal to the row rank. Hence the row and column ranks are equal.

Definition 3.8. The rank of a matrix is the common value of its row and column rank.

EXAMPLE 2 Compute the rank of the following matrix:

$$A = \begin{bmatrix} 1 & 0 & 1 & 0 \\ 2 & 1 & 1 & 1 \\ 0 & 2 & 3 & 2 \end{bmatrix}$$

Solution. This matrix is easily seen to be row equivalent to the matrix.

$$B = \begin{bmatrix} 1 & 0 & 1 & 0 \\ 0 & 1 & -1 & 1 \\ 0 & 0 & 5 & 0 \end{bmatrix}$$

This last matrix has rank equal to 3. Since the rows of B were obtained from linear combinations of the rows of A and we can also obtain the rows of A from linear combinations of the rows of B, the row spaces of A and B must be the same. Hence, A and B have the same row rank and thus the same rank, namely, 3. ■

In the preceding example we computed the row rank of A by first finding the row rank of a matrix that was row equivalent to A and then used the fact that their row ranks were equal. We formalize this in the next theorem.

Theorem 3.6. If A and B are two matrices that are row or column equivalent, then the rank of A is equal to the rank of B.

Proof. See problem 10 at the end of this section.

We now need to relate these ideas to the problem of describing the solution space of a system of equations. Consider the linear system of equations $A\mathbf{x} = \mathbf{b}$, where $A = [a_{jk}]$ is an $m \times n$ matrix. The matrix A defines a linear transformation L from $\mathbb{R}^n$ to $\mathbb{R}^m$. Since $L(\mathbf{e}_k)$ equals the kth column of A, it is clear that the column space of A is the same as the range of L. In fact if we let $\mathbf{C}_k$ be the kth column of A, and $\mathbf{x} = (x_1, \ldots, x_n)^T$, then $A\mathbf{x} = x_1\mathbf{C}_1 + \cdots + x_n\mathbf{C}_n$; that is, $A\mathbf{x}$ is just a linear combination of the columns of A. Thus, if A is

$$\begin{bmatrix} 2 & 1 & 3 \\ 4 & -2 & 8 \end{bmatrix}$$

we may write $A[x_1, x_2, x_3]^T$ as

$$\begin{bmatrix} 2 & 1 & 3 \\ 4 & -2 & 8 \end{bmatrix}\begin{bmatrix} x_1 \\ x_2 \\ x_3 \end{bmatrix} = x_1\begin{bmatrix} 2 \\ 4 \end{bmatrix} + x_2\begin{bmatrix} 1 \\ -2 \end{bmatrix} + x_3\begin{bmatrix} 3 \\ 8 \end{bmatrix}$$

Clearly this is a linear combination of the columns of A.

These remarks make it clear that the column rank of A (the dimension of the column space) is equal to the dimension of the range of L. Hence we have

$$\dim(\ker(L)) = n - \dim(\mathrm{Rg}(L)) = n - \text{rank of } A$$

Remembering that $\ker(L)$ is just the solution space of the homogeneous system $A\mathbf{x} = \mathbf{0}$, we have our promised algorithm for computing the size of the solution space.

What can be said about nonhomogeneous equations? Essentially the same thing, as the following theorem indicates.

Theorem 3.7. Suppose $\mathbf{x}_0$ satisfies $A\mathbf{x}_0 = \mathbf{b}$. Then the solution space of $A\mathbf{x} = \mathbf{b}$ equals $\mathbf{x}_0 + \ker(A)$; that is, every solution of $A\mathbf{x} = \mathbf{b}$ is equal to $\mathbf{x}_0$ plus some vector in the kernel. Note that this is just Theorem 1.10 restated.

Proof. Suppose $A\mathbf{x} = \mathbf{b}$. Then

$$\mathbf{0} = \mathbf{b} - \mathbf{b} = A\mathbf{x} - A\mathbf{x}_0 = A(\mathbf{x} - \mathbf{x}_0)$$

Thus $\mathbf{x} - \mathbf{x}_0$ is in the kernel of A, and $\mathbf{x} = \mathbf{x}_0 + (\mathbf{x} - \mathbf{x}_0)$. Conversely if $\mathbf{x} = \mathbf{x}_0 + \mathbf{z}$ for some $\mathbf{z}$ in the kernel, then

$$A\mathbf{x} = A(\mathbf{x}_0 + \mathbf{z}) = A\mathbf{x}_0 + A\mathbf{z} = \mathbf{b} + \mathbf{0} = \mathbf{b}$$

EXAMPLE 3 Describe the solution set of the following system of equations:

$$\begin{aligned} 2x_1 \qquad\quad + 4x_3 + \ 5x_4 &= 8 \\ x_1 + 2x_2 \qquad\quad + \ 5x_4 &= 4 \\ x_1 + 6x_2 + 3x_3 + 10x_4 &= 11 \end{aligned}$$

Solution. The coefficient matrix A, which equals

$$\begin{bmatrix} 2 & 0 & 4 & 5 \\ 1 & 2 & 0 & 5 \\ 1 & 6 & 3 & 10 \end{bmatrix}$$

is row equivalent to the matrix

$$\begin{bmatrix} 1 & 0 & 0 & \frac{5}{2} \\ 0 & 1 & 0 & \frac{5}{4} \\ 0 & 0 & 1 & 0 \end{bmatrix}$$

Hence, $\dim(\mathrm{Rg}(A)) = \text{rank of } A = 3$. Since $4 - 3 = 1$, $\dim(\ker A) = 1$. A basis for the kernel is $(-\frac{5}{2}, -\frac{5}{4}, 0, 1)$. Thus, the solution space of the equation $A\mathbf{x} =$

$(8,4,11)^T$ is of the form $\mathbf{x} = \mathbf{x}_0 + c(-\frac{5}{2},-\frac{5}{4},0,1)$, for any constant c, assuming of course that there is at least one particular solution $\mathbf{x}_0$. There is a solution if and only if (8,4,11) is in the range of A and equivalently if and only if $(8,4,11)^T$ is in the column space of A. A basis of the column space is $\{(2,1,1)^T, (0,2,6)^T, (4,0,3)^T\}$. One easily sees that $(8,4,11)^T = 2(2,1,1)^T + (0,2,6)^T + (4,0,3)^T$; that is, b equals 2(first column) + (second column) + (third column) + 0(fourth column). Thus, $\mathbf{x}_0 = (2,1,1,0)^T$ is a particular solution, and every solution is of the form

$$\mathbf{x} = (2,1,1,0) + c(-\tfrac{5}{2},-\tfrac{5}{4},0,1)$$ ■

A common problem has to do with data fitting. In its simplest form, this problem can be viewed in the following manner. Assume we have a collection of n points (x_k, y_k), $1 \le k \le n$, where $x_j \ne x_k$ if $j \ne k$. We wish to find a polynomial $\mathbf{p}(t)$ of degree m such that $\mathbf{p}(x_k) = y_k$ for each k. In fact, we would like to find the smallest value of m such that regardless of the values y_k such a polynomial will exist. Since there are n data points and a polynomial of degree m has $m + 1$ arbitrary coefficients we might conjecture m equal to $n - 1$ is the smallest value of m for which we are guaranteed a solution. We now recast this problem in terms of linear transformations. Thus, let (x_j, y_j), $j = 1, 2, \ldots, n$, be given. Let $\mathbf{p}(t)$ be any polynomial in P_m. Define $L: P_m \to \mathbb{R}^n$ by

$$L(\mathbf{p}) = (\mathbf{p}(x_1), \mathbf{p}(x_2), \ldots, \mathbf{p}(x_n))$$

That is, we evaluate our polynomial at the n numbers x_j. For example, if we had the three points $(-1,-2)$, $(0,6)$, and $(1,0)$ and $\mathbf{p}(t) = 8 + 2t - 8t^3 + t^4$, then

$$L(\mathbf{p}) = (\mathbf{p}(-1), \mathbf{p}(0), \mathbf{p}(1)) = (15,8,3)$$

The question then, as to whether or not a polynomial $\mathbf{p}$ of degree m can be picked so that $\mathbf{p}(x_j) = y_j$, amounts to deciding if the point $(y_1, y_2, \ldots, y_n)$ lies in the range of the linear transformation L. We next find a matrix representation for this linear transformation using the natural basis $\{1, t, t^2, \ldots, t^m\}$ in P_m and the standard basis $\{\mathbf{e}_1, \ldots, \mathbf{e}_n\}$ in $\mathbb{R}^n$.

$$\begin{aligned} L(1) &= (1,1,\ldots,1) \\ L(t) &= (x_1, x_2, \ldots, x_n) \\ L(t^k) &= (x_1^k, \ldots, x_n^k) \qquad k = 1, 2, \ldots, m \end{aligned}$$

Thus, we have the $n \times (m+1)$ matrix

$$A = \begin{bmatrix} 1 & x_1 & x_1^2 & \cdots & x_1^m \\ 1 & x_2 & x_2^2 & \cdots & x_2^m \\ \cdots & \cdots & \cdots & \cdots & \cdots \\ 1 & x_n & x_n^2 & \cdots & x_n^m \end{bmatrix}$$

If all the x_j's are different, it can be shown that the rank of A is the smaller of the two numbers $m + 1$ and n. Thus, if we wish to always be able to solve $L(\mathbf{p}) = (y_1, \ldots, y_n)$ we need rank $A = n$. Clearly, the smallest value of m that works is $m = n - 1$; i.e., A is a square matrix. Another way of stating this is that for any n distinct numbers $x_1, \ldots, x_n$ the linear transformation L maps P_{n-1} (polynomials of degree $n - 1$ or less) onto $\mathbb{R}^n$ in a one-to-one fashion.

EXAMPLE 4 Given the three points $(1,-6)$, $(2,0)$, and $(3,6)$, we know from the above discussion that there is a polynomial $\mathbf{p}(t)$, of degree $3 - 1 = 2$, such that $\mathbf{p}(1) = -6$, $\mathbf{p}(2) = 0$, and $\mathbf{p}(3) = 6$. Find this polynomial.

Solution. The transformation $L\colon P_2 \to \mathbb{R}^3$ defined by the abscissas of these three points satisfies

$$L(\mathbf{1}) = (1,1,1) \qquad L(\mathbf{t}) = (1,2,3) \qquad L(\mathbf{t}^2) = (1,4,9)$$

The matrix representation for this transformation is

$$A = \begin{bmatrix} 1 & 1 & 1 \\ 1 & 2 & 4 \\ 1 & 3 & 9 \end{bmatrix}$$

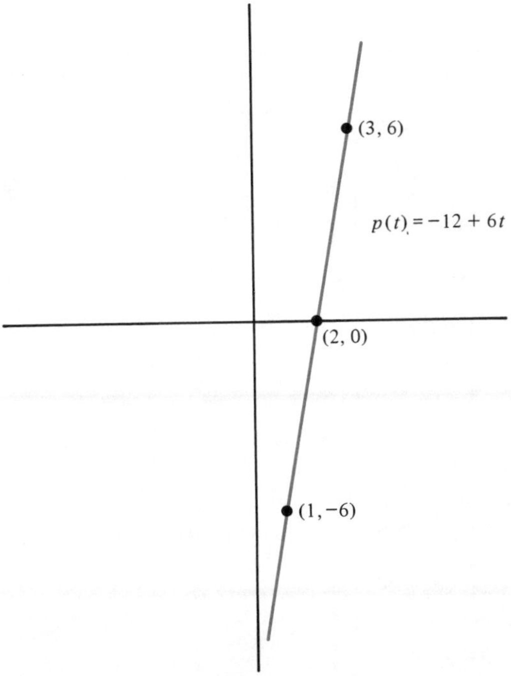

Figure 3.4

This matrix has rank equal to 3. We wish to find $\mathbf{p}(t)$ such that $L(\mathbf{p}) = (-6,0,6)$. In terms of the coefficients a_j of $\mathbf{p}(t) = a_0 + a_1 t + a_2 t^2$, we want a solution to the equation

$$A\begin{bmatrix} a_0 \\ a_1 \\ a_2 \end{bmatrix} = \begin{bmatrix} -6 \\ 0 \\ 6 \end{bmatrix}$$

The unique solution to this equation is $a_0 = -12$, $a_1 = 6$, $a_2 = 0$. Thus, $\mathbf{p}(t) = -12 + 6t$ is the unique polynomial of degree 2 or less that fits the data. We know that the polynomial is unique, since the matrix A has rank 3, which implies that $\dim(\ker(L)) = 3 - 3 = 0$. See Figure 3.4. ■

PROBLEM SET 3.4

1. Calculate the rank of each of the following matrices:

a. $[1 \quad 0 \quad 1]$ **b.** $\begin{bmatrix} 1 \\ 0 \\ 1 \end{bmatrix}$ **c.** $\begin{bmatrix} 1 & 1 \\ 0 & 1 \end{bmatrix}$ **d.** $\begin{bmatrix} 1 & 1 & 1 \\ 1 & 0 & 1 \\ 1 & 1 & 1 \\ 1 & 0 & -1 \end{bmatrix}$

2. Calculate the rank of each of the following matrices:

a. $[1 \quad 2]$ **b.** $\begin{bmatrix} 1 & 2 \\ -1 & 0 \\ 0 & 1 \end{bmatrix}$ **c.** $\begin{bmatrix} 7 \\ 5 \end{bmatrix}$

3. Each of the matrices below represents a linear transformation from $\mathbb{R}^n$ to $\mathbb{R}^m$. Determine the values of n and m for each matrix. Then determine the dimensions of the range and kernel of L.

a. $\begin{bmatrix} 1 & 0 & -1 \\ 0 & 0 & 4 \\ 1 & 0 & 0 \\ 0 & 0 & 1 \end{bmatrix}$ **b.** $\begin{bmatrix} 1 & 2 & -1 & 3 \\ 1 & -1 & 1 & -1 \\ 0 & 1 & 0 & 1 \\ 1 & 0 & 1 & 0 \\ 1 & 1 & 0 & 0 \end{bmatrix}$

4. For each matrix below, determine the dimensions of the range and kernel. Then decide if the linear transformation it represents is onto and/or one-to-one.

a. $[1 \quad 2]$ **b.** $\begin{bmatrix} 1 & 2 \\ -1 & 0 \\ 0 & 1 \end{bmatrix}$ **c.** $\begin{bmatrix} 7 \\ 5 \end{bmatrix}$

5. For each matrix below determine the dimensions of the range and kernel. Then decide if the linear transformation it represents is onto and/or one-to-one.

a. $\begin{bmatrix} 1 & 1 & -1 \\ 5 & 0 & 4 \\ 2 & 1 & 0 \\ 1 & 1 & 1 \end{bmatrix}$ **b.** $\begin{bmatrix} 2 & 3 & 5 & 7 \\ 1 & -1 & 1 & -1 \\ 5 & 0 & 5 & 1 \\ 1 & 2 & 3 & 4 \\ 1 & 1 & 0 & 0 \end{bmatrix}$

6. Compute the row rank and column rank of each of the following matrices:

a. $\begin{bmatrix} 1 & 6 & 0 & 3 \\ 2 & -1 & 1 & 0 \end{bmatrix}$ **b.** $\begin{bmatrix} -3 & 6 & 4 & 1 \\ 2 & 8 & 4 & 3 \\ -4 & 1 & 0 & 0 \end{bmatrix}$ **c.** $\begin{bmatrix} 1 & 2 & 1 \\ 4 & 5 & 6 \\ 8 & -1 & 2 \\ 6 & 1 & 0 \end{bmatrix}$

7. Consider the following system of linear equations:

$$\begin{aligned} 2x_1 \qquad - 6x_3 &= -6 \\ x_2 + x_3 &= 1 \end{aligned}$$

Let A be the coefficient matrix of this system.

a. Compute the rank of A.
b. $\dim(\ker(A)) = ?$ Find a basis for $\ker(A)$.
c. $\dim(\text{Rg}(A)) = ?$ Find a basis for $\text{Rg}(A)$.
d. Is A a one-to-one linear transformation?
e. Is A onto?
f. Does the above system of equations have a solution? If yes, characterize the solution set in terms of the kernel of A and one particular solution.

8. Let $L: \mathbb{R}^3 \to \mathbb{R}^2$ have the matrix representation

$$A = \begin{bmatrix} -2 & 1 & 3 \\ 4 & -1 & 0 \end{bmatrix}$$

Show that the range of L and the column space of A are the same subspace of $\mathbb{R}^2$.

9. Consider the following system of linear equations:

$$\begin{aligned} 4x_1 \qquad + 2x_3 + x_4 &= 0 \\ 2x_1 - x_2 + x_3 + 3x_4 &= 1 \\ -8x_1 - 2x_2 - 4x_3 + 3x_4 &= 2 \end{aligned}$$

Let A be the coefficient matrix of this system.

a. Compute the rank of A.
b. $\dim(\ker(A)) = ?$ Find a basis for $\ker(A)$.
c. $\dim(\text{Rg}(A)) = ?$ Find a basis for $\text{Rg}(A)$.
d. Is A a one-to-one linear transformation?
e. Is A onto?
f. Does the above system of equations have a solution? If yes, characterize the solution set in terms of the kernel of A and a particular solution.

10. Let $\mathbf{x}_k$ for $k = 1, 2, \ldots, p$ be vectors in a vector space. Let $V = S[\mathbf{x}_1, \ldots, \mathbf{x}_p]$. If c is any nonzero constant show that

a. $V = S[c\mathbf{x}_1, \mathbf{x}_2, \ldots, \mathbf{x}_p]$.

b. $V = S[\mathbf{x}_1, c\mathbf{x}_1 + \mathbf{x}_2, \ldots, \mathbf{x}_p]$

The reader should note that the result proved in this problem is what is needed to verify Theorem 3.6, the vectors then representing the rows or columns of a matrix.

11. Given data points (1,0) and (2,1), define $L: P_2 \to \mathbb{R}^2$ by $L(\mathbf{p}) = (\mathbf{p}(1), \mathbf{p}(2))$. Here $\mathbf{p} = \mathbf{p}(t)$ is any polynomial of degree 2 or less. Find the matrix representation of L with respect to the standard bases in P_2 and $\mathbb{R}^2$. Show that the rank of A equals 2. What does this say about fitting polynomials of degree at most 2 through the data points (1,0) and (2,1)?

12. Let $\mathbf{p}$ be any polynomial in P_2. Define $L\colon P_2 \to \mathbb{R}^3$ by $L(\mathbf{p}) = (\mathbf{p}(-2),\mathbf{p}(0),\mathbf{p}(1))$. Find the matrix representation A of L with respect to the standard bases in P_2 and $\mathbb{R}^3$. Show that A has rank equal to 3. What does this say about fitting polynomials of degree 2 or less through three points in the plane with x coordinates -2, 0, and 1?

13. Find a polynomial in P_1, if possible, that fits the following data:
a. (2,6), (3,6) **b.** (2,0), (−1,4) **c.** (2,6), (3,6), (4,7)

14. Find a polynomial in P_2, if possible, that fits the following data:
a. (−2,6), (3,7) **b.** (−2,6), (3,7), (4,7)
c. (−2,6), (3,7), (4,7), (5,8)

15. Let $V = C[0,1]$, the vector space of real-valued continuous functions defined on [0,1]. Define the mapping L by

$$L[\mathbf{f}] = \left[\frac{\mathbf{f}(0)}{2} + \mathbf{f}\left(\frac{1}{2}\right) + \frac{\mathbf{f}(1)}{2}\right]\left(\frac{1}{2}\right)$$

Thus, if $\mathbf{f}(t) = \sin t$, we have $L[\mathbf{f}] = [(\sin 0)/2 + \sin \frac{1}{2} + (\sin 1)/2](\frac{1}{2})$. This formula is just the trapezoid rule for the approximate evaluation of integrals over the interval [0,1] using the points 0, $\frac{1}{2}$, and 1.

a. Show that L is a linear transformation, after deciding of course what W should be. Characterize the range and kernel of L.

b. Let $L_1[\mathbf{f}] = L[\mathbf{f}] - \int_0^1 \mathbf{f}(t)dt$. Show that L_1 is also a linear transformation. How would you describe its kernel?

16. For each of the following matrices A compute the rank of A, A^T, AA^T, and A^TA. You should get rank A^TA equals rank A and rank AA^T equals rank A^T. Hence all four numbers are equal.

a. $[1 \quad 1 \quad 2]$ **b.** $\begin{bmatrix} 1 & 0 & 2 \\ 0 & 1 & 3 \end{bmatrix}$ **c.** $\begin{bmatrix} 1 & -2 \\ 3 & 2 \end{bmatrix}$

17. Let A be any $m \times n$ matrix. How are the row space of A and the range of A^T related?

18. Let x_1, x_2, and x_3 be three different numbers.

a. Show rank $\begin{bmatrix} 1 & x_1 \\ 1 & x_2 \end{bmatrix} = 2$.

b. Show rank $\begin{bmatrix} 1 & x_1 & x_1^2 \\ 1 & x_2 & x_2^2 \\ 1 & x_3 & x_3^2 \end{bmatrix} = 3$.

Hint: The matrix in a is row equivalent to $\begin{bmatrix} 1 & x_1 \\ 0 & x_2 - x_1 \end{bmatrix}$.

19. Define $L\colon P_2 \to \mathbb{R}^3$ by $L[\mathbf{p}] = (\mathbf{p}(x_1),\mathbf{p}(x_2),\mathbf{p}(x_3))$, where the x_j's are all different. Show that L is one-to-one and onto.

20. Let $x_1, x_2, \ldots, x_n$ be n pairwise distinct numbers. Let

$$A = \begin{bmatrix} 1 & x_1 & x_1^2 & \cdots & x_1^{n-1} \\ 1 & x_2 & x_2^2 & \cdots & x_2^{n-1} \\ \cdot & \cdot & \cdot & \cdots & \cdot \\ 1 & x_n & x_n^2 & \cdots & x_n^{n-1} \end{bmatrix}$$

Show that rank $A = n$.

3.5 CHANGE OF BASIS FORMULAS

In Section 3.3 we learned how to represent a linear transformation as a matrix, and in the last section we saw how the matrix can be used to tell us some facts about the kernel and range of the linear transformation. Clearly the "simpler" the matrix representation the easier it is to understand the linear transformation. Since Gaussian elimination is a method used to obtain a "simple" matrix representation for a system of equations, we have already seen the utility of finding nice representations.

Once we have a fixed linear transformation, that maps a vector space V into itself, the only variable in determining its matrix representation is our choice of basis.

Before studying how to pick such a basis (cf. Chapter 5), we need to learn how the matrix representations of the same linear transformation are related to each other. Thus, let L be a linear transformation from V into V, where $\dim(V) = n$. Let $F = \{\mathbf{f}_j : j = 1, \ldots, n\}$ and $G = \{\mathbf{g}_j : j = 1, \ldots, n\}$ be two bases of V. Let $A = [a_{jk}]$ be the matrix representation of L with respect to F and let $B = [b_{jk}]$ be the matrix representation of L with respect to G. This means that the following equations hold:

$$L[\mathbf{f}_j] = \sum_{k=1}^{n} a_{kj}\mathbf{f}_k \tag{3.7}$$

$$L[\mathbf{g}_j] = \sum_{k=1}^{n} b_{kj}\mathbf{g}_k \tag{3.8}$$

Let $P = [p_{jk}]$ be the change of basis matrix that satisfies $[\mathbf{x}]_G^T = P[\mathbf{x}]_F^T$, that is

$$\mathbf{f}_j = \sum_{k=1}^{n} p_{kj}\mathbf{g}_k \tag{3.9}$$

and if $P^{-1} = [q_{jk}]$, then

$$\mathbf{g}_j = \sum_{k=1}^{n} q_{kj}\mathbf{f}_k \tag{3.10}$$

We refer the reader to Theorem 2.11 and the material preceding it. Computing $L[\mathbf{f}_j]$, we have

$$\begin{aligned} L[\mathbf{f}_j] &= L\left[\sum_{k=1}^{n} p_{kj}\mathbf{g}_k\right] = \sum_{k=1}^{n} p_{kj}L[\mathbf{g}_k] \\ &= \sum_{k=1}^{n} p_{kj}\left(\sum_{s=1}^{n} b_{sk}\mathbf{g}_s\right) = \sum_{k=1}^{n}\sum_{s=1}^{n} p_{kj}b_{sk}\left(\sum_{m=1}^{n} q_{ms}\mathbf{f}_m\right) \\ &= \sum_{m=1}^{n}\left(\sum_{k=1}^{n}\sum_{s=1}^{n}(q_{ms}b_{sk}p_{kj})\right)\mathbf{f}_m \end{aligned}$$

Comparing this expression with (3.7) and equating coefficients, we have

$$a_{mj} = \sum_{k=1}^{n} \sum_{s=1}^{n} q_{ms} b_{sk} p_{kj} \tag{3.11}$$

for m and j varying from 1 through n. However, the right-hand side of (3.11) is the m, j entry of the matrix $P^{-1}BP$. Thus we have

$$A = P^{-1}BP \quad \text{and} \quad B = PAP^{-1} \tag{3.12}$$

A perhaps simpler way to remember how the matrix representations A and B are related is to utilize formulas (3.3), that is,

$$[L(\mathbf{x})]_F^T = A[\mathbf{x}]_F^T \qquad [L(\mathbf{x})]_G^T = B[\mathbf{x}]_G^T$$

Thus,

$$\begin{aligned} [L(\mathbf{x})]_G^T &= P[L(\mathbf{x})]_F^T = P(A[\mathbf{x}]_F^T) \\ &= (PA)(P^{-1}[\mathbf{x}]_G^T) = (PAP^{-1})[\mathbf{x}]_G^T \end{aligned}$$

Hence, we must have $B = PAP^{-1}$. The reader is referred to problem 11 in Section 2.6 for a justification of this last step.

EXAMPLE 1 Let L be a linear transformation from $\mathbb{R}^2$ to $\mathbb{R}^2$ defined by $L(x_1, x_2) = (2x_1 + x_2, 3x_1 - x_2)$. Verify formulas (3.12) for $F = \{(1,1), (-1,2)\}$ and $G = \{(-1,-1), (2,0)\}$.

Solution. Writing the vectors in F as linear combinations of the vectors in G, we have

$$\begin{aligned} (1,1) &= -(-1,-1) + 0(2,0) \\ (-1,2) &= -2(-1,-1) + (-\tfrac{3}{2})(2,0) \end{aligned}$$

Thus,

$$P = \begin{bmatrix} -1 & -2 \\ 0 & -\frac{3}{2} \end{bmatrix} \qquad P^{-1} = \begin{bmatrix} -1 & \frac{4}{3} \\ 0 & -\frac{2}{3} \end{bmatrix}$$

To determine A, we compute

$$\begin{aligned} L(1,1) &= (3,1) = \tfrac{7}{3}(1,1) - \tfrac{2}{3}(-1,2) \\ L(-1,2) &= (0,-7) = -\tfrac{7}{3}(1,1) - \tfrac{7}{3}(-1,2) \end{aligned}$$

Thus,

$$A = \begin{bmatrix} \frac{7}{3} & -\frac{7}{3} \\ -\frac{2}{3} & -\frac{7}{3} \end{bmatrix}$$

To determine B, we compute

$$L(-1,-1) = (-3,-1) = (-1,-1) - (2,0)$$
$$L(2,0) = (4,6) = -6(-1,-1) - (2,0)$$

Thus,

$$B = \begin{bmatrix} 1 & -6 \\ -1 & -1 \end{bmatrix}$$

Computing $P^{-1}BP$ we have

$$\begin{bmatrix} -1 & \frac{4}{3} \\ 0 & -\frac{2}{3} \end{bmatrix}\begin{bmatrix} 1 & -6 \\ -1 & -1 \end{bmatrix}\begin{bmatrix} -1 & -2 \\ 0 & -\frac{3}{2} \end{bmatrix} = \begin{bmatrix} \frac{7}{3} & -\frac{7}{3} \\ -\frac{2}{3} & -\frac{7}{3} \end{bmatrix} = A$$

which is formula (3.12). ■

EXAMPLE 2 Let $L: \mathbb{R}^4 \to \mathbb{R}^4$ be a linear transformation defined by

$$L(x_1,x_2,x_3,x_4) = (2x_3 + x_4, -3x_1 + x_2 - x_4, x_1 - x_3 + 6x_4, x_2 - x_3)$$

Let S be the standard basis of $\mathbb{R}^4$ and let $G = \{(-1,0,1,1,), (0,1,-1,0), (0,0,1,1), (1,0,1,0)\}$. Find the matrix representations of L with respect to S, and then G, by employing (3.12).

Solution. Let A be the matrix representation of L with respect to S. By inspection we have

$$A = \begin{bmatrix} 0 & 0 & 2 & 1 \\ -3 & 1 & 0 & -1 \\ 1 & 0 & -1 & 6 \\ 0 & 1 & -1 & 0 \end{bmatrix}$$

If P^{-1} is the change of basis matrix that satisfies $[x]_S^T = P^{-1}[x]_G^T$ then,

$$P^{-1} = \begin{bmatrix} -1 & 0 & 0 & 1 \\ 0 & 1 & 0 & 0 \\ 1 & -1 & 1 & 1 \\ 1 & 0 & 1 & 0 \end{bmatrix}$$

An easy computation gives us

$$P = \begin{bmatrix} -1 & 1 & 1 & -1 \\ 0 & 1 & 0 & 0 \\ 1 & -1 & -1 & 2 \\ 0 & 1 & 1 & -1 \end{bmatrix}$$

Using (3.12), where B is the matrix representation of L with respect to the basis G, we have

$$B = PAP^{-1}$$

$$= \begin{bmatrix} -1 & 1 & 1 & -1 \\ 0 & 1 & 0 & 0 \\ 1 & -1 & -1 & 2 \\ 0 & 1 & 1 & -1 \end{bmatrix} \begin{bmatrix} 0 & 0 & 2 & 1 \\ -3 & 1 & 0 & -1 \\ 1 & 0 & -1 & 6 \\ 0 & 1 & -1 & 0 \end{bmatrix} \begin{bmatrix} -1 & 0 & 0 & 1 \\ 0 & 1 & 0 & 0 \\ 1 & -1 & 1 & 1 \\ 1 & 0 & 1 & 0 \end{bmatrix}$$

$$= \begin{bmatrix} 4 & 2 & 2 & -1 \\ 2 & 1 & -1 & -3 \\ -5 & 0 & -3 & 3 \\ 7 & 0 & 5 & -2 \end{bmatrix}$$

As a check on our computations we compute $L(-1,0,1,1)$ using the matrix B, and then compare this with our original definition of L. The vector $(-1,0,1,1)$ is the first vector in the basis G. Thus, the first column of B contains the coordinates of $L(-1,0,1,1)$ with respect to the basis G. Hence,

$$\begin{aligned} L(-1,0,1,1) &= 4(-1,0,1,1) + 2(0,1,-1,0) - 5(0,0,1,1) + 7(1,0,1,0) \\ &= (3,2,4,-1) \end{aligned}$$

This is exactly what we get when we compute $L(-1,0,1,1)$ directly. ■

The preceding calculations have shown us that if $L: V \to V$ is a linear transformation, and A and B are two matrix representations of L with respect to two different bases of V, then there is a matrix P such that $A = P^{-1}BP$.

Definition 3.9. Let A and B be two $n \times n$ matrices. We say that A is similar to B, if there is a nonsingular matrix P such that $A = P^{-1}BP$.

As we have seen, the matrix P is nothing more than the matrix relating two different bases of our vector space. Moreover, we understand that similar matrices are just different matrix representations of the same linear transformation.

The change of basis discussion assumed that the linear transformation L mapped V into V. What happens when $L: V \to W$, and we change bases in V and W? Without going through the calculations, we state the appropriate theorem.

Theorem 3.8. Let $L: V \to W$ be a linear transformation. Let F and $\mathcal{F}$ be two bases of V. Let G and $\mathcal{G}$ be two bases of W. Let A be the matrix representation of L using the bases F and G while B is the representation using the bases $\mathcal{F}$ and $\mathcal{G}$. Let $P = [p_{jk}]$ be the change of basis matrix that satisfies $[\mathbf{x}]_{\mathcal{F}}^T = P[\mathbf{x}]_F^T$. Let $Q = [q_{jk}]$ be the matrix that satisfies $[\mathbf{y}]_{\mathcal{G}}^T = Q[\mathbf{y}]_G^T$. Then

$$A = Q^{-1}BP \tag{3.13}$$

The reader should look at problem 9 at the end of this section for one method of proving this result.

PROBLEM SET 3.5

1. Let $L: \mathbb{R}^2 \to \mathbb{R}^2$ be defined by $L(x_1, x_2) = (2x_1, 2x_2)$. Let $F = \{(1,-1), (2,5)\}$.
   a. Find the matrix representation A of L with respect to the standard basis S.
   b. Let P be the change of basis matrix that satisfies $[\mathbf{x}]_F^T = P[\mathbf{x}]_S^T$. Find P and P^{-1}.
   c. Find the matrix representation B of L with respect to the basis F by using (3.12).
2. Let A be any 2×2 scalar matrix, that is

$$A = \begin{bmatrix} c & 0 \\ 0 & c \end{bmatrix} = cI_2$$

   Let P be any 2×2 nonsingular matrix. Show that PAP^{-1} equals A; cf. problem 1.
3. Let $L: \mathbb{R}^3 \to \mathbb{R}^3$. Let $F = \{(1,2,-3), (1,0,0), (0,1,0)\}$. Suppose

$$A = \begin{bmatrix} 0 & 1 & -2 \\ 2 & 1 & 0 \\ 5 & 0 & 1 \end{bmatrix}$$

   is the matrix representation of L with respect to the basis F. Use (3.12) to find the matrix representation of L with respect to the standard basis; cf. problem 4 of Section 3.2.
4. Suppose $A = \begin{bmatrix} 2 & 1 \\ -1 & 2 \end{bmatrix}$ is the matrix representation of a linear transformation with respect to the standard basis. Let $P = \begin{bmatrix} 0 & 1 \\ 1 & 0 \end{bmatrix}$ and set $B = PAP^{-1}$. Then B can be thought of as the matrix representation of L with respect to some other basis. What is this basis?
5. Let A, B, and C be three $n \times n$ matrices. Show that
   a. A is similar to itself.
   b. If A is similar to B, then B is similar to A.
   c. If A is similar to B and B is similar to C, then A is similar to C.
6. Let A be any 2×2 matrix. Show that if A is similar to I_2, then $A = I_2$.
7. Let $F = \{(1,2),(1,0)\}$ and $G = \{(1,-1), (0,1)\}$. Let $L(x_1, x_2) = (x_1 + x_2, 2x_1 - x_2)$. Find the matrix representation of L with respect to bases F and G. Verify (3.12).
8. Let $F = \{(1,0,1),(0,1,1),(1,1,0)\}$ and $G = \{(1,0,-1),(-1,1,0),(0,0,1)\}$. Let $L(x_1, x_2, x_3) = (x_1 + x_2 + x_3, x_2 + x_3, x_3)$. Verify (3.12) for the above bases and linear transformation L.
9. Prove Theorem 3.8. Hint: Use problem 11 in Section 2.6 and formula (3.3).
10. Let $F_1 = \left\{ \begin{bmatrix} 1 & 0 \\ 0 & 0 \end{bmatrix} \begin{bmatrix} 0 & 1 \\ 0 & 0 \end{bmatrix} \begin{bmatrix} 0 & 0 \\ 1 & 0 \end{bmatrix} \begin{bmatrix} 0 & 0 \\ 0 & 1 \end{bmatrix} \right\}$

    Let $F_2 = \left\{ \begin{bmatrix} 0 & 1 \\ 1 & 1 \end{bmatrix} \begin{bmatrix} 1 & 0 \\ 1 & 1 \end{bmatrix} \begin{bmatrix} 1 & 1 \\ 0 & 1 \end{bmatrix} \begin{bmatrix} 1 & 1 \\ 1 & 0 \end{bmatrix} \right\}$

Let $G_1 = \left\{ \begin{bmatrix} 1 & 0 & 0 \\ 0 & 0 & 0 \end{bmatrix} \begin{bmatrix} 0 & 1 & 0 \\ 0 & 0 & 0 \end{bmatrix} \begin{bmatrix} 0 & 0 & 1 \\ 0 & 0 & 0 \end{bmatrix} \begin{bmatrix} 0 & 0 & 0 \\ 1 & 0 & 0 \end{bmatrix} \begin{bmatrix} 0 & 0 & 0 \\ 0 & 1 & 0 \end{bmatrix} \begin{bmatrix} 0 & 0 & 0 \\ 0 & 0 & 1 \end{bmatrix} \right\}$

Let $G_2 = \left\{ \begin{bmatrix} 0 & 1 & 1 \\ 1 & 1 & 1 \end{bmatrix} \begin{bmatrix} 1 & 0 & 1 \\ 1 & 1 & 1 \end{bmatrix} \begin{bmatrix} 1 & 1 & 0 \\ 1 & 1 & 1 \end{bmatrix} \begin{bmatrix} 1 & 1 & 1 \\ 0 & 1 & 1 \end{bmatrix} \begin{bmatrix} 1 & 1 & 1 \\ 1 & 0 & 1 \end{bmatrix} \begin{bmatrix} 1 & 1 & 1 \\ 1 & 1 & 0 \end{bmatrix} \right\}$

Let ${}_jA_k$ be the matrix representation of a linear transformation $L\colon M_{22} \to M_{23}$ with respect to the bases G_j and F_k. That is,

$$[L(\mathbf{x})]^T_{G_j} = {}_jA_k[\mathbf{x}]^T_{F_k}$$

Suppose that

$${}_1A_1 = \begin{bmatrix} -2 & 0 & -3 & 0 \\ 1 & 0 & 1 & 0 \\ 0 & 1 & 4 & 1 \\ 3 & 1 & -5 & 0 \\ 2 & -2 & 0 & 1 \\ 1 & 2 & 3 & 4 \end{bmatrix}$$

Find the other three matrix representations.

11. Suppose A and B are similar $n \times n$ matrices, i.e., there is a matrix P such that $A = PBP^{-1}$.

a. How are $\ker(A)$ and $\ker(B)$ related?

b. How are $\mathrm{Rg}(A)$ and $\mathrm{Rg}(B)$ related?

12. Let $L\colon P_2 \to P_2$ be a linear transformation. Let $F = \{t^2 + t - 1, t^2 + 2, t - 6\}$. Suppose the matrix representation of L with respect to F is

$$A = \begin{bmatrix} -14 & -2 & -18 \\ 23 & 11 & 18 \\ 11 & 2 & 15 \end{bmatrix}$$

Find the matrix representation of L with respect to the standard basis of P_2.

SUPPLEMENTARY PROBLEMS

1. Define and give examples of each of the following:

a. Linear transformation

b. Range and kernel of a linear transformation

c. Rank of a matrix

d. Column (row) rank

2. Let $A = \begin{bmatrix} 2 & 6 \\ 1 & -2 \\ 3 & 5 \end{bmatrix}$ be the matrix representation of a linear transformation from $\mathbb{R}^k$ to $\mathbb{R}^p$.

a. $k = ?$ $p = ?$

b. Determine the dimensions of $\ker(A)$ and $\mathrm{Rg}(A)$.

3. For each of the following matrices determine the rank and find a basis for the kernel:

a. $\begin{bmatrix} 3 & 1 & 2 \\ 1 & 0 & 1 \\ -1 & 1 & 1 \end{bmatrix}$ **b.** $\begin{bmatrix} -1 & 2 & 0 & 6 \\ 4 & 3 & -1 & 0 \\ 1 & 0 & 0 & 1 \end{bmatrix}$ **c.** $\begin{bmatrix} 4 & 2 & 4 & 6 \\ 6 & 3 & 6 & 9 \\ 2 & 1 & 2 & 1 \end{bmatrix}$

4. Let A and B be two matrices. Show that

$$\text{Rank}(AB) \leq \text{minimum}\{\text{rank}(A), \text{rank}(B)\}$$

Hint: Rank (A) equals $\dim(\text{Rg}(A))$.

5. Define $L: \mathbb{R}^2 \to \mathbb{R}^2$ by $L(x_1, x_2) = (x_1 + x_2, -x_2)$. Show that L maps the straight line $y = mx$ onto the straight line $y = -[m/(m+1)]x$. What happens to the line $y = -x$?

6. Describe, geometrically, the following linear transformations, and in each case determine the kernel and range:
a. $L(x_1, x_2) = (x_1 - x_2, x_1 + x_2)$
b. $L(x_1, x_2) = (2x_1 - x_2, 4x_2 - 2x_1)$

7. Let V_1, V_2, and V_3 be vector spaces with bases F_1, F_2, and F_3, respectively. Suppose $L_1: V_1 \to V_2$ and $L_2: V_2 \to V_3$ are linear transformations. Let A_1 and A_2 be their corresponding matrix representations with respect to the given bases. Define $L: V_1 \to V_3$ by $L(\mathbf{x}) = L_2(L_1(\mathbf{x}))$.
a. Show that L is a linear transformation from V_1 to V_3.
b. Show that the matrix representation of L with respect to the bases F_1 and F_3 is A_2A_1.

8. A linear transformation $L: \mathbb{R}^2 \to \mathbb{R}^2$ is said to be positive if it maps the first quadrant into the first quadrant; that is, if x_1 and x_2 are both positive, then so are y_1 and y_2, where $L(x_1, x_2) = (y_1, y_2)$.
a. Let $L(x_1, x_2) = (ax_1 + bx_2, cx_1 + dx_2)$. Show that L is positive if and only if a, b, c, and d are all positive. Thus, L is a positive linear transformation if and only if its matrix representation with respect to the standard basis has only positive entries.
b. Find a positive linear transformation whose matrix representation with respect to some basis has at least one negative entry.
c. Find an example of a positive linear transformation whose kernel has dimension equal to 1.

9. A mapping T from a vector space V into V is said to be an affine transformation if

$$T(\mathbf{x}) = L(\mathbf{x}) + \mathbf{a}$$

where L is a linear transformation from V into V and $\mathbf{a}$ is any fixed vector in V.
a. Show that $T(\mathbf{x}) = L(\mathbf{x}) + \mathbf{a}$ is a linear transformation if and only if $\mathbf{a}$ equals the zero vector.
b. Given any straight line in $\mathbb{R}^2$, show that there is an affine transformation that maps the line $x_2 = 0$ onto the given line. Hint: Rotate and then translate.

10. Two vector spaces V_1 and V_2 are said to be isomorphic if there is a linear transformation $L: V_1 \to V_2$ that is both one-to-one and onto.
a. Suppose $\dim(V_1) = m$ and V_1 is isomorphic to V_2. Show $\dim(V_2) = m$.
b. Show that two finite-dimensional vector spaces are isomorphic if and only if they have the same dimension.

c. Let $V_1 = \mathbb{R}^1$ and let V_2 be the vector space defined in Example 5 of Section 2.2. Define $L: V_1 \to V_2$ by $L(x) = e^x$. Show that L is a linear transformation that is one-to-one and onto.

11. Let $V = \{\Sigma_{k=1}^{n} a_k \sin(kx)e^{k^2t}$, n any positive integer and a_k arbitrary numbers$\}$.

a. Under ordinary addition and multiplication show that V is a vector space.

Define $L: V \to V$ by $L[u] = \dfrac{\partial u}{\partial t} - \dfrac{\partial^2 u}{\partial x^2}$.

b. Show L is a linear transformation.

c. Find the kernel and range of L.

12. If V is a finite-dimensional space and L is a linear transformation from V into V, then L is one-to-one if and only if L is onto. This result, as the following shows, may not be true if V is infinite-dimensional. Let V be the vector space of all polynomials in the variable t.

a. Define $L[\mathbf{p}](t) = t\mathbf{p}(t)$. Thus, $L[\mathbf{t}] = \mathbf{t}^2$, $L[\mathbf{t}^2] = \mathbf{t}^3$, and $L[\mathbf{t}^n] = \mathbf{t}^{n+1}$. Show that L is one-to-one but not onto.

b. Define $L[\mathbf{p}](t) = \mathbf{p}'(t)$. Thus, $L[\mathbf{t}^n] = n\mathbf{t}^{n-1}$. Show that L is onto but not one-to-one. What is the kernel of L?

13. Let L be a linear transformation from V into V. A subspace W of V is said to be invariant under L if, for every $\mathbf{x}$ in W, $L[\mathbf{x}]$ is also in W. Thus, we may also consider L to be a linear transformation from W to W. Show that for any linear transformation L each of the following is an invariant subspace:

a. $W = \{0\}$, $W = V$.

b. For any constant λ, show $W_\lambda = \{\mathbf{x}: L[\mathbf{x}] = \lambda x\}$ is an invariant subspace.

14. Let $A = \begin{bmatrix} 2 & 1 & 0 \\ -2 & 3 & 0 \\ 0 & 0 & 4 \end{bmatrix}$ be the matrix representation of a linear transformation L with respect to the standard basis. Show that $S[\mathbf{e}_1]$ is not an invariant subspace but that $S[\mathbf{e}_1, \mathbf{e}_2]$ and $S[\mathbf{e}_3]$ are invariant subspaces; cf. problem 13.

15. Let $F = \{\mathbf{x}_1, \ldots, \mathbf{x}_n\}$ be a basis for a vector space V. Let L be a linear transformation from V into V for which $S[\mathbf{x}_1, \ldots, \mathbf{x}_k]$ and $S[\mathbf{x}_{k+1}, \ldots, \mathbf{x}_n]$ are both invariant subspaces of L. Let A be the matrix representation of L with respect to the basis F. Show that

$$A = \begin{bmatrix} B_1 & 0 \\ 0 & B_2 \end{bmatrix}$$

where B_1 is a $k \times k$ matrix and B_2 is an $(n-k) \times (n-k)$ matrix. Conversely, suppose that A has the above form. Show that the above two subspaces must be invariant.

4 DETERMINANTS

Chapter 3 entailed a discussion of linear transformations and how to identify them with matrices. When we study a particular linear transformation we would like its matrix representation to be simple, diagonal if possible. We therefore need some way of deciding if we can simplify the matrix representation and then how to do so. This problem has a solution, and in order to implement it, we need to talk about something called the determinant of a matrix.

The determinant of a square matrix is a number. It turns out that this number is nonzero if and only if the matrix is invertible. In the first section of this chapter, different ways of computing the determinant of a matrix are presented. Few proofs are given; in fact no attempt has been made to even give a precise definition of a determinant. Those readers interested in a more rigorous discussion are encouraged to read Appendices C and D.

4.1 PROPERTIES OF THE DETERMINANT

The first thing to note is that the determinant of a matrix is defined only if the matrix is square. Thus, if A is a 2×2 matrix, it has a determinant, but if A is a 2×3 matrix it does not. The determinant of a 2×2 matrix is now defined.

Definition 4.1. Determinant(A) = det(A)

$$= \det \begin{bmatrix} a & b \\ c & d \end{bmatrix} = ad - bc.$$

EXAMPLE 1 Compute the determinants of the following matrices:

a. $\det \begin{bmatrix} 1 & 6 \\ -2 & 3 \end{bmatrix} = 3 - (-12) = 15$

b. $\det \begin{bmatrix} 2 & 0 \\ 1 & 4 \end{bmatrix} = 8 - 0 = 8$

c. $\det \begin{bmatrix} 2 & 6 \\ 1 & 3 \end{bmatrix} = 6 - 6 = 0$ ■

To compute the determinant of a 3×3 or $n \times n$ matrix, we need to introduce some notation.

Definition 4.2. Let $A = [a_{jk}]$ be an $n \times n$ matrix. Let M_{jk} be that $(n-1) \times (n-1)$ matrix obtained from A by deleting its jth row and kth column. This submatrix of A is referred to as the j,k minor of A.

EXAMPLE 2 Let $A = \begin{bmatrix} -1 & 6 & 3 \\ 2 & 0 & 9 \\ 4 & 8 & 7 \end{bmatrix}$. Find M_{11}, M_{23}, and M_{32}.

$$M_{11} = \begin{bmatrix} 0 & 9 \\ 8 & 7 \end{bmatrix} \quad M_{23} = \begin{bmatrix} -1 & 6 \\ 4 & 8 \end{bmatrix} \quad M_{32} = \begin{bmatrix} -1 & 3 \\ 2 & 9 \end{bmatrix}$$ ■

Using minors we demonstrate one way to compute the determinant of a 3×3 matrix. The technique is called expansion by cofactors.

Let A be any 3×3 matrix:

$$A = \begin{bmatrix} a_{11} & a_{12} & a_{13} \\ a_{21} & a_{22} & a_{23} \\ a_{31} & a_{32} & a_{33} \end{bmatrix}$$

Then

$$\det(A) = a_{11}\det(M_{11}) - a_{12}\det(M_{12}) + a_{13}\det(M_{13})$$

Note that any minor of a 3×3 matrix is a 2×2 matrix, and hence its determinant is defined. We also wish to stress that we did not have to expand across the first row. We could have used any row or column.

EXAMPLE 3 Compute the determinant of the matrix below by expanding across the first row and also by expanding down the second column.

$$A = \begin{bmatrix} -1 & 2 & 4 \\ 6 & 3 & 5 \\ -3 & 7 & 0 \end{bmatrix}$$

1. Expanding across the first row we have

$$\det(A) = a_{11}\det(M_{11}) - a_{12}\det(M_{12}) + a_{13}\det(M_{13})$$

$$= (-1)\det\begin{bmatrix} 3 & 5 \\ 7 & 0 \end{bmatrix} - (2)\det\begin{bmatrix} 6 & 5 \\ -3 & 0 \end{bmatrix} + (4)\det\begin{bmatrix} 6 & 3 \\ -3 & 7 \end{bmatrix}$$

$$= -(-35) - 2(15) + 4(42 + 9) = 209$$

2. Expanding down the second column we have

$$\det(A) = -a_{12}\det(M_{12}) + a_{22}\det(M_{22}) - a_{32}\det(M_{32})$$

$$= -(2)\det\begin{bmatrix} 6 & 5 \\ -3 & 0 \end{bmatrix} + (3)\det\begin{bmatrix} -1 & 4 \\ -3 & 0 \end{bmatrix} - (7)\det\begin{bmatrix} -1 & 4 \\ 6 & 5 \end{bmatrix}$$

$$= (-2)(15) + 3(12) - 7(-29) = 209$$ ■

It seems from the above two computations that minus signs creep in at random. That is not true. There is a rule for deciding whether or not a minus sign should appear, and it is given in the following theorem.

Theorem 4.1. Let $A = [a_{jk}]$ be any $n \times n$ matrix. Then

$$\det(A) = \sum_{j=1}^{n} (-1)^{j+k} a_{jk} \det(M_{jk}) \qquad k = 1, 2, \ldots, n$$
expansion down the kth column

(4.1)

$$\det(A) = \sum_{k=1}^{n} (-1)^{j+k} a_{jk} \det(M_{jk}) \qquad j = 1, 2, \ldots, n$$
expansion across the jth row

Figure 4.1 should clarify whether or not a minus sign precedes the term $\det(M_{jk})$. Notice that the 1,1 entry has a + sign, and whenever we move one space horizontally or vertically the sign changes. Since it takes three moves to go from 1, 1 to 2, 3, the coefficient of a_{23} in (4.1) equals $-\det(M_{23})$.

The terms $(-1)^{j+k}\det(M_{jk})$ are called the cofactors of a_{jk}, hence the phrase expansion by cofactors. Notice that this theorem reduces the problem of computing the determinant of an $n \times n$ matrix to the problem of calculating the determinant of an $(n-1) \times (n-1)$ matrix. Continued use of this procedure

$$\begin{bmatrix} + & - & + & - & + & \cdots \\ - & + & - & + & - & \cdots \\ + & - & & & & \\ - & & & & & \\ + & & & & & \\ - & + & - & + & \cdots & \\ \vdots & & & & & \end{bmatrix}$$

Figure 4.1

will reduce the original problem to one of calculating the determinants of 2×2 matrices.

It is clear that computing the determinant of a matrix, especially a large one, is painful. It's also clear that the more zeros in a matrix the easier the chore. The following theorems enable us to increase the number of zeros in a matrix and at the same time keep track of how the value of the determinant changes.

Theorem 4.2. Let A be a square matrix. If A_1 is a matrix obtained from A by interchanging any two rows or columns, then $\det(A_1) = -\det(A)$.

EXAMPLE 4

$$\det \begin{bmatrix} 1 & 2 & 4 \\ 0 & 3 & 2 \\ 1 & 0 & 5 \end{bmatrix} = -\det \begin{bmatrix} 0 & 3 & 2 \\ 1 & 2 & 4 \\ 1 & 0 & 5 \end{bmatrix} \text{: rows one and two interchanged}$$

$$\det \begin{bmatrix} 1 & 2 & 4 \\ 0 & 3 & 2 \\ 1 & 0 & 5 \end{bmatrix} = -\det \begin{bmatrix} 1 & 4 & 2 \\ 0 & 2 & 3 \\ 1 & 5 & 0 \end{bmatrix} \text{: columns two and three interchanged}$$

■

Corollary 4.1. If an $n \times n$ matrix has two identical rows or columns, its determinant must equal zero.

Proof. The preceding theorem says that if you interchange any two rows or columns, the determinant changes sign. But if the two rows interchanged are identical, the determinant must remain unchanged. Since zero is the only number equal to its negative, we have $\det(A) = 0$.

EXAMPLE 5

a. $\det \begin{bmatrix} 1 & 2 \\ 1 & 2 \end{bmatrix} = 2 - 2 = 0$

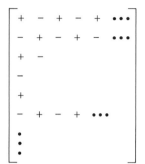

b. $\det\begin{bmatrix} 1 & -6 & 1 \\ 2 & 3 & 8 \\ 1 & -6 & 1 \end{bmatrix} = 0$: rows one and three are identical ■

Theorem 4.3. If any row or column of a square matrix A is multiplied by a constant c to get a matrix A_1, then $\det(A_1) = c[\det(A)]$.

Corollary 4.2. If a square matrix A has a row or column of zeros, then $\det(A) = 0$.

EXAMPLE 6

$$\det\begin{bmatrix} 3 & 4 & 12 \\ 6 & 16 & 30 \\ 9 & 8 & 21 \end{bmatrix} = \det\begin{bmatrix} 3(1) & 4 & 12 \\ 3(2) & 16 & 30 \\ 3(3) & 8 & 21 \end{bmatrix} = 3\det\begin{bmatrix} 1 & 4 & 12 \\ 2 & 16 & 30 \\ 3 & 8 & 21 \end{bmatrix}$$

$$= 3\det\begin{bmatrix} 1 & 4 & 12 \\ 2(1) & 2(8) & 2(15) \\ 3 & 8 & 21 \end{bmatrix}$$

$$= 6\det\begin{bmatrix} 1 & 4 & 3(4) \\ 1 & 8 & 3(5) \\ 3 & 8 & 3(7) \end{bmatrix} = 18\det\begin{bmatrix} 1 & 4(1) & 4 \\ 1 & 4(2) & 5 \\ 3 & 4(2) & 7 \end{bmatrix}$$

$$= 72\det\begin{bmatrix} 1 & 1 & 4 \\ 1 & 2 & 5 \\ 3 & 2 & 7 \end{bmatrix}$$ ■

Theorem 4.4. If any multiple of a row (column) is added to another row (column) of a square matrix A to get another matrix A_1, then $\det(A_1) = \det(A)$.

EXAMPLE 7

$$\det\begin{bmatrix} 1 & 1 & 4 \\ 1 & 2 & 5 \\ 3 & 2 & 7 \end{bmatrix} = \det\begin{bmatrix} 1 & 1 & 4 \\ 0 & 1 & 1 \\ 3 & 2 & 7 \end{bmatrix} = \det\begin{bmatrix} 1 & 1 & 4 \\ 0 & 1 & 1 \\ 0 & -1 & -5 \end{bmatrix}$$

$$= \det\begin{bmatrix} 1 & 1 & 4 \\ 0 & 1 & 1 \\ 0 & 0 & -4 \end{bmatrix} = (1)\det\begin{bmatrix} 1 & 1 \\ 0 & -4 \end{bmatrix} = -4$$ ■

This last example illustrates perhaps the easiest way to evaluate the determinant of a matrix. That is, use the elementary row or column operations to get a row or column with at most one nonzero entry and then use Theorem 4.1.

Our next example also demonstrates this idea. The reader might try computing the determinant in this example by using Theorem 4.1 directly and then comparing the two techniques.

EXAMPLE 8

$$\det\begin{bmatrix} 2 & 1 & -1 & 0 \\ 1 & 1 & 0 & 4 \\ 0 & 0 & 1 & -2 \\ 1 & 0 & 1 & 1 \end{bmatrix} = \det\begin{bmatrix} 2 & 1 & -1 & -2 \\ 1 & 1 & 0 & 4 \\ 0 & 0 & 1 & 0 \\ 1 & 0 & 1 & 3 \end{bmatrix} = \det\begin{bmatrix} 2 & 1 & -2 \\ 1 & 1 & 4 \\ 1 & 0 & 3 \end{bmatrix}$$

$$= \det\begin{bmatrix} 0 & 1 & -8 \\ 0 & 1 & 1 \\ 1 & 0 & 3 \end{bmatrix} = \det\begin{bmatrix} 1 & -8 \\ 1 & 1 \end{bmatrix} = 9$$

■

Theorem 4.5. Let A and B be two $n \times n$ matrices. Then

$$\det(AB) = \det(A)\det(B)$$

EXAMPLE 9 Let $A = \begin{bmatrix} 3 & 4 \\ 1 & -2 \end{bmatrix}$, $B = \begin{bmatrix} 1 & 3 \\ -2 & 8 \end{bmatrix}$. Verify Theorem 4.5 for these two matrices.

Solution

$$\det(A) = \det\begin{bmatrix} 3 & 4 \\ 1 & -2 \end{bmatrix} = -10 \qquad \det(B) = \det\begin{bmatrix} 1 & 3 \\ -2 & 8 \end{bmatrix} = 14$$

$$\det(AB) = \det\begin{bmatrix} -5 & 41 \\ 5 & -13 \end{bmatrix} = -140 = (-10)(14) = \det(A)\det(B)$$

■

Theorem 4.6. Let A be any square matrix and let A^T be its transpose, then

$$\det(A) = \det(A^T)$$

EXAMPLE 10 Verify Theorem 4.6 for the matrix $A = \begin{bmatrix} 6 & 4 \\ -2 & 3 \end{bmatrix}$.

Solution

$$\det(A) = \det\begin{bmatrix} 6 & 4 \\ -2 & 3 \end{bmatrix} = 18 - (-8) = 26$$

$$\det(A^T) = \det\begin{bmatrix} 6 & -2 \\ 4 & 3 \end{bmatrix} = 18 - (-8) = 26 = \det(A)$$

■

Theorem 4.7. A square matrix A is invertible if and only if $\det(A)$ is nonzero.

This last theorem is one that we use repeatedly in the remainder of this text. For example, in the next section we discuss how to compute the inverse of a matrix in terms of the determinants of its minors, and in Chapter 5 we use an equivalent version of Theorem 4.7 that says, if ker(A) has nonzero vectors in it, then det(A) = 0.

PROBLEM SET 4.1

1. Compute the determinants of each of the following matrices:

a. $\begin{bmatrix} -6 & 0 \\ 1 & 2 \end{bmatrix}$ **b.** $\begin{bmatrix} -2 & 1 \\ -4 & 2 \end{bmatrix}$ **c.** $\begin{bmatrix} 1 & 2 & 3 \\ 4 & 6 & 2 \\ -1 & 4 & 1 \end{bmatrix}$ **d.** $\begin{bmatrix} -1 & 2 & 6 & 4 \\ 1 & 0 & 2 & 8 \\ 0 & 3 & 9 & 6 \\ 2 & 7 & 5 & 6 \end{bmatrix}$

2. Compute the determinants of each of the following matrices:

a. $\begin{bmatrix} 1 & 3 \\ 2 & 4 \end{bmatrix}$ **b.** $\begin{bmatrix} 1 & 2 \\ 3 & 4 \end{bmatrix}$ **c.** $\begin{bmatrix} 1 & 0 & -2 \\ 4 & 6 & 0 \\ 1 & 1 & 0 \end{bmatrix}$ **d.** $\begin{bmatrix} 1 & 4 & 1 \\ 0 & 6 & 1 \\ -2 & 0 & 0 \end{bmatrix}$

3. a. $\det\begin{bmatrix} 1 & 2 & 3 \\ 4 & 5 & 6 \\ 7 & 8 & 9 \end{bmatrix}$ **b.** $\det\begin{bmatrix} a_1 & a_2 & a_3 \\ a_1+k & a_2+k & a_3+k \\ a_1+2k & a_2+2k & a_3+2k \end{bmatrix}$

4. a. $\det\begin{bmatrix} 2 & 3 & 0 \\ 3 & 4 & 1 \\ 0 & 1 & -2 \end{bmatrix}$ **b.** $\det\begin{bmatrix} 1 & 0 & 2 \\ 0 & -2 & 1 \\ 2 & 1 & 4 \end{bmatrix}$

5. If E is an elementary row matrix associated with adding a multiple of one row to another, then det(E) = ?

6. Let A be any upper or lower triangular matrix. Show that $\det(A) = a_{11}a_{22} \cdots a_{nn}$; that is, the determinant of A equals the product of the diagonal entries of A.

7. Let a_j, $j = 1, \ldots, n$ be arbitrary numbers and let k be any number. We assume below that $n > 2$.

a. Show that

$$\det\begin{bmatrix} a_1 & a_2 & \cdots & a_n \\ a_1+k & a_2+k & \cdots & a_n+k \\ a_1+2k & a_2+2k & \cdots & a_n+2k \\ \cdots & \cdots & \cdots & \cdots \\ a_1+(n-2)k & a_2+(n-2)k & \cdots & a_n+(n-2)k \\ a_1+(n-1)k & a_2+(n-1)k & \cdots & a_n+(n-1)k \end{bmatrix} = 0$$

b. Show that

$$\det\begin{bmatrix} 1 & 2 & \cdots & n \\ n+1 & n+2 & \cdots & 2n \\ 2n+1 & 2n+2 & \cdots & 2n+n \\ \cdots & \cdots & \cdots & \cdots \\ n(n-1)+1 & & \cdots & n^2 \end{bmatrix} = 0$$

c. What happens if $n = 2$?

8. Let $A_2 = \begin{bmatrix} 0 & a_2 \\ a_1 & 0 \end{bmatrix}$. Let $A_3 = \begin{vmatrix} 0 & 0 & a_3 \\ 0 & a_2 & 0 \\ a_1 & 0 & 0 \end{vmatrix}$, and

$$\text{let } A_n = \begin{bmatrix} 0 & 0 & \cdots & & a_n \\ 0 & 0 & \cdots & a_{n-1} & 0 \\ \cdots & \cdots & \cdots & \cdots & \cdots \\ 0 & a_2 & 0 & \cdots & 0 \\ a_1 & 0 & 0 & \cdots & 0 \end{bmatrix}$$

Thus, A_n is an $n \times n$ matrix whose only possible nonzero entries occur in the $(n-i+1)$st column of the ith row. Show that $\det(A_n) = (-1)^{n(n-1)/2} a_1 a_2 \cdots a_n$.

9. Let $x_1, x_2, \ldots, x_n$ be any n numbers. Let

$$\begin{bmatrix} 1 & 1 \\ x_1 & x_2 \end{bmatrix} = A_2 \qquad \begin{bmatrix} 1 & 1 & 1 \\ x_1 & x_2 & x_3 \\ x_1^2 & x_2^2 & x_3^2 \end{bmatrix} = A_3$$

$$\begin{bmatrix} 1 & 1 & \cdots & 1 \\ x_1 & x_2 & \cdots & x_n \\ \cdots & \cdots & \cdots & \cdots \\ x_1^{n-1} & & \cdots & x_n^{n-1} \end{bmatrix} = A_n$$

The reader should note that these matrices (actually their transposes) appeared in Section 3.4, when we discussed fitting polynomials to prescribed data. The determinants in this problem are called Vandermonde determinants.

a. Show that $\det(A_2) = x_2 - x_1$.
b. $\det(A_3) = (x_3 - x_2)(x_3 - x_1)(x_2 - x_1)$.
c. Find a similar formula for $\det(A_n)$.

10. Let A be any $n \times n$ matrix. Show that $\det(A)$ equals zero if and only if the rows (columns) of A form a linearly dependent set of vectors in $\mathbb{R}^n$. (Hint: Use Theorems 4.4 and 4.7.)

11. Let A be a nonsingular matrix. Show that $\det(A^{-1}) = [\det(A)]^{-1}$. (Hint: $AA^{-1} = I_n$. Now use Theorem 4.5.)

12. Let S be any scalar matrix, that is, $S = cI_n$ for some number c. Show that $\det(S) = c^n$.

13. Let A and B be any two similar matrices; that is, there is a nonsingular matrix P such that $A = PBP^{-1}$. Show that $\det(A) = \det(B)$.

14. Compute the determinants of the following matrices:

a. $\begin{bmatrix} 2 & 3 & 4 & 5 \\ 3 & 4 & 5 & 6 \\ 4 & 5 & 6 & 7 \\ 5 & 6 & 7 & 8 \end{bmatrix}$ **b.** $\begin{bmatrix} -8 & 7 & 6 & -1 \\ 2 & 5 & 4 & 3 \\ 1 & -5 & 7 & 11 \\ 0 & 2 & -3 & 6 \end{bmatrix}$

15. A matrix is said to be skew symmetric if $A = -A^T$. Thus $\begin{bmatrix} 0 & -1 \\ 1 & 0 \end{bmatrix}$ is skew symmetric, while $\begin{bmatrix} 0 & 1 \\ 1 & 0 \end{bmatrix}$ is not.

a. Show that all the diagonal elements of a skew symmetric matrix are equal to zero.
b. Let A be a 3×3 skew symmetric matrix. Show that $\det(A) = 0$.

c. If A is a 2×2 skew symmetric matrix, must $\det(A) = 0$?
d. What happens if A is an $n \times n$ skew symmetric matrix?

16. Let P be an $n \times n$ permutation matrix. Show that $\det(P) = \pm 1$. For the definition of a permutation matrix see problem 13 in Problem Set 1.5.

4.2 THE ADJOINT MATRIX AND A^{-1}

Theorem 4.7 states that A is invertible if and only if $\det(A)$ is nonzero. In this section we show how to compute the inverse of a matrix by using the determinants of its $(n-1) \times (n-1)$ minors (cf. Definition 4.2). Since these determinants will appear quite frequently, we introduce a special notation for them.

Definition 4.3. Let $A = [a_{jk}]$ be any $n \times n$ matrix. The cofactor of a_{jk}, denoted by A_{jk}, is defined to be

$$A_{jk} = (-1)^{j+k} \det(M_{jk})$$

where M_{jk} is the j,k minor of A

EXAMPLE 1 Let $A = \begin{bmatrix} -1 & 6 & 3 \\ 2 & 0 & 9 \\ 4 & 8 & 7 \end{bmatrix}$.

$$\begin{aligned} &A_{11} = (-1)^2(-72) && A_{12} = (-1)^3(14 - 36) && A_{13} = (-1)^4(16) \\ &A_{21} = -(42 - 24) && A_{22} = (-7 - 12) && A_{23} = -(-8 - 24) \\ &A_{31} = (54) && A_{32} = -(-9 - 6) && A_{33} = (-12) \end{aligned}$$ ■

Using the cofactors of a matrix, we construct its adjoint matrix.

Definition 4.4. Let $A = [a_{jk}]$ be an $n \times n$ matrix. The adjoint matrix of A is the following $n \times n$ matrix:

$$\operatorname{adj}(A) = [A_{jk}]^T$$

where A_{jk} are the cofactors of A.

Note that the transpose of the matrix $[A_{jk}]$ must be taken.

EXAMPLE 2 Let $A = \begin{bmatrix} -1 & 6 & 3 \\ 2 & 0 & 9 \\ 4 & 8 & 7 \end{bmatrix}$.

From Example 1 and the definition of adj(A) we have

$$\operatorname{adj}(A) = \begin{bmatrix} -72 & 22 & 16 \\ -18 & -19 & 32 \\ 54 & 15 & -12 \end{bmatrix}^T$$
$$= \begin{bmatrix} -72 & -18 & 54 \\ 22 & -19 & 15 \\ 16 & 32 & -12 \end{bmatrix}$$

■

EXAMPLE 3 Let A be the matrix in the preceding example. Compute det(A) and A[adj(A)].

$$\det(A) = \det \begin{bmatrix} -1 & 6 & 3 \\ 2 & 0 & 9 \\ 4 & 8 & 7 \end{bmatrix} = \det \begin{bmatrix} -1 & 6 & 3 \\ 0 & 12 & 15 \\ 0 & 32 & 19 \end{bmatrix} = 252$$

$$A[\operatorname{adj}(A)] = \begin{bmatrix} -1 & 6 & 3 \\ 2 & 0 & 9 \\ 4 & 8 & 7 \end{bmatrix} \begin{bmatrix} -72 & -18 & 54 \\ 22 & -19 & 15 \\ 16 & 32 & -12 \end{bmatrix} = \begin{bmatrix} 252 & 0 & 0 \\ 0 & 252 & 0 \\ 0 & 0 & 252 \end{bmatrix}$$
$$= 252 I_3 = \det(A) I_3$$

■

EXAMPLE 4 Let $A = \begin{bmatrix} a & b \\ c & d \end{bmatrix}$. Compute det($A$) and A adj(A).

Solution. We exhibit the minors of A first, and then compute the adjoint of A.

$$M_{11} = d \qquad M_{12} = c$$
$$M_{21} = b \qquad M_{22} = a$$

Thus,

$$\operatorname{adj}(A) = \operatorname{adj} \begin{bmatrix} a & b \\ c & d \end{bmatrix} = \begin{bmatrix} d & -c \\ -b & a \end{bmatrix}^T = \begin{bmatrix} d & -b \\ -c & a \end{bmatrix}$$

$$A \operatorname{adj}(A) = \begin{bmatrix} a & b \\ c & d \end{bmatrix} \begin{bmatrix} d & -b \\ -c & a \end{bmatrix} = \begin{bmatrix} ad-bc & 0 \\ 0 & ad-bc \end{bmatrix}$$
$$= (ad-bc) \begin{bmatrix} 1 & 0 \\ 0 & 1 \end{bmatrix} = \det(A) I_2$$

■

The preceding two examples showed that A adj(A) = det(A)I for any 2×2 matrix and one particular 3×3 matrix. The next theorem states that this formula is true for any square matrix.

Theorem 4.8. Let A be any $n \times n$ matrix. Then

$$A[\text{adj}(A)] = [\text{adj}(A)]A = [\det(A)]I_n$$

Corollary 4.3. Let A be any square matrix with nonzero determinant. Then A is nonsingular and

$$A^{-1} = (\det(A))^{-1}\text{adj}(A)$$

EXAMPLE 5 Let $A = \begin{bmatrix} -1 & 6 & 3 \\ 2 & 0 & 9 \\ 4 & 8 & 7 \end{bmatrix}$. In Examples 2 and 3, we computed the determinant and adjoint of this matrix. Using those results, we have

$$A^{-1} = (\det(A))^{-1}\text{adj}(A) = (252)^{-1}\begin{bmatrix} -72 & -18 & 54 \\ 22 & -19 & 15 \\ 16 & 32 & -12 \end{bmatrix}$$ ■

EXAMPLE 6 Let $A = \begin{bmatrix} 2 & 1 & 0 & 1 \\ -1 & 0 & -1 & 2 \\ 0 & 0 & 1 & 3 \\ 2 & -1 & 0 & 0 \end{bmatrix}$. Compute the adjoint of A and verify Theorem 4.8.

$$\text{adj}(A) = \begin{bmatrix} 5 & 10 & -3 & 1 \\ -1 & -2 & -12 & 4 \\ -1 & -2 & 9 & 4 \\ 5 & -11 & -3 & 1 \end{bmatrix}^T = \begin{bmatrix} 5 & -1 & -1 & 5 \\ 10 & -2 & -2 & -11 \\ -3 & -12 & 9 & -3 \\ 1 & 4 & 4 & 1 \end{bmatrix}.$$

A quick computation shows that

$$A\,\text{adj}(A) = \begin{bmatrix} 21 & 0 & 0 & 0 \\ 0 & 21 & 0 & 0 \\ 0 & 0 & 21 & 0 \\ 0 & 0 & 0 & 21 \end{bmatrix} = \det(A)I_4$$ ■

PROBLEM SET 4.2

1. Compute the adjoint matrix of each of the following matrices and verify that $A\,\text{adj}(A) = \det(A)I_n$:

a. $\begin{bmatrix} 2 & 0 \\ 0 & 3 \end{bmatrix}$ **b.** $\begin{bmatrix} a & 0 \\ 0 & b \end{bmatrix}$ **c.** $\begin{bmatrix} 6 & 1 \\ 3 & 4 \end{bmatrix}$ **d.** $\begin{bmatrix} 2 & 4 \\ -1 & -2 \end{bmatrix}$

2. Compute the adjoint matrix of each of the following matrices and verify that $A\,\text{adj}(A) = \det(A)I_n$:

a. $\begin{bmatrix} -2 & 3 & 4 \\ 0 & 1 & 2 \\ 0 & 0 & -2 \end{bmatrix}$ b. $\begin{bmatrix} 2 & 6 & 0 \\ 4 & 0 & 8 \\ -2 & 3 & 0 \end{bmatrix}$ c. $\begin{bmatrix} 2 & -1 & -2 \\ 1 & 0 & -4 \\ 0 & -1 & 6 \end{bmatrix}$

3. Using Corollary 4.3, compute the inverse of each of the nonsingular matrices in problem 1.
4. Using Corollary 4.3, compute the inverse of each of the nonsingular matrices in problem 2.
5. Consider the following system of equations:

$$\begin{aligned} x_1 - x_2 + x_3 - x_4 &= 1 \\ x_1 + x_2 - x_3 + x_4 &= 2 \\ x_1 + x_2 + x_3 - x_4 &= 3 \\ x_1 + x_2 + x_3 + x_4 &= 4 \end{aligned}$$

   a. Solve this system using Gaussian elimination.
   b. Using Corollary 4.3, compute the inverse of the coefficient matrix A.
   c. Using A^{-1}, solve this system.
6. Suppose A is a 2×2 upper triangular matrix. Show that adj(A) is also upper triangular. Is this also true for 3×3 matrices?
7. Let A be a square matrix.
   a. If A is invertible, show that adj(A) is also invertible. Find a formula for det(adj(A)) and for the inverse of adj(A).
   b. If A is not invertible, show that adj(A) is not invertible. (Hint: Theorem 4.8.) Thus, $\det(A) = 0$ if and only if $\det(\text{adj}(A)) = 0$.
8. How are the adjoints of A and A^T related?

4.3 CRAMER'S RULE

Cramer's rule is a formula for computing the solution to a system of linear equations when the coefficient matrix A is nonsingular. This formula is just the component version of the equation

$$A^{-1} = (\det(A))^{-1}\text{adj}(A) \tag{4.2}$$

We derive Cramer's rule for a system of two equations with two unknowns before stating the rule for n equations with n unknowns.

EXAMPLE 1 Find the solution to

$$\begin{aligned} a_{11}x_1 + a_{12}x_2 &= b_1 \\ a_{21}x_1 + a_{22}x_2 &= b_2 \end{aligned} \tag{4.3}$$

Assuming that $\det(A) = a_{11}a_{22} - a_{12}a_{21} \neq 0$,

$$A^{-1} = (\det(A))^{-1}\text{adj}(A) = (\det A)^{-1}\begin{bmatrix} a_{22} & -a_{12} \\ -a_{21} & a_{11} \end{bmatrix}$$

Thus,

$$\begin{bmatrix} x_1 \\ x_2 \end{bmatrix} = A^{-1}\begin{bmatrix} b_1 \\ b_2 \end{bmatrix} = (\det A)^{-1}\begin{bmatrix} a_{22} & -a_{12} \\ -a_{21} & a_{11} \end{bmatrix}\begin{bmatrix} b_1 \\ b_2 \end{bmatrix}$$

and

$$\begin{bmatrix} x_1 \\ x_2 \end{bmatrix} = \begin{bmatrix} b_1a_{22} - b_2a_{12} \\ b_2a_{11} - b_1a_{21} \end{bmatrix}(\det A)^{-1}$$

Thus,

$$x_1 = \frac{b_1a_{22} - b_2a_{12}}{\det(A)} = \frac{\det\begin{bmatrix} b_1 & a_{12} \\ b_2 & a_{22} \end{bmatrix}}{\det(A)}$$

$$x_2 = \frac{b_2a_{11} - b_1a_{21}}{\det(A)} = \frac{\det\begin{bmatrix} a_{11} & b_1 \\ a_{21} & b_2 \end{bmatrix}}{\det(A)}$$

■

Thus, we see that x_1 can be expressed as the ratio of two determinants. The matrix in the numerator is obtained by replacing the first column of A with the right-hand side of (4.3). To calculate x_2, the matrix in the numerator is obtained by replacing the second column of A with the right-hand side of (4.3).

Consider the general system with n equations and n unknowns

$$A\mathbf{x} = \mathbf{b} \tag{4.4}$$

where $A = [a_{jk}]$ is an $n \times n$ nonsingular matrix $\mathbf{x} = [x_1, \ldots, x_n]^T$, and $\mathbf{b} = [b_1, \ldots, b_n]^T$. We know that

$$\mathbf{x} = A^{-1}\mathbf{b} = (\det A)^{-1}\text{adj}(A)\mathbf{b} \tag{4.5}$$

Using (4.5), we write x_k in terms of $\mathbf{b}$:

$$x_k = (\det A)^{-1}(A_{1k}b_1 + A_{2k}b_2 + \cdots + A_{nk}b_n) \tag{4.6}$$

remember $\text{adj}(A) = [A_{jk}]^T$, where A_{jk} is the cofactor of a_{jk}. The last factor in (4.6) can be thought of as the expansion by minors (going down the kth column) of the determinant of the matrix obtained by replacing the kth column of A with the column $[b_1, \ldots, b_n]^T$. That is,

$$x_k = \frac{\det\begin{bmatrix} a_{11} & \cdots & a_{1(k-1)} & b_1 & a_{1(k+1)} & \cdots & a_{1n} \\ \cdot & \cdot & \cdot & \cdot & \cdot & \cdot & \cdot \\ a_{j1} & \cdots & \cdots & b_j & \cdots & \cdots & a_{jn} \\ \cdot & \cdot & \cdot & \cdot & \cdot & \cdot & \cdot \\ a_{n1} & \cdots & a_{n(k-1)} & b_n & a_{n(k+1)} & \cdots & a_{nn} \end{bmatrix}}{\det(A)} \tag{4.7}$$

Formula (4.7) is Cramer's rule. Note that it is valid only for systems whose coefficient matrix is nonsingular.

EXAMPLE 2 Find the value of x_4 for which x_1, x_2, x_3, x_4, and x_5 solves the following system:

$$\begin{aligned}
2x_1 + x_2 \qquad\qquad\qquad + 2x_5 &= 2\\
x_1 + \qquad\qquad x_4 \qquad &= 2\\
3x_2 + x_3 + 2x_4 \qquad &= -1\\
x_1 + x_2 + \qquad x_4 + x_5 &= -1\\
x_3 + x_4 + x_5 &= -1
\end{aligned}$$

An easy computation shows that the coefficient matrix A has a determinant equal to -7. Thus, A is nonsingular and we have

$$x_4 = \frac{\det\begin{bmatrix} 2 & 1 & 0 & 2 & 2\\ 1 & 0 & 0 & 2 & 0\\ 0 & 3 & 1 & -1 & 0\\ 1 & 1 & 0 & -1 & 1\\ 0 & 0 & 1 & -1 & 1\end{bmatrix}}{\det(A)} = -\left(\frac{13}{7}\right)$$

We now have three techniques for solving systems of equations: Gaussian elimination, which always works; inversion of the coefficient matrix A, i.e., compute A^{-1}; and Cramer's rule. The last two methods assume, of course, that A is invertible. For small systems, $n = 2, 3$, or 4, any of these methods would be fine, at least if A is invertible. For large n, however, the most efficient method to use in solving a system of equations is usually Gaussian elimination. It is for this reason that Cramer's rule is a theoretical rather than a problem-solving tool.

PROBLEM SET 4.3

1. Use Cramer's rule to solve each of the following systems of equations:

a. $\begin{aligned} 2x_1 + 3x_2 &= 7\\ 8x_1 + x_2 &= -2 \end{aligned}$ **b.** $\begin{aligned} -8x_1 + 6x_2 &= 4\\ 3x_1 + 2x_2 &= 6 \end{aligned}$

2. Use Gaussian elimination to solve each of the systems in problem 1.

3. Use Cramer's rule to solve the following system:

$$\begin{aligned}
2x_1 - 6x_2 + x_3 &= 2\\
x_2 + x_3 &= 1\\
x_1 - x_2 - x_3 &= 0
\end{aligned}$$

4. Consider the following system of equations:

$$\begin{aligned}
2x_1 + x_2 + x_3 - x_4 &= 1\\
3x_1 - x_2 + x_3 + x_4 &= 0\\
-x_1 + x_2 - x_3 + x_4 &= 0\\
6x_1 - x_2 \qquad + x_4 &= 0
\end{aligned}$$

a. Use Cramer's rule to solve for x_1.
b. Use Gaussian elimination to solve for x_1.

5. Consider the following system of equations:

$$\begin{aligned} -x_1 + 3x_2 + x_3 &= 7 \\ x_1 - x_2 - x_3 &= -2 \\ -x_1 + 6x_2 + 2x_3 &= 3 \end{aligned}$$

a. Solve this system using Cramer's rule.
b. Solve using Gaussian elimination.
c. Solve by finding the inverse of the coefficient matrix.

6. Solve the following system by using Cramer's rule.

$$\begin{aligned} 2x_1 - \qquad x_3 &= 1 \\ 2x_1 + 4x_2 - x_3 &= 0 \\ x_1 - 8x_2 - 3x_3 &= -2 \end{aligned}$$

7. Solve using Cramer's rule.

$$\begin{aligned} x_1 + x_2 + x_3 &= a \\ x_1 + (1+a)x_2 + x_3 &= 2a \\ x_1 + x_2 + (1+a)x_3 &= 0 \end{aligned}$$

8. Solve using Cramer's rule.

$$\begin{aligned} x_1 + x_2 + x_3 + x_4 &= 4 \\ x_1 + 2x_2 + 2x_3 + 2x_4 &= 1 \\ x_1 + x_2 + 2x_3 + 2x_4 &= 2 \\ x_1 + x_2 + x_3 + 2x_4 &= 3 \end{aligned}$$

4.4 AREA AND VOLUME

Given any two vectors in $\mathbb{R}^2$ that are not parallel, they determine a parallelogram; cf. problems 9 and 10 in Section 2.1. With a few manipulations it is possible to express the area of this parallelogram as the absolute value of the determinant of a matrix constructed from the coordinates of these vectors.

EXAMPLE 1 Compute the area of the parallelogram determined by the vectors (1,2) and (8,0) (see Figure 4.2).

The area of this parallelogram is the base length 8 times the height 2.

$$\text{Area} = 8 \times 2 = \det \begin{bmatrix} 8 & 1 \\ 0 & 2 \end{bmatrix} \tag{4.8}$$

where the columns of the matrix $\begin{bmatrix} 8 & 1 \\ 0 & 2 \end{bmatrix}$ are the coordinates of the two vectors (8,0) and (1,2). ■

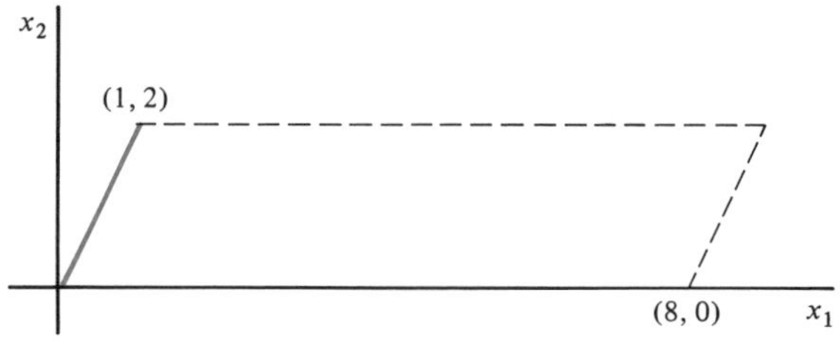

Figure 4.2

We derive a similar formula for any two vectors (a_1, a_2) and (b_1, b_2). Our proof and picture assume that both vectors lie in the first quadrant, but the other cases can be handled in a similar manner.

Consider the parallelogram determined by the two vectors (a_1, a_2) and (b_1, b_2) (see Figure 4.3).

$$\begin{aligned}\text{Area of parallelogram } OBCA &= \text{area of triangle}(OBP_2)\\ &\quad + \text{area of trapezoid}(P_2BCP_3)\\ &\quad - \text{area of triangle}(OAP_1)\\ &\quad - \text{area of trapezoid}(P_1ACP_3)\\ &= \tfrac{1}{2}(b_1b_2) + \tfrac{1}{2}(a_1)(b_2 + a_2 + b_2)\\ &\quad - \tfrac{1}{2}(a_1a_2) - \tfrac{1}{2}(b_1)(a_2 + a_2 + b_2)\\ &= a_1b_2 - a_2b_1 = \det\begin{bmatrix} a_1 & b_1 \\ a_2 & b_2 \end{bmatrix}\end{aligned}$$

Since we do not wish to worry about the order in which we list the columns of the above matrix, we write the area as

$$\text{Area} = \left|\det\begin{bmatrix} a_1 & b_1 \\ a_2 & b_2 \end{bmatrix}\right|$$

We formally state this in the following theorem.

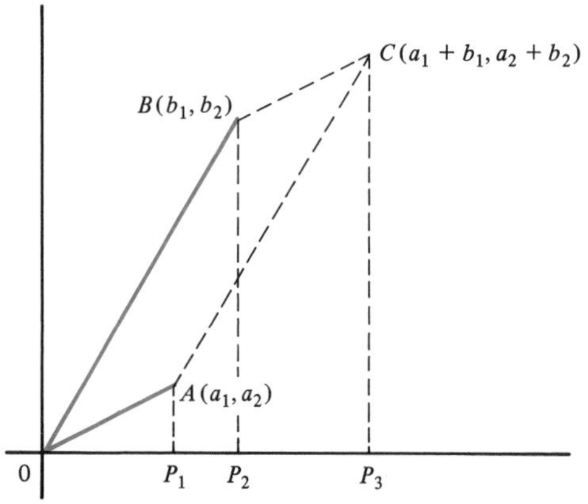

Figure 4.3

Theorem 4.9. Let $\mathbf{a} = (a_1, a_2)$ and $\mathbf{b} = (b_1, b_2)$ be any two vectors in $\mathbb{R}^2$. Let P be the parallelogram generated by these two vectors, i.e., $P = \{\mathbf{x}: \mathbf{x} = t_1\mathbf{a} + t_2\mathbf{b},\ t_1$ and t_2 arbitrary scalars between 0 and 1$\}$. Then

$$\text{Area}(P) = \left|\det\begin{bmatrix} a_1 & b_1 \\ a_2 & b_2 \end{bmatrix}\right|$$

EXAMPLE 2

a. Sketch the parallelogram determined by the vectors $(-6,4)$ and $(3,-2)$ and calculate its area.

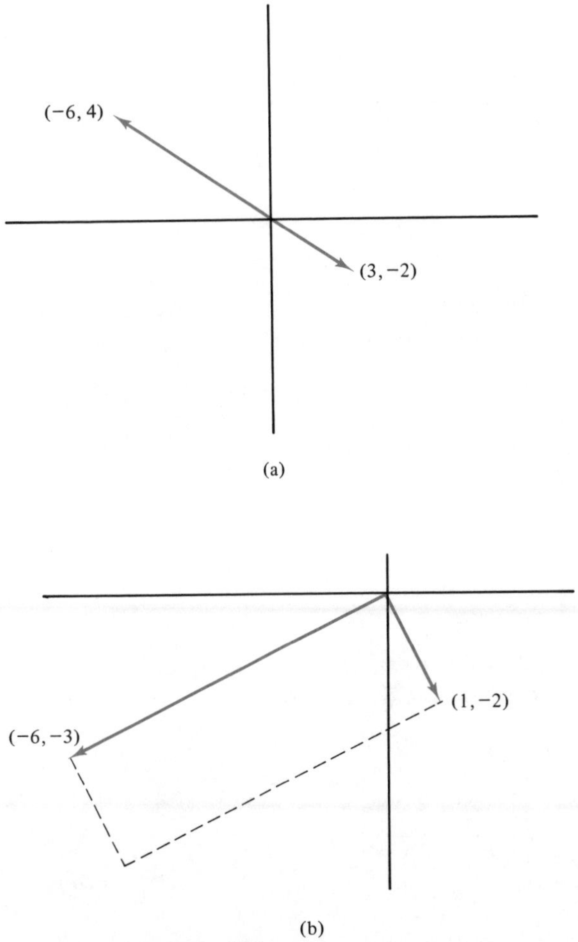

Figure 4.4

We see from Figure 4.4a that the parallelogram is just the straight-line segment joining the two points $(-6,4)$ and $(3,-2)$. Thus the area should be zero. Indeed, we have

$$\text{Area}(P) = \left|\det \begin{bmatrix} -6 & 3 \\ 4 & -2 \end{bmatrix}\right| = |12 - 12| = 0$$

b. Repeat the above, using the vectors $(1,-2)$ and $(-6,-3)$ (see Figure 4.4b).

$$\text{Area}(P) = \left|\det \begin{bmatrix} 1 & -6 \\ -2 & -3 \end{bmatrix}\right| = |(-3 - 12)| = 15$$

■

As one would expect, there is a generalization of this formula to higher dimensions, and we state it in the next theorem.

Theorem 4.10. Let $\{\mathbf{x}_1, \ldots, \mathbf{x}_n\}$ be any n vectors in $\mathbb{R}^n$. Let P be the n-dimensional parallelepiped generated by these vectors, that is,

$$P = \{\mathbf{y}: \mathbf{y} = t_1\mathbf{x}_1 + \cdots + t_n\mathbf{x}_n: 0 \leq t_j \leq 1\}$$

Then the n-dimensional volume of P equals

$$\text{Vol}(P) = \left|\det \begin{bmatrix} \mathbf{x}_1 & \mathbf{x}_2 & \cdots & \mathbf{x}_n \\ \cdot & \cdot & & \cdot \\ \cdot & \cdot & & \cdot \\ \cdot & \cdot & & \cdot \end{bmatrix}\right| \tag{4.9}$$

The matrix in (4.9) is obtained by using the coordinates of the vector $\mathbf{x}_j$ (with respect to the standard basis) as the jth column. Thus, if we had the four vectors

$$\mathbf{x}_1 = (1,1,2,1) \qquad \mathbf{x}_2 = (-1,-1,3,4) \qquad \mathbf{x}_3 = (8,9,1,1) \qquad \mathbf{x}_4 = (10,11,1,0)$$

then the matrix would be

$$\begin{bmatrix} 1 & -1 & 8 & 10 \\ 1 & -1 & 9 & 11 \\ 2 & 3 & 1 & 1 \\ 1 & 4 & 1 & 0 \end{bmatrix}$$

EXAMPLE 3 Sketch the parallelepiped generated by the three vectors $\mathbf{a} = (1,1,0)$, $\mathbf{b} = (0,1,1)$, and $\mathbf{c} = (1,0,1)$ (see Figure 4.5), and determine its volume.

$$\begin{aligned} \text{Vol}(P) &= \left|\det \begin{bmatrix} 1 & 0 & 1 \\ 1 & 1 & 0 \\ 0 & 1 & 1 \end{bmatrix}\right| \\ &= \left|\det \begin{bmatrix} 1 & 0 \\ 1 & 1 \end{bmatrix} - \det \begin{bmatrix} 0 & 1 \\ 1 & 1 \end{bmatrix}\right| \\ &= |1 + 1| = 2 \end{aligned}$$

■

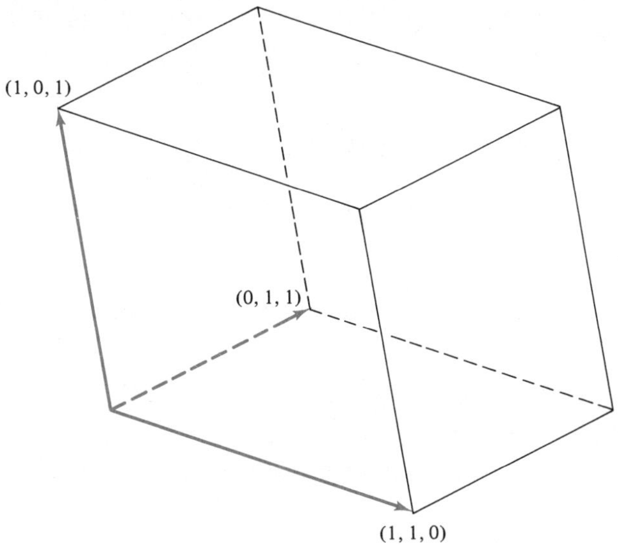

Figure 4.5

We remark that these determinants will be zero if and only if the n vectors used to form the matrices are linearly dependent. In that case, the solid they generate will lie in an $(n-1)$-dimensional plane, and hence should have n-dimensional volume equal to zero. See Example 2a.

If L is a linear transformation from $\mathbb{R}^2$ to $\mathbb{R}^2$, it has a matrix representation A, where

$$A = \begin{bmatrix} a_{11} & a_{12} \\ a_{21} & a_{22} \end{bmatrix}$$

As we saw in Chapter 3, $L(\mathbf{e}_1) = (a_{11}, a_{21})$ and $L(\mathbf{e}_2) = (a_{12}, a_{22})$. Thus, geometrically we can picture L as transforming the parallelogram generated by $\mathbf{e}_1$ and $\mathbf{e}_2$ into the parallelogram generated by (a_{11}, a_{21}) and (a_{12}, a_{22}) (see Figure 4.6). Moreover, we have

$$\text{Area}(L(P)) = \left| \det \begin{bmatrix} a_{11} & a_{12} \\ a_{21} & a_{22} \end{bmatrix} \right| = |\det(A)|$$

Since in this particular case we have area$(P) = 1$, we may rewrite this formula as

$$\text{Area}(L(P)) = |\det(A)| \text{area}(P)$$

To see that this formula is true in general let $\mathbf{x} = (x_1, x_2)$ and $\mathbf{y} = (y_1, y_2)$. Let P be the parallelogram generated by $\mathbf{x}$ and $\mathbf{y}$, that is,

$$P = \{t_1\mathbf{x} + t_2\mathbf{y}: 0 \le t_1 \le 1, 0 \le t_2 \le 1\}$$

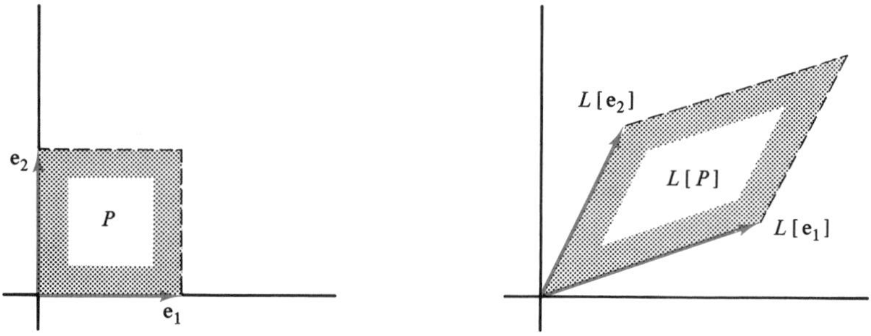

Figure 4.6

Then

$$L(P) = \{t_1L(\mathbf{x}) + t_2L(\mathbf{y}): 0 \leq t_1 \leq 1, 0 \leq t_2 \leq 1\}$$

where

$$L(\mathbf{x}) = (a_{11}x_1 + a_{12}x_2, a_{21}x_1 + a_{22}x_2)$$

and

$$L(\mathbf{y}) = (a_{11}y_1 + a_{12}y_2, a_{21}y_1 + a_{22}y_2).$$

We have from (4.9) that

$$\begin{aligned}\text{Area}(L(P)) &= \left|\det \begin{bmatrix} a_{11}x_1 + a_{12}x_2 & a_{11}y_1 + a_{12}y_2 \\ a_{21}x_1 + a_{22}x_2 & a_{21}y_1 + a_{22}y_2 \end{bmatrix}\right| \\ &= \left|\det \left(\begin{bmatrix} a_{11} & a_{12} \\ a_{21} & a_{22} \end{bmatrix} \begin{bmatrix} x_1 & y_1 \\ x_2 & y_2 \end{bmatrix} \right)\right| \\ &= \left|\det \begin{bmatrix} a_{11} & a_{12} \\ a_{21} & a_{22} \end{bmatrix} \det \begin{bmatrix} x_1 & y_1 \\ x_2 & y_2 \end{bmatrix}\right| \\ &= |\det(A)|[\text{area}(P)]\end{aligned}$$

Thus, a linear transformation from $\mathbb{R}^2$ to $\mathbb{R}^2$ maps parallelograms into parallelograms (rank $A = 2$), or line segments (rank $A = 1$), or a point (rank $A = 0$). Moreover, the change in area depends only on the linear transformation and not on the particular parallelogram. Naturally there is a generalization of this to higher dimensions which we state below.

Theorem 4.11. Let A be the $n \times n$ matrix representation of $L:\mathbb{R}^n \to \mathbb{R}^n$. Let P be an n-dimensional parallelogram generated by the n vectors $\{\mathbf{x}_1, \ldots, \mathbf{x}_n\}$. Then $L(P)$ is generated by the n vectors $\{L(\mathbf{x}_1), \ldots, L(\mathbf{x}_n)\}$ and their volumes are related by the formula

$$\text{Vol}(L(P)) = |\det(A)|\text{Vol}(P) \tag{4.10}$$

PROBLEM SET 4.4

1. Calculate the area of the triangles whose vertices are:
a. (0,0), (1,6), (−2,3) **b.** (8,17), (9,2), (4,6)

2. Calculate the volume of the tetrahedron whose vertices are:
a. (0,0,0), (1,−1,2), (−3,6,7), (1,1,1)
b. (1,1,1), (−1,−1,−1), (0,4,8), (−3,0,2)

3. Find the area of the parallelograms determined by the following vectors; cf. problem 1.
a. (1,6), (−2,3) **b.** (1,−15), (−4,−11)

4. Sketch the parallelograms P generated by the following pairs of vectors. Let O be the parallelogram generated by the standard basis vectors. For each of the parallelograms P find a linear transformation L such that $P = L(O)$.
a. (1,0), (1,2) **b.** (1,−1), (3,6)

5. In $\mathbb{R}^2$, show that the straight line passing through the two points (a_1,a_2) and (b_1,b_2) has equation

$$\det \begin{bmatrix} x_1 & x_2 & 1 \\ a_1 & a_2 & 1 \\ b_1 & b_2 & 1 \end{bmatrix} = 0$$

6. Let L, a linear transformation from $\mathbb{R}^2$ to $\mathbb{R}^2$, have matrix representation

$$A = \begin{bmatrix} 1 & 2 \\ -1 & -2 \end{bmatrix}$$

Let P be the parallelogram generated by the vectors (−1,2) and (1,3). Sketch P and $L(P)$ and compute their areas. Then verify (4.10).

7. Let $A = \begin{bmatrix} -1 & -2 \\ -2 & 1 \end{bmatrix}$ be the matrix representation of a linear transformation L. Let P be the parallelogram generated by the following pairs of vectors:
a. (1,0), (0,1)
b. (1,1), (−1,1)
In each case sketch P, $L(P)$, and verify (4.10).

8. Let $A = \begin{bmatrix} 1 & 0 & 2 \\ -2 & 1 & 1 \\ 0 & 1 & 0 \end{bmatrix}$ be the matrix representation of a linear transformation L.

Let P be the solid generated by the following vectors:
a. (1,0,0), (0,1,0), (0,0,1)
b. (1,1,0), (1,0,1), (0,1,1)
In each case sketch P, $L(P)$, and verify (4.10).

SUPPLEMENTARY PROBLEMS

1. Let $A = [a_{jk}]$ be an $n \times n$ matrix. Define each of the following:
a. M_{jk} minor of A
b. Cofactor of a_{jk}
c. Adj(A)

2. Compute the determinant and adjoint of each of the following matrices:

a. $\begin{bmatrix} 3 & 4 \\ 2 & 5 \end{bmatrix}\begin{bmatrix} 4 & 2 \\ 6 & 3 \end{bmatrix}$

b. $\begin{bmatrix} -2 & 3 & 4 \\ 1 & 0 & -2 \\ 5 & 1 & 6 \end{bmatrix}\begin{bmatrix} 5 & -3 & 1 \\ 0 & 1 & 1 \\ -5 & 5 & 1 \end{bmatrix}$

3. Let $A = \begin{bmatrix} -2 & x & 1 \\ x & 1 & 1 \\ 2 & 3 & -1 \end{bmatrix}$. Find all values of x for which $\det(A) = 0$.

4. Find all values of λ for which the following system of equations has a nontrivial solution:

$$\begin{aligned} 2x_1 - 7x_2 &= \lambda x_1 \\ -4x_1 + \ x_2 &= \lambda x_2 \end{aligned}$$

5. Let $f(t) = \det\begin{bmatrix} a(t) & b(t) \\ c(t) & d(t) \end{bmatrix}$, where $a(t)$, $b(t)$, $c(t)$, and $d(t)$ are differentiable functions of t.

a. Show that $f'(t) = \det\begin{bmatrix} a' & b' \\ c & d \end{bmatrix} + \det\begin{bmatrix} a & b \\ c' & d' \end{bmatrix}$.

b. Derive a similar formula for the determinant of a 3×3 matrix of functions.

6. Let $y_1(t)$ and $y_2(t)$ be two solutions of the differential equation

$$ay'' + by' + cy = 0$$

Define the Wronskian $W(t)$ of the solutions $y_j(t)$ by

$$W(t) = \det\begin{bmatrix} y_1(t) & y_2(t) \\ y_1'(t) & y_2'(t) \end{bmatrix}$$

Show that the Wronskian satisfies the differential equation

$$aW' + bW = 0$$

7. Using the fact that a matrix has nonzero determinant if and only if its rows (columns) are linearly independent, determine which of the following sets are linearly independent:

a. $(2,1)$, $(-3,4)$

b. $(2,1,-2,3)$, $(0,1,0,1)$, $(1,1,1,0)$, $(3,3,-1,4)$

8. If $\mathbf{A} = (a_1,a_2,a_3)$ and $\mathbf{B} = (b_1,b_2,b_3)$, then the cross product of **A** and **B** is defined as

$$\begin{aligned} \mathbf{A} \times \mathbf{B} &= \det\begin{bmatrix} \mathbf{i} & \mathbf{j} & \mathbf{k} \\ a_1 & a_2 & a_3 \\ b_1 & b_2 & b_3 \end{bmatrix} \\ &= (a_2b_3 - a_3b_2, a_3b_1 - a_1b_3, a_1b_2 - a_2b_1) \end{aligned}$$

The reader should realize that the term involving the determinant is merely a mnemonic, useful in remembering the last expression. Show that $\mathbf{A} \times \mathbf{B} = -\mathbf{B} \times \mathbf{A}$ and that $\mathbf{A} \times \mathbf{A} = \mathbf{0}$.

9. Verify the following formulas:

a. $\det \begin{bmatrix} a & b & 0 \\ c & d & 0 \\ 0 & 0 & e \end{bmatrix} = \det \begin{bmatrix} a & b \\ c & d \end{bmatrix} e$

b. $\det \begin{bmatrix} a & b & 0 & 0 \\ c & d & 0 & 0 \\ 0 & 0 & e & f \\ 0 & 0 & g & h \end{bmatrix} = \det \begin{bmatrix} a & b \\ c & d \end{bmatrix} \det \begin{bmatrix} e & f \\ g & h \end{bmatrix}$

10. Let $A = \begin{bmatrix} A_1 & 0 \\ 0 & A_2 \end{bmatrix}$ be an $n \times n$ matrix, where A_1 is an $m \times m$ matrix and A_2 is an $(n-m) \times (n-m)$ matrix, $1 \leq m \leq n-1$. Show that

$$\det(A) = \det(A_1)\det(A_2)$$

11. A group $(\mathscr{G}, \cdot)$ is a mathematical system that consists of a collection of objects $\mathscr{G}$ and an operation $\cdot$ between two elements of $\mathscr{G}$, which gives an element of $\mathscr{G}$. Thus, if x and y are elements of $\mathscr{G}$, then $x \cdot y$ (we drop the dot in the future) is also in $\mathscr{G}$. We also suppose that this operation, called multiplication, satisfies the following properties:

1. $(xy)z = x(yz)$.
2. There is an identity element e in $\mathscr{G}$ such that $ex = xe = x$ for every x in $\mathscr{G}$.
3. For each x in $\mathscr{G}$, there is an x^{-1} in $\mathscr{G}$ such that $xx^{-1} = x^{-1}x = e$.

a. Show that the set of vectors in a vector space forms a group, the group operation being vector addition.

b. Let $Gl(n)$ denote the set of invertible $n \times n$ matrices, with the group operation being matrix multiplication. Show that $Gl(n)$ forms a group. This group is called the general linear group.

c. Let $Sl(n)$ denote the set of invertible $n \times n$ matrices with determinant equal to 1. Show that $Sl(n)$, with the group operation being matrix multiplication, is a group. This group is called the special linear group.

d. Show that the set of all $n \times n$ matrices under matrix multiplication does not form a group.

12. An $n \times n$ matrix P is said to be orthogonal if $P^{-1} = P^T$.

a. Show that if P is orthogonal, then $\det(P) = \pm 1$.

b. Deduce from part a that if P is orthogonal and if K is some n-dimensional parallelepiped, then $\mathrm{vol}(PK) = \mathrm{vol}(K)$.

c. Find a 2×2 matrix P that is not orthogonal, and for which $\det(P)$ equals 1.

d. Show that $\mathscr{O}(n)$, the set of $n \times n$ orthogonal matrices, forms a group under the operation of matrix multiplication; cf. problem 11.

13. A matrix B is said to be similar to the matrix A if there is a nonsingular matrix P such that $B = PAP^{-1}$.

a. Show that if B is similar to A, then $\det(B) = \det(A)$.

b. Find a pair of 2×2 matrices A and B for which $\det(A) = \det(B)$ and B is not similar to A. Hint: If B is similar to a scalar matrix, then $B = A$.

14. Find two 2×2 matrices A and B for which $\det(A + B) \neq \det(A) + \det(B)$.

5
EIGENVALUES AND EIGENVECTORS

In this chapter we return to the study of linear transformations that we started in Chapter 3. The ideas presented here are related to finding the "simplest" matrix representation for a fixed linear transformation. As you recall, a matrix representation is determined once the bases for the two vector spaces are picked. Thus our problem is how to pick these bases.

5.1 WHAT IS AN EIGENVECTOR?

Before defining eigenvectors and eigenvalues let us look at the linear transformation L, from $\mathbb{R}^2$ to $\mathbb{R}^2$, whose matrix representation is

$$A = \begin{bmatrix} 2 & 0 \\ 0 & 3 \end{bmatrix}$$

We cannot compute $L(x_1, x_2)$ until we specify which basis G we used. Let's assume that $G = \{\mathbf{g}_1, \mathbf{g}_2\}$. Then we know that $L(\mathbf{g}_1) = 2\mathbf{g}_1 + 0\mathbf{g}_2 = 2\mathbf{g}_1$ and

$L(\mathbf{g}_2) = 0\mathbf{g}_1 + 3\mathbf{g}_2 = 3\mathbf{g}_2$. Thus, $L(\mathbf{g}_k)$ just multiplies $\mathbf{g}_k$ by the corresponding element in the main diagonal of A. Figure 5.1 illustrates this. If $\mathbf{x} = x_1\mathbf{g}_1 + x_2\mathbf{g}_2$, then $L(\mathbf{x}) = 2x_1\mathbf{g}_1 + 3x_2\mathbf{g}_2$. Since L multiplies each basis vector by some constant, it is extremely easy to compute and visualize what the linear transformation does to $\mathbb{R}^2$. In fact, since scalar multiplication is the simplest linear transformation possible, we would like to be able to do the following.

Given a linear transformation L from $\mathbb{R}^n$ to $\mathbb{R}^n$, find a basis $F = \{\mathbf{f}_1, \ldots, \mathbf{f}_n\}$ such that $L(\mathbf{f}_k) = \lambda_k \mathbf{f}_k$ for $k = 1, 2, \ldots, n$, that is, find n linearly independent vectors upon which L acts as scalar multiplication. Unfortunately, it is not always possible to do this. There are, however, large classes of linear transformations, that we mention later, for which it is possible to find such a set of vectors.

These vectors are called eigenvectors and the scalar multipliers λ_k are called the eigenvalues of L. The reader should note that the terms characteristic vector and characteristic value are also used; sometimes the word "proper" is substituted for characteristic. We formalize this discussion with the following:

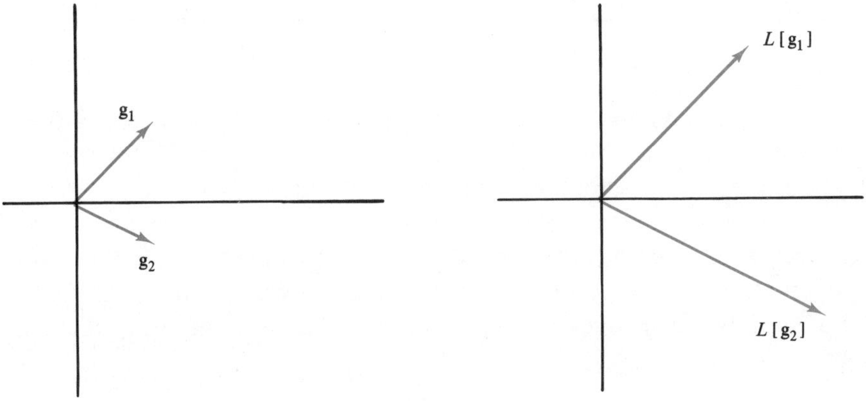

Figure 5.1

Definition 5.1. Let L be a linear transformation that maps a vector space into itself. A nonzero vector $\mathbf{x}$ is called an eigenvector of L if there is a scalar λ such that $L(\mathbf{x}) = \lambda\mathbf{x}$. The scalar λ is called an eigenvalue of L and the eigenvector is said to belong to, or correspond to, λ.

OK, we know what we want, eigenvectors. How do we find them? Let's examine a 2×2 matrix. Let

$$A = \begin{bmatrix} 2 & 6 \\ 1 & 3 \end{bmatrix}$$

and suppose that A is the matrix representation of a linear transformation L with respect to the standard basis. Thus, for any $\mathbf{x} = (x_1, x_2)$ we have

$$L(\mathbf{x}) = \begin{bmatrix} 2 & 6 \\ 1 & 3 \end{bmatrix} \begin{bmatrix} x_1 \\ x_2 \end{bmatrix} = \begin{bmatrix} 2x_1 + 6x_2 \\ x_1 + 3x_2 \end{bmatrix}$$

We want to find those numbers λ for which there is a nonzero vector $\mathbf{x}$ such that $L(\mathbf{x}) = \lambda\mathbf{x}$. Thus,

$$A \begin{bmatrix} x_1 \\ x_2 \end{bmatrix} = \lambda \begin{bmatrix} x_1 \\ x_2 \end{bmatrix}$$

or

$$(A - \lambda I_2) \begin{bmatrix} x_1 \\ x_2 \end{bmatrix} = \begin{bmatrix} 0 \\ 0 \end{bmatrix}$$

Hence, we are looking for those numbers λ for which the equation $(A - \lambda I_2)\mathbf{x} = \mathbf{0}$ has a nontrivial solution. But this happens if and only if $\det(A - \lambda I_2) = 0$. For this particular A we have

$$\det(A - \lambda I_2) = \det \begin{bmatrix} 2-\lambda & 6 \\ 1 & 3-\lambda \end{bmatrix} = (2-\lambda)(3-\lambda) - 6$$
$$= \lambda^2 - 5\lambda = \lambda(\lambda - 5)$$

The only values of λ that satisfy the equation $\det(A - \lambda I_2) = 0$ are $\lambda = 0$ and $\lambda = 5$. Thus the eigenvalues of L are 0 and 5. An eigenvector of 5, for example, will be any nonzero vector $\mathbf{x}$ in the kernel of $A - 5I_2$.

In the following pages when we talk about finding the eigenvalues and eigenvectors of some $n \times n$ matrix A, what we mean is that A is the matrix representation, with respect to the standard basis in $\mathbb{R}^n$, of a linear transformation L, and the eigenvalues and eigenvectors of A are just the eigenvalues and eigenvectors of L.

EXAMPLE 1 Find the eigenvalues and eigenvectors of the matrix

$$\begin{bmatrix} 2 & 6 \\ 1 & 3 \end{bmatrix}$$

From the above discussion we know that the only possible eigenvalues of A are 0 and 5.

$\lambda = 0$: We want $\mathbf{x} = (x_1, x_2)$ such that

$$\left(\begin{bmatrix} 2 & 6 \\ 1 & 3 \end{bmatrix} - 0 \begin{bmatrix} 1 & 0 \\ 0 & 1 \end{bmatrix} \right) \begin{bmatrix} x_1 \\ x_2 \end{bmatrix} = \begin{bmatrix} 0 \\ 0 \end{bmatrix}$$

The coefficient matrix of this system is $\begin{bmatrix} 2 & 6 \\ 1 & 3 \end{bmatrix}$ and it is row equivalent to the matrix $\begin{bmatrix} 1 & 3 \\ 0 & 0 \end{bmatrix}$. The solutions to this homogeneous equation satisfy $x_1 = -3x_2$. Therefore, $\ker(A - 0I_2) = S[(-3,1)]$, and any eigenvector of A corresponding to the eigenvalue 0 is a nonzero multiple of $(-3,1)$. As a check we compute

$$A\begin{bmatrix} -3 \\ 1 \end{bmatrix} = \begin{bmatrix} 2 & 6 \\ 1 & 3 \end{bmatrix}\begin{bmatrix} -3 \\ 1 \end{bmatrix} = \begin{bmatrix} 0 \\ 0 \end{bmatrix} = 0\begin{bmatrix} -3 \\ 1 \end{bmatrix}$$

$\lambda = 5$: We want to find those vectors $\mathbf{x}$ such that $(A - 5I_2)\mathbf{x} = \mathbf{0}$. This leads to the equation

$$\left(\begin{bmatrix} 2 & 6 \\ 1 & 3 \end{bmatrix} - 5\begin{bmatrix} 1 & 0 \\ 0 & 1 \end{bmatrix}\right)\begin{bmatrix} x_1 \\ x_2 \end{bmatrix} = \begin{bmatrix} 0 \\ 0 \end{bmatrix}$$

The coefficient matrix of this system, $\begin{bmatrix} -3 & 6 \\ 1 & -2 \end{bmatrix}$, is row equivalent to the matrix $\begin{bmatrix} 1 & -2 \\ 0 & 0 \end{bmatrix}$. Any solution of this system satisfies $x_1 = 2x_2$. Hence, $\ker(A - 5I_2) = S[(2,1)]$. All eigenvectors corresponding to the eigenvalue $\lambda = 5$ must be nonzero multiples of $(2,1)$. Checking to see that $(2,1)$ is indeed an eigenvector corresponding to 5, we have

$$A\begin{bmatrix} 2 \\ 1 \end{bmatrix} = \begin{bmatrix} 2 & 6 \\ 1 & 3 \end{bmatrix}\begin{bmatrix} 2 \\ 1 \end{bmatrix} = \begin{bmatrix} 10 \\ 5 \end{bmatrix} = 5\begin{bmatrix} 2 \\ 1 \end{bmatrix}$$ ■

We summarize the above discussion with the following definition and theorem.

Definition 5.2. Let A be any $n \times n$ matrix. Let λ be any scalar. Then the $n \times n$ matrix $A - \lambda I_n$ is called the characteristic matrix of A and the nth degree polynomial $p(\lambda) = \det(A - \lambda I_n)$ is called the characteristic polynomial of A.

The characteristic polynomial is sometimes defined as $\det(\lambda I - A) = \det[-(A - \lambda I)] = (-1)^n \det(A - \lambda I) = (-1)^n p(\lambda)$. Thus, the two versions differ by at most a minus sign.

Theorem 5.1. Let A be any $n \times n$ matrix. Then λ_0 is an eigenvalue of A with corresponding eigenvector $\mathbf{x}_0$ if and only if $\det(A - \lambda_0 I_n) = 0$ and $\mathbf{x}_0$ is a nonzero vector in $\ker(A - \lambda_0 I_n)$.

Proof. $\mathbf{x}_0$ is an eigenvector corresponding to λ_0 if and only if $A\mathbf{x}_0 = \lambda_0 \mathbf{x}_0$ and $\mathbf{x}_0$ is nonzero. But this is equivalent to saying that $\mathbf{x}_0$ is in $\ker(A - \lambda_0 I_n)$ and $\mathbf{x}_0$ is nonzero. But if $\ker(A - \lambda_0 I_n)$ has a nonzero vector in it, then $\det(A - \lambda_0 I_n) = 0$. Thus a necessary condition for λ_0 to be an eigenvalue of A is that it is a root of the characteristic polynomial of A.

A mistake that is sometimes made when trying to calculate the characteristic polynomial of a matrix is to first find a matrix B, in row echelon form, that is row equivalent to A and then compute the characteristic polynomial of B. There is usually no relationship whatsoever between the characteristic polynomials of A and B.

EXAMPLE 2 Let $A = \begin{bmatrix} 1 & 1 & -2 \\ -1 & 2 & 1 \\ 0 & 1 & -1 \end{bmatrix}$. Compute the characteristic matrix and polynomial of A. Determine the eigenvalues and eigenvectors of A. The characteristic matrix of A is

$$A - \lambda I_3 = \begin{bmatrix} 1-\lambda & 1 & -2 \\ -1 & 2-\lambda & 1 \\ 1 & 1 & -1-\lambda \end{bmatrix}$$

The characteristic polynomial of A is $\det(A - \lambda I_3)$ and equals

$$\det\begin{bmatrix} 1-\lambda & 1 & -2 \\ -1 & 2-\lambda & 1 \\ 0 & 1 & -1-\lambda \end{bmatrix} = \det\begin{bmatrix} 1-\lambda & 1 & -1-\lambda \\ -1 & 2-\lambda & 0 \\ 0 & 1 & -1-\lambda \end{bmatrix}$$

$$= \det\begin{bmatrix} 1-\lambda & 0 & 0 \\ -1 & 2-\lambda & 0 \\ 0 & 1 & -1-\lambda \end{bmatrix}$$

$$= -(\lambda+1)(\lambda-1)(\lambda-2)$$

Thus, the eigenvalues of A are -1, 1, and 2. We determine the eigenvectors of A by finding the nonzero vectors in $\ker(A - \lambda_0 I_3)$, for $\lambda_0 = -1$, 1, and 2.

$\lambda = -1$: The matrix $A - (-1)I_3$ equals $\begin{bmatrix} 2 & 1 & -2 \\ -1 & 3 & 1 \\ 0 & 1 & 0 \end{bmatrix}$ and it is row equivalent to the matrix $\begin{bmatrix} 1 & 0 & -1 \\ 0 & 1 & 0 \\ 0 & 0 & 0 \end{bmatrix}$. This implies that $\ker(A + I_3)$ equals $S[(1,0,1)]$. One easily checks that $A(1,0,1)^T = (-1)(1,0,1)^T$.

$\lambda = 1$: $A - I_3 = \begin{bmatrix} 0 & 1 & -2 \\ -1 & 1 & 1 \\ 0 & 1 & -2 \end{bmatrix}$, and this matrix is row equivalent to $\begin{bmatrix} 0 & 1 & -2 \\ 1 & 0 & -3 \\ 0 & 0 & 0 \end{bmatrix}$. Clearly $\ker(A - I_3) = S[(3,2,1)]$, and a quick calculation shows that $A(3,2,1)^T = (3,2,1)^T$.

$$\lambda = 2\text{: } A - 2I_3 = \begin{bmatrix} -1 & 1 & -2 \\ -1 & 0 & 1 \\ 0 & 1 & -3 \end{bmatrix}.$$ This matrix is row equivalent to

$$\begin{bmatrix} 1 & 0 & -1 \\ 0 & 1 & -3 \\ 0 & 0 & 0 \end{bmatrix}.$$ Thus, $\ker(A - 2I_3) = S[(1,3,1)]$. Computing $A(1,3,1)^T$, we see that it equals $2(1,3,1)^T$. ■

If A is an $n \times n$ matrix, we've defined the characteristic polynomial $p(\lambda)$ of A to be $\det(A - \lambda I_n)$. If $\lambda_1, \lambda_2, \ldots, \lambda_q$ are the distinct roots of $p(\lambda)$, then we may write

$$p(\lambda) = (-1)^n(\lambda - \lambda_1)^{m_1}(\lambda - \lambda_2)^{m_2} \ldots (\lambda - \lambda_q)^{m_q}$$

where $m_1 + m_2 + \cdots + m_q = n$. The multiplicity of an eigenvalue is the exponent corresponding to that eigenvalue. Thus, the multiplicity of λ_1 is m_1, that of λ_2 being m_2, etc. For example, if A equals $\begin{bmatrix} 2 & 1 & 0 & 0 \\ 0 & 2 & 1 & 0 \\ 0 & 0 & 2 & 2 \\ 0 & 0 & 0 & 4 \end{bmatrix}$, then $p(\lambda) = (\lambda - 2)^3(\lambda - 4)$. The eigenvalue 2 has multiplicity 3 while the eigenvalue 4 has multiplicity 1.

One last bit of terminology: When we wish to talk about all the eigenvectors associated with one eigenvalue we use the term eigenspace. Thus, in Example 2 the eigenspace of the eigenvalue (-1) is just $\ker(A - (-1)I)$. In general, if L is any linear transformation from a vector space into itself and λ_0 is an eigenvalue of L, the eigenspace of λ_0 is $\ker(L - \lambda_0 I)$. That is, the eigenspace of λ_0 consists of all its eigenvectors plus the zero vector. Note that the zero vector is never an eigenvector.

We've seen how to compute the eigenvalues of a linear transformation if the linear transformation is matrix multiplication. What do we do in the more abstract setting when $L\colon V \to V$? Well, in Chapter 3 we saw that once we fix a basis $F = \{\mathbf{f}_1, \ldots, \mathbf{f}_n\}$ of V, we have a matrix representation, A, of L. Moreover, $L(\mathbf{x}) = \lambda\mathbf{x}$ for some nonzero vector $\mathbf{x}$ and scalar λ if and only if $A[\mathbf{x}]_F = \lambda[\mathbf{x}]_F$. Thus, λ is an eigenvalue of L if and only if $\det(A - \lambda I) = 0$, and $\ker(A - \lambda I)$ has a nonzero vector and the nonzero vectors in $\ker(A - \lambda I)$ are the coordinates, with respect to the basis F, of the nonzero vectors in $\ker(L - \lambda I)$. What this means is that we may calculate the eigenvalues and eigenvectors of L by calculating the eigenvalues and eigenvectors of any one of its matrix representations.

PROBLEM SET 5.1

1. Compute the characteristic matrix and polynomial of each of the following matrices:

a. $\begin{bmatrix} 1 & 4 \\ 3 & 8 \end{bmatrix}$ **b.** $\begin{bmatrix} 7 & 8 \\ 0 & 4 \end{bmatrix}$ **c.** $\begin{bmatrix} 1 & 3 & 0 \\ 1 & 2 & 1 \\ 4 & -5 & 8 \end{bmatrix}$

2. Calculate the characteristic polynomial of the following matrices:

a. $\begin{bmatrix} 2 & -6 & 1 \\ 4 & 0 & -5 \\ 0 & -1 & 0 \end{bmatrix}$ **b.** $\begin{bmatrix} -1 & 3 & 2 \\ 6 & 1 & 0 \\ 4 & 5 & 0 \end{bmatrix}$

3. Find the eigenvalues, their multiplicities, and the dimensions of the eigenspaces of the following matrices:

a. $\begin{bmatrix} 3 & 0 \\ 0 & 3 \end{bmatrix}$ **b.** $\begin{bmatrix} 3 & 1 \\ 0 & 3 \end{bmatrix}$ **c.** $\begin{bmatrix} 3 & 1 \\ 1 & 3 \end{bmatrix}$ **d.** $\begin{bmatrix} 3 & 0 \\ 0 & 2 \end{bmatrix}$

4. Find the eigenvalues, their multiplicities, and the dimensions of the eigenspaces of the following matrices:

a. $\begin{bmatrix} -1 & 0 & 0 \\ 0 & -1 & 0 \\ 0 & 0 & -1 \end{bmatrix}$ **b.** $\begin{bmatrix} -1 & 1 & 0 \\ 0 & -1 & 0 \\ 0 & 0 & -1 \end{bmatrix}$ **c.** $\begin{bmatrix} -1 & 1 & 0 \\ 0 & -1 & 1 \\ 0 & 0 & -1 \end{bmatrix}$

5. Let V be a vector space. Let L: $V \rightarrow V$ be a linear transformation. If λ_0 is an eigenvalue of L, show that the eigenspace of V corresponding to λ_0 is a subspace of V and has dimension at least 1. The eigenspace of λ_0 is defined to be the set of vectors $\mathbf{x}$ such that $L(\mathbf{x}) = \lambda_0\mathbf{x}$.

6. Let A and B be two similar matrices. Thus, there is a matrix P such that $A = P^{-1}BP$.
a. Show that A and B have the same eigenvalues with the same multiplicities.
b. If W is the eigenspace of A corresponding to λ, what is the eigenspace of B corresponding to λ?

7. Let A be an $n \times n$ matrix. Let $p(\lambda) = \det(A - \lambda I_n)$ be its characteristic polynomial. Then if $\lambda_1, \lambda_2, \ldots, \lambda_n$ are the roots of $p(\lambda)$, we may write

$$\begin{aligned} p(\lambda) &= (-1)^n(\lambda - \lambda_1)(\lambda - \lambda_2) \cdots (\lambda - \lambda_n) \\ &= (-1)^n[\lambda^n - c_1\lambda^{n-1} + \cdots + (-1)^n c_n] \end{aligned}$$

a. Assume A is a 2×2 matrix, $A = [a_{jk}]$. Then $p(\lambda) = \lambda^2 - c_1\lambda + c_2$. Show that $c_2 = \det(A) = \lambda_1\lambda_2$ and $c_1 = \lambda_1 + \lambda_2 = a_{11} + a_{22}$.
b. Assume $A = [a_{jk}]$ is a 3×3 matrix. Then $p(\lambda) = -(\lambda^3 - c_1\lambda^2 + c_2\lambda - c_3)$. Show that $c_1 = \lambda_1 + \lambda_2 + \lambda_3 = a_{11} + a_{22} + a_{33}$ and that $c_3 = \det(A) = \lambda_1\lambda_2\lambda_3$.
c. Generalize parts a and b to $n \times n$ matrices.

8. If A is an $n \times n$ matrix, $A = [a_{jk}]$, we define the trace of $A = \mathrm{Tr}(A) = a_{11} + a_{22} + \cdots + a_{nn}$. Show that the following are true for any two $n \times n$ matrices A and B:
a. $\mathrm{Tr}(A + B) = \mathrm{Tr}(A) + \mathrm{Tr}(B)$
b. $\mathrm{Tr}(AB) = \mathrm{Tr}(BA)$
c. $\mathrm{Tr}(A) = \mathrm{Tr}(A^T)$
d. Show that if A and B are similar matrices, $Tr(A) = Tr(B)$.

9. Let A be an $n \times n$ matrix and let $p(\lambda) = \det(A - \lambda I_n)$, the characteristic polynomial of A. Then if $\lambda_1, \lambda_2, \ldots, \lambda_n$ are the roots of $p(\lambda)$, we may write

$$\begin{aligned} p(\lambda) &= (-1)^n(\lambda - \lambda_1)(\lambda - \lambda_2) \cdots (\lambda - \lambda_n) \\ &= (-1)^n[\lambda^n - c_1\lambda^{n-1} + \cdots + (-1)^n c_n] \end{aligned}$$

By $p(A)$ we mean

$$p(A) = (-1)^n[A^n - c_1 A^{n-1} + \cdots + (-1)^n c_n I_n]$$

That is, wherever λ appears in $p(\lambda)$, it is replaced by A. For each of the matrices in problem 1 compute $p(A)$.

10. Let $A = \begin{bmatrix} 0 & 2 \\ 2 & 0 \end{bmatrix}$. Show that the eigenvalues of A are ± 2. Let $\mathbf{x}$ and $\mathbf{y}$ be two eigenvectors corresponding to 2 and -2, respectively.

a. Show that $A^n\mathbf{x} = 2^n\mathbf{x}$ and $A^n\mathbf{y} = (-2)^n\mathbf{y}$.

b. Let $g(\lambda) = \lambda^n + c_1\lambda^{n-1} + \cdots + c_n$ be an arbitrary polynomial. Define the matrix $g(A)$ by $g(A) = A^n + c_1 A^{n-1} + \cdots + c_n I$. Show that $g(A)\mathbf{x} = g(2)\mathbf{x}$ and $g(A)\mathbf{y} = g(-2)\mathbf{y}$.

11. Let A be any $n \times n$ matrix. Suppose λ_0 is an eigenvalue of A with $\mathbf{x}_0$ any eigenvector corresponding to λ_0. Let $g(\lambda)$ be any polynomial in λ; define $g(A)$ as in problem 10. Show that $g(A)\mathbf{x}_0 = g(\lambda_0)\mathbf{x}_0$.

12. Let $A = \begin{bmatrix} 0 & -1 \\ 1 & 0 \end{bmatrix}$. As we saw in Chapter 3, this matrix represents a rotation of 90 degrees about the origin. As such, we should not expect any eigenvectors. Why? Compute $\det(A - \lambda I)$ and find all the roots of this polynomial. Show that there is no vector $\mathbf{x}$ in $\mathbb{R}^2$ such that $A\mathbf{x} = \lambda\mathbf{x}$ except for the zero vector. What happens if $\mathbb{R}^2$ is replaced by $\mathbb{C}^2$? For which rotations in $\mathbb{R}^2$, if any, are there eigenvectors?

13. Find the eigenvalues and eigenspaces of the matrices below:

$$\begin{bmatrix} 0 & 0 & -3 & 4 \\ -2 & 2 & -2 & 4 \\ -2 & 0 & 1 & 2 \\ -3 & 0 & -3 & 7 \end{bmatrix} \qquad \begin{bmatrix} \frac{4}{3} & \frac{4}{3} & 0 & 2 \\ \frac{2}{3} & \frac{2}{3} & 0 & 1 \\ -1 & -1 & 0 & 3 \\ -\frac{1}{3} & -\frac{1}{3} & 0 & 0 \end{bmatrix}$$

14. Let $A = \begin{bmatrix} d_1 & 0 \\ 0 & d_2 \end{bmatrix}$.

a. Find the eigenvalues and eigenspaces of A.

b. Let $p(\lambda) = \det(A - \lambda I)$. Show that $p(A) = 0_{22}$, the zero 2×2 matrix; cf. problem 10.

c. Assume that $\det(A) = d_1 d_2 \neq 0$. Use the equation $p(A) = 0_{22}$ to compute A^{-1}, where $p(\lambda) = \det(A - \lambda I)$.

15. Let $A = [a_{jk}]$ be any 2×2 matrix. Let $p(\lambda)$ be its characteristic polynomial. Show that $p(A) = 0_{22}$. Assume that $\det(A) \neq 0$ and show how to compute A^{-1} using this information.

16. Let A be any $n \times n$ matrix and let $p(\lambda)$ be its characteristic polynomial. The Hamilton-Cayley theorem states that $p(A) = 0_{nn}$. Assuming that $\det(A) \neq 0$, explain how one could compute A^{-1} by using the equation $p(A) = 0_{nn}$.

17. Let $A = \begin{bmatrix} 6 & 1 \\ 0 & 9 \end{bmatrix}$.

a. Find the eigenvalues of A.
b. For any constant c, find the eigenvalues of the matrix $A - cI$.
c. For any constant c, find the eigenvalues of the matrix cA.

18. Let A be any $n \times n$ matrix. Suppose the eigenvalues of A are $\{\lambda_1, \ldots, \lambda_n\}$. Let c be any constant.
a. Show that the eigenvalues of $A - cI$ are $\{\lambda_1 - c, \ldots, \lambda_n - c\}$.
b. What are the eigenvalues of cA?

19. Let $A = [a_{jk}]$ be an invertible $n \times n$ matrix. Suppose that $\{\lambda_1, \ldots, \lambda_n\}$ is the set of eigenvalues of A. Show that $\lambda_j \neq 0$ for $j = 1, \ldots, n$. Then show that the eigenvalues of A^{-1} are the reciprocals of the λ_j. That is, if 2 is an eigenvalue of A, then $\frac{1}{2}$ is an eigenvalue of A^{-1}. How are the eigenvectors of A and A^{-1} related?

20. Let A be an $n \times n$ matrix. Suppose $\mathbf{x}_0$ is an eigenvector of A.
a. Show that $\mathbf{x}_0$ is an eigenvector of A^n for any n.
b. What is the eigenvalue of A^n for which $\mathbf{x}_0$ is the corresponding eigenvector?
c. Show that $\mathbf{x}_0$ is an eigenvector of $A - kI$ for any constant k.
d. What is the eigenvalue of $A - kI$ for which $\mathbf{x}_0$ is the corresponding eigenvector?

21. Show that $\det(A - \lambda I) = \det(A^T - \lambda I)$ for any constant λ. Thus, A and A^T have the same characteristic polynomial, and hence the same eigenvalues.

22. Find a 2×2 matrix A for which A and A^T do not have the same eigenspaces.

23. Find the eigenvalues, their multiplicities, and the dimensions of the corresponding eigenspaces for each of the following matrices:

a. $\begin{bmatrix} 1 & 1 & 0 & 0 \\ 0 & 1 & 0 & 0 \\ 0 & 0 & 1 & 0 \\ 0 & 0 & 0 & 1 \end{bmatrix}$ **b.** $\begin{bmatrix} 1 & 1 & 0 & 0 \\ 0 & 1 & 1 & 0 \\ 0 & 0 & 1 & 0 \\ 0 & 0 & 0 & 1 \end{bmatrix}$ **c.** $\begin{bmatrix} 1 & 1 & 0 & 0 \\ 0 & 1 & 1 & 0 \\ 0 & 0 & 1 & 1 \\ 0 & 0 & 0 & 1 \end{bmatrix}$

24. Let $A = \begin{bmatrix} -1 & 3 \\ 1 & 1 \end{bmatrix}$. Define $L: M_{22} \to M_{22}$ by $L[B] = AB$ for any B in M_{22}. Find the eigenvalues of L, their multiplicities, and the dimensions of the eigenspaces.

25. Let A be any $n \times n$ matrix. Define $L: M_{nn} \to M_{nn}$ by $L[B] = AB$. Show that λ is an eigenvalue of L if and only if λ is an eigenvalue of the matrix A. Remember that λ is an eigenvalue of A only if there exists a nonzero $\mathbf{x}$ in $\mathbb{R}^n$ such that $A\mathbf{x} = \lambda\mathbf{x}$.

26. Define $L: P_2 \to P_2$ by $L[\mathbf{p}](t) = (1/t) \int_0^t \mathbf{p}(s)\, ds$. Find the eigenvalues, their multiplicities, and the dimensions of the eigenspaces.

27. Let $L: P_2 \to P_2$ be defined as $L[\mathbf{p}] = \mathbf{p}'$. Find the eigenvalues, their multiplicities, and the dimensions of the eigenspaces.

28. Let $A = \begin{bmatrix} 1 & 3 \\ 0 & 1 \end{bmatrix}$. Let $B = \begin{bmatrix} 1 & 0 \\ 0 & 6 \end{bmatrix}$. Define $L: M_{22} \to M_{22}$ by $L[\mathbf{x}] = A\mathbf{x} + \mathbf{x}B$. Show that L is a linear transformation, and then find its eigenvalues, their multiplicities, and the corresponding eigenspaces.

5.2 DIAGONALIZATION OF MATRICES

In the last section we stated that the eigenvalues of a matrix A are those roots of the characteristic polynomial $p(\lambda) = \det(A - \lambda I)$ for which $\ker(A - \lambda I)$ contains more than just the zero vector. The fundamental theorem of algebra states that every polynomial has at least one root. Thus, every matrix should have at least one eigenvalue and a corresponding eigenvector. This argument erroneously leads us to believe that every linear transformation L, from a vector space V into V, has at least one eigenvector and eigenvalue. This need not be true when V is a real vector space, since multiplication by complex numbers is not allowed, and the root of $p(\lambda) = 0$ that is guaranteed by the fundamental theorem of algebra may be a complex number. The matrix $\begin{bmatrix} 0 & -1 \\ 1 & 0 \end{bmatrix}$ is one such example; cf. problem 12 in Section 5.1. The trouble with a complex number being an eigenvalue is that our vector space may only allow multiplication by real numbers. For example, if i represents the square root of -1 and $\mathbf{x} = (1,2)$, a vector in $\mathbb{R}^2$, then $i\mathbf{x} = (i,2i)$ makes sense, but it is no longer in $\mathbb{R}^2$. If our vector space is $\mathbb{C}^2$, then, since multiplication by complex scalars is allowed, the above problem does not arise.

Let's step back a second and review what it is that we want, simple representations for linear transformations. From this, we are led to the idea of eigenvectors and the realization that we want not just one eigenvector, but a basis of eigenvectors. In fact, we have the following theorem.

Theorem 5.2. Let $L: V \to V$ be a linear transformation of a vector space V into itself. Suppose there is a basis of V that consists of eigenvectors of L. Then the matrix representation of L with respect to this basis will be diagonal and the diagonal elements will be the eigenvalues of L.

Proof. Let $F = \{\mathbf{f}_1, \ldots, \mathbf{f}_n\}$ be a basis of V such that $L(\mathbf{f}_k) = \lambda_k \mathbf{f}_k$, i.e., $\mathbf{f}_k$ is an eigenvector of L and λ_k is the corresponding eigenvalue. Then if $A = [a_{jk}]$ is the matrix representation of L with respect to the basis F, we have

$$\lambda_k \mathbf{f}_k = L(\mathbf{f}_k) = \sum_{j=1}^{n} a_{jk} \mathbf{f}_j$$

Thus, $a_{jk} = 0$ if $j \neq k$ and $a_{kk} = \lambda_k$. In other words A is a diagonal matrix whose diagonal elements are precisely the eigenvalues of L.

Notice, there is no problem with complex numbers. We avoid the difficulty by assuming a basis of eigenvectors. The next question is, how can we tell if there is such a basis? The following lemma helps answer this question.

Lemma 5.1. Let $L: V \to V$ be a linear transformation. Suppose $\lambda_1, \lambda_2, \ldots, \lambda_p$ are distinct eigenvalues of L. Let $\mathbf{f}_1, \ldots, \mathbf{f}_p$ be eigenvectors corresponding to them. Then the set of vectors $\{\mathbf{f}_1, \ldots, \mathbf{f}_p\}$ is linearly independent.

Proof. We prove that this set is linearly independent by an inductive process; that is, we show the theorem is true when $p = 1$ and 2 and then show

how to go from $p-1$ to p. Suppose $p=1$: then the set in question is just $\{\mathbf{f}_1\}$, and since $\mathbf{f}_1 \neq 0$, it is linearly independent. Now suppose there are constants c_1 and c_2 such that

$$\mathbf{0} = c_1\mathbf{f}_1 + c_2\mathbf{f}_2 \tag{5.1}$$

Then we also have

$$\begin{aligned}\mathbf{0} &= L(\mathbf{0}) = L(c_1\mathbf{f}_1 + c_2\mathbf{f}_2)\\ &= c_1 L(\mathbf{f}_1) + c_2 L(\mathbf{f}_2)\\ &= c_1\lambda_1\mathbf{f}_1 + c_2\lambda_2\mathbf{f}_2\end{aligned} \tag{5.2}$$

Multiplying (5.1) by λ_1, and subtracting the resulting equation from (5.2), we have

$$\mathbf{0} = c_2(\lambda_2 - \lambda_1)\mathbf{f}_2$$

Since $\mathbf{f}_2 \neq 0$ and $\lambda_2 \neq \lambda_1$, we must have $c_2 = 0$. This and (5.1) imply $c_1 = 0$. Hence, $\{\mathbf{f}_1, \mathbf{f}_2\}$ is linearly independent. Assume now that the set $\{\mathbf{f}_1, \ldots, \mathbf{f}_{p-1}\}$ is linearly independent and

$$\mathbf{0} = c_1\mathbf{f}_1 + c_2\mathbf{f}_2 + \cdots + c_{p-1}\mathbf{f}_{p-1} + c_p\mathbf{f}_p \tag{5.3}$$

for some constants $c_1, \ldots, c_p$. Then

$$\begin{aligned}\mathbf{0} &= L[c_1\mathbf{f}_1 + \cdots + c_p\mathbf{f}_p]\\ &= c_1 L(\mathbf{f}_1) + \cdots + c_p L(\mathbf{f}_p)\\ &= c_1\lambda_1\mathbf{f}_1 + \cdots + c_p\lambda_p\mathbf{f}_p\end{aligned} \tag{5.4}$$

Multiplying (5.3) by λ_p, and then subtracting from (5.4), we have

$$\mathbf{0} = c_1(\lambda_1 - \lambda_p)\mathbf{f}_1 + c_2(\lambda_2 - \lambda_p)\mathbf{f}_2 + \cdots + c_{p-1}(\lambda_{p-1} - \lambda_p)\mathbf{f}_{p-1}$$

Since the $\mathbf{f}_k$'s, $1 \le k \le p-1$, are linearly independent we must have $c_k(\lambda_k - \lambda_p) = 0$ for $k = 1, 2, \ldots, p-1$. But $\lambda_p \neq \lambda_k$; hence $c_k = 0$, $k = 1, 2, \ldots, p-1$. Thus, (5.3) reduces to $c_p\mathbf{f}_p = \mathbf{0}$ and we conclude that $c_p = 0$ also. This of course means that our set of vectors is linearly independent.

Theorem 5.3. Let V be a real n-dimensional vector space. Let L be a linear transformation from V into V. Suppose that the characteristic polynomial of L has n distinct real roots. Then L has a diagonal matrix representation.

Proof. Let the roots of the characteristic polynomial be $\lambda_1, \lambda_2, \ldots, \lambda_n$. By hypothesis they are all different and real. Thus, for each j, $\ker(L - \lambda_j I)$ will contain more than just the zero vector. Hence, each of these roots is an eigenvalue, and if $\mathbf{f}_1, \ldots, \mathbf{f}_n$ is a set of associated eigenvectors, the previous lemma ensures that they form a linearly independent set. Since $\dim(V) = n$, they also form a basis for V. Theorem 5.2 guarantees that L does indeed have a matrix representation that is diagonal.

We note that if V in the above theorem is an n-dimensional complex vector space, we would not need to insist that the roots of the characteristic polynomial be real.

EXAMPLE 1 Let L be a linear transformation from $\mathbb{R}^4$ to $\mathbb{R}^4$ whose matrix representation A with respect to the standard basis is

$$\begin{bmatrix} -3 & 0 & 2 & -4 \\ -6 & 2 & 2 & -5 \\ 4 & 0 & -1 & 4 \\ 6 & 0 & -2 & 7 \end{bmatrix}$$

Find a diagonal representation for L.

Solution. The characteristic polynomial $p(\lambda)$ of L is

$$\begin{aligned} p(\lambda) &= \det\begin{bmatrix} -3-\lambda & 0 & 2 & -4 \\ -6 & 2-\lambda & 2 & -5 \\ 4 & 0 & -1-\lambda & 4 \\ 6 & 0 & -2 & 7-\lambda \end{bmatrix} \\ &= (2-\lambda)\det\begin{bmatrix} -3-\lambda & 2 & -4 \\ 4 & -1-\lambda & 4 \\ 6 & -2 & 7-\lambda \end{bmatrix} \\ &= (2-\lambda)\det\begin{bmatrix} 3-\lambda & 0 & 3-\lambda \\ 4 & -1-\lambda & 4 \\ 6 & -2 & 7-\lambda \end{bmatrix} \\ &= (2-\lambda)\det\begin{bmatrix} 3-\lambda & 0 & 0 \\ 4 & -1-\lambda & 0 \\ 6 & -2 & 1-\lambda \end{bmatrix} \\ &= (-1-\lambda)(1-\lambda)(2-\lambda)(3-\lambda) \end{aligned}$$

Thus, the roots of $p(\lambda)$ are -1, 1, 2, and 3. Since they are real and distinct, Theorem 5.3 guarantees that $\mathbb{R}^4$ will have a basis that consists of eigenvectors of L. We next find one eigenvector for each of the eigenvalues.

$\lambda = -1$:

$$A-(-1)I = \begin{bmatrix} -2 & 0 & 2 & -4 \\ -6 & 3 & 2 & -5 \\ 4 & 0 & 0 & 4 \\ 6 & 0 & -2 & 8 \end{bmatrix}$$

is row equivalent to $\begin{bmatrix} 1 & 0 & 0 & 1 \\ 0 & 1 & 0 & 1 \\ 0 & 0 & 1 & -1 \\ 0 & 0 & 0 & 0 \end{bmatrix}$

Thus, $\ker(A+I) = S[(-1,-1,1,1)]$, and $\mathbf{f}_1 = (-1,-1,1,1)$ is an eigenvector corresponding to -1.

$\lambda = 1$:

$$A - I = \begin{bmatrix} -4 & 0 & 2 & -4 \\ -6 & 1 & 2 & -5 \\ 4 & 0 & -2 & 4 \\ 6 & 0 & -2 & 6 \end{bmatrix}$$

is row equivalent to the matrix $\begin{bmatrix} 1 & 0 & 0 & 1 \\ 0 & 1 & 0 & 1 \\ 0 & 0 & 1 & 0 \\ 0 & 0 & 0 & 0 \end{bmatrix}$

Thus, $\mathbf{f}_2 = (-1,-1,0,1)$ is an eigenvector corresponding to the eigenvalue 1.

$\lambda = 2$: The matrix $A - 2I$ is row equivalent to the matrix

$$\begin{bmatrix} 1 & 0 & 0 & 1 \\ 0 & 0 & 1 & -1 \\ 0 & 0 & 0 & 1 \\ 0 & 0 & 0 & 0 \end{bmatrix}$$

Hence, $\mathbf{f}_3 = (0,1,0,0)$ is an eigenvector for the eigenvalue 2.

$\lambda = 3$: The matrix $A - 3I$ is row equivalent to the matrix

$$\begin{bmatrix} 1 & 0 & 0 & \frac{1}{2} \\ 0 & 1 & 0 & 1 \\ 0 & 0 & 1 & -\frac{1}{2} \\ 0 & 0 & 0 & 0 \end{bmatrix}$$

Hence, $\mathbf{f}_4 = (-1,-2,1,2)$ is an eigenvector corresponding to the eigenvalue 3. Lemma 5.1 guarantees that the four vectors $\{\mathbf{f}_1, \mathbf{f}_2, \mathbf{f}_3, \mathbf{f}_4\}$ are linearly independent, and since $\dim(\mathbb{R}^4) = 4$, they form a basis. Since $L(\mathbf{f}_1) = -\mathbf{f}_1$, $L(\mathbf{f}_2) = \mathbf{f}_2$, $L(\mathbf{f}_3) = 2\mathbf{f}_3$, and $L(\mathbf{f}_4) = 3\mathbf{f}_4$, the matrix representation of L with respect to this basis is

$$\begin{bmatrix} -1 & 0 & 0 & 0 \\ 0 & 1 & 0 & 0 \\ 0 & 0 & 2 & 0 \\ 0 & 0 & 0 & 3 \end{bmatrix}$$ ■

It is clear from some of our previous examples that rather than having distinct eigenvalues it is possible that some eigenvalues will appear with multiplicity greater than 1. In this case, Theorem 5.3 is not applicable, and we use Theorem 5.4, whose proof is omitted.

Theorem 5.4. Let V be an n-dimensional real vector space. Let $L: V \to V$ be a linear transformation. Let $p(\lambda)$, the characteristic polynomial of L, equal $(\lambda - \lambda_1)^{m_1}(\lambda - \lambda_2)^{m_2} \cdots (\lambda - \lambda_p)^{m_p}$. Assume each of the roots λ_j, $1 \le j \le p$ is real. Then L has a diagonal matrix representation if and only if $\dim(\ker(A - \lambda_j I)) = m_j$ for each of the eigenvalues λ_j; that is, the number of linearly independent solutions to the homogeneous equation $(A - \lambda_j I)\mathbf{x} = \mathbf{0}$ must equal the multiplicity of the eigenvalue λ_j.

We illustrate this theorem in the next example.

EXAMPLE 2 Determine which of the following linear transformations has a diagonal representation. The matrices that are given are the representations of the transformations with respect to the standard basis of $\mathbb{R}^3$.

a. $A = \begin{bmatrix} 3 & 1 & 0 \\ 0 & 3 & 0 \\ 0 & 0 & 4 \end{bmatrix} \qquad p(\lambda) = (3 - \lambda)^2(4 - \lambda)$

The eigenvalues of A are 3, with multiplicity 2, and 4 with multiplicity 1. The matrix $A - 3I$ equals

$$\begin{bmatrix} 0 & 1 & 0 \\ 0 & 0 & 0 \\ 0 & 0 & 1 \end{bmatrix}$$

Clearly the kernel of this matrix has dimension 1. Thus, $\dim(\ker(A - 3I))$ equals 1, which is less than the multiplicity of the eigenvalue 3. Hence, the linear transformation L, represented by A, cannot be diagonalized.

b. $A = \begin{bmatrix} 3 & 1 & 0 \\ 1 & 3 & 0 \\ 0 & 0 & 4 \end{bmatrix} \qquad p(\lambda) = -(\lambda - 4)^2(\lambda - 2)$

The eigenvalue 4 has multiplicity 2 and the eigenvalue 2 has multiplicity 1. Hence, this linear transformation will have a diagonal representation if and only if $\dim(\ker(A - 4I)) = 2$ and $\dim(\ker(A - 2I)) = 1$. Since the last equality is obvious, we will only check the first one. The matrix $(A - 4I)$ is row equivalent to the matrix

$$\begin{bmatrix} 1 & -1 & 0 \\ 0 & 0 & 0 \\ 0 & 0 & 0 \end{bmatrix}$$

Since this matrix has rank equal to 1, its kernel must have dimension equal to 2. In fact the eigenspace $\ker(A - 4I)$ has $\{(1,1,0),(0,0,1)\}$ as a basis. A routine calculation shows that the vector $(1,-1,0)$ is an eigenvector corresponding to the eigenvalue 2. Thus, a basis of eigenvectors is $\{(1,1,0),(0,0,1),(1,-1,0)\}$, and L with respect to this basis has the diagonal representation

$$\begin{bmatrix} 4 & 0 & 0 \\ 0 & 4 & 0 \\ 0 & 0 & 2 \end{bmatrix}$$

■

This last example is a special case of a very important and useful result. Before stating it, we remind the reader that a matrix A is said to be symmetric if $A = A^T$; cf. Example 2b.

Theorem 5.5. Let L be a linear transformation from $\mathbb{R}^n$ into $\mathbb{R}^n$. Let A be the matrix representation of L with respect to the standard basis of $\mathbb{R}^n$. If A is a symmetric matrix, then L has a diagonal representation.

We omit the proof of this result and content ourselves with one more example.

EXAMPLE 3 Let L be a linear transformation from $\mathbb{R}^3$ to $\mathbb{R}^3$ whose matrix representation A with respect to the standard basis is given below. Find the eigenvalues of L and a basis of eigenvectors.

$$A = \begin{bmatrix} 1 & 3 & -3 \\ 3 & 1 & -3 \\ -3 & -3 & 1 \end{bmatrix}$$

Solution. We note that A is symmetric and hence Theorem 5.5 guarantees that there is a basis of $\mathbb{R}^3$ that consists of eigenvectors of A. The characteristic polynomial of A, $p(\lambda)$, equals $(2+\lambda)^2(7-\lambda)$. Thus, L has two real eigenvalues -2 and 7; -2 has multiplicity 2 and 7 has multiplicity 1. Computing the eigenvectors we have:

$\lambda = -2$: $A + 2I$ is row equivalent to the matrix

$$\begin{bmatrix} 1 & 1 & -1 \\ 0 & 0 & 0 \\ 0 & 0 & 0 \end{bmatrix}$$

Two linearly independent eigenvectors corresponding to -2 are $(1,0,1)$ and $(0,1,1)$.

$\lambda = 7$: $A - 7I$ is row equivalent to

$$\begin{bmatrix} 1 & 0 & 1 \\ 0 & 1 & 1 \\ 0 & 0 & 0 \end{bmatrix}$$

Thus, $(1,1,-1)$ is a basis for $\ker(A - 7I)$ and the set $\{(1,1,-1),(1,0,1),(0,1,1)\}$ is a basis for $\mathbb{R}^3$. Clearly the matrix representation of L with respect to this basis is

$$\begin{bmatrix} 7 & 0 & 0 \\ 0 & -2 & 0 \\ 0 & 0 & -2 \end{bmatrix}$$

■

There is one more idea to discuss before we conclude this section on diagonalization. Suppose we start out with a matrix A (the representation of L with respect to the standard basis of $\mathbb{R}^n$), and after calculating the eigenvalues and eigenvectors we see that this particular matrix has n linearly independent eigenvectors; i.e., $\mathbb{R}^n$ has a basis of eigenvectors of A. Suppose the eigenvalues and eigenvectors are $\{\lambda_j: j = 1, \ldots, n\}$ and $F = \{\mathbf{f}_j: j = 1, \ldots, n\}$ respectively. Then the matrix representation of L with respect to the basis F is $[\lambda_j \delta_{jk}] = D$. How are A and D related? The answer to this has already been given in Section 3.5; for if P is the change of basis matrix that satisfies $[\mathbf{x}]_F^T = P[\mathbf{x}]_S^T$, then P^{-1} is easy to write out, for its columns are the coordinates of the eigenvectors with respect to the standard basis. Moreover, we have

$$A = P^{-1}DP \tag{5.5}$$

or

$$D = PAP^{-1}$$

Note that in the formula $D = PAP^{-1}$, the matrix that multiplies A on the right is the matrix whose columns consist of the eigenvectors of A. Note also that A and D are similar; cf. Section 3.5.

EXAMPLE 4 Let A be the matrix

$$\begin{bmatrix} 1 & 3 & -3 \\ 3 & 1 & -3 \\ -3 & -3 & 1 \end{bmatrix}$$

Since A is symmetric, we know that it can be diagonalized. In fact, in Example 3, we computed the eigenvalues and eigenvectors of A and got

$$\begin{aligned} \lambda_1 &= -2 & \mathbf{f}_1 &= (1,0,1) \\ \lambda_2 &= -2 & \mathbf{f}_2 &= (0,1,1) \\ \lambda_3 &= 7 & \mathbf{f}_3 &= (1,1,-1) \end{aligned}$$

Thus,

$$D = \begin{bmatrix} -2 & 0 & 0 \\ 0 & -2 & 0 \\ 0 & 0 & 7 \end{bmatrix} \quad \text{and} \quad P^{-1} = \begin{bmatrix} 1 & 0 & 1 \\ 0 & 1 & 1 \\ 1 & 1 & -1 \end{bmatrix}$$

Computing $P = (P^{-1})^{-1}$ we have

$$D = PAP^{-1}$$

$$= \frac{1}{3}\begin{bmatrix} 2 & -1 & 1 \\ -1 & 2 & 1 \\ 1 & 1 & -1 \end{bmatrix}\begin{bmatrix} 1 & 3 & -3 \\ 3 & 1 & -3 \\ -3 & -3 & 1 \end{bmatrix}\begin{bmatrix} 1 & 0 & 1 \\ 0 & 1 & 1 \\ 1 & 1 & -1 \end{bmatrix}$$

$$= \begin{bmatrix} -2 & 0 & 0 \\ 0 & -2 & 0 \\ 0 & 0 & 7 \end{bmatrix}$$

■

Formula (5.5) and the following calculations enable us to compute fairly easily the various powers of A, once we know the eigenvalues and eigenvectors of this matrix.

$$A^2 = [P^{-1}DP][P^{-1}DP] = P^{-1}D(PP^{-1})DP = PD^2P^{-1}$$

In a similar fashion, we have

$$A^n = P^{-1}D^nP \tag{5.6}$$

The advantage in using (5.6) to compute A^n is that D is a diagonal matrix and hence its powers are easy to calculate.

EXAMPLE 5 Let $A = \begin{bmatrix} 1 & 3 \\ 0 & 4 \end{bmatrix}$; cf. Example 4 in Section 1.4. Compute A^n for arbitrary n.

Solution. Computing the characteristic polynomial of A we get

$$p(\lambda) = \det\begin{bmatrix} 1-\lambda & 3 \\ 0 & 4-\lambda \end{bmatrix} = (\lambda - 1)(\lambda - 4)$$

Computing the eigenvectors of A, we have

$$A - I = \begin{bmatrix} 0 & 3 \\ 0 & 3 \end{bmatrix} \qquad \text{thus } \mathbf{f}_1 = (1,0)$$

$$A - 4I = \begin{bmatrix} -3 & 3 \\ 0 & 0 \end{bmatrix} \qquad \text{thus } \mathbf{f}_2 = (1,1)$$

A basis of eigenvectors is $\{(1,0),(1,1)\}$ and

$$D = \begin{bmatrix} 1 & 0 \\ 0 & 4 \end{bmatrix} \qquad P = \begin{bmatrix} 1 & -1 \\ 0 & 1 \end{bmatrix} \qquad P^{-1} = \begin{bmatrix} 1 & 1 \\ 0 & 1 \end{bmatrix}$$

Using (5.6) we calculate A^n:

$$\begin{bmatrix} 1 & 3 \\ 0 & 4 \end{bmatrix}^n = \begin{bmatrix} 1 & 1 \\ 0 & 1 \end{bmatrix}\begin{bmatrix} 1 & 0 \\ 0 & 4 \end{bmatrix}^n\begin{bmatrix} 1 & -1 \\ 0 & 1 \end{bmatrix}$$
$$= \begin{bmatrix} 1 & 1 \\ 0 & 1 \end{bmatrix}\begin{bmatrix} 1 & 0 \\ 0 & 4^n \end{bmatrix}\begin{bmatrix} 1 & -1 \\ 0 & 1 \end{bmatrix}$$
$$= \begin{bmatrix} 1 & 4^n \\ 0 & 4^n \end{bmatrix}\begin{bmatrix} 1 & -1 \\ 0 & 1 \end{bmatrix} = \begin{bmatrix} 1 & 4^n - 1 \\ 0 & 4^n \end{bmatrix}$$

■

PROBLEM SET 5.2

1. For each of the following linear transformations find a basis in which the matrix representation of L is diagonal:
a. $L(x_1,x_2) = (-x_1, 2x_1 + 3x_2)$
b. $L(x_1,x_2) = (8x_1 + 2x_2, x_1 + 7x_2)$

2. For each of the following matrices find a diagonal matrix that is similar to the given matrix:

a. $\begin{bmatrix} -1 & 0 \\ 2 & 3 \end{bmatrix}$ **b.** $\begin{bmatrix} 8 & 2 \\ 1 & 7 \end{bmatrix}$ **c.** $\begin{bmatrix} 4 & 2 \\ 2 & 1 \end{bmatrix}$ **d.** $\begin{bmatrix} 1 & 2 \\ 0 & 0 \end{bmatrix}$

3. For each of the matrices A in problem 2 compute A^n for n an arbitrary positive integer. If A is invertible, compute A^n for n an arbitrary integer. [If A is invertible, (5.6) is valid for $n = 0, \pm 1, \pm 2, \ldots$.]

4. Let A be the matrix

$$A = \begin{bmatrix} 4 & 1 & -1 \\ 1 & 4 & -1 \\ -1 & -1 & 4 \end{bmatrix}$$

a. Find a matrix P such that $PAP^{-1} = D$ is a diagonal matrix.
b. Compute A^{10}.
c. Compute A^{-10}.

5. Let A be the matrix

$$A = \begin{bmatrix} 3 & 0 & 1 & -2 \\ 2 & 3 & 2 & -4 \\ 0 & 0 & 2 & 0 \\ 1 & 0 & 1 & 0 \end{bmatrix}$$

a. Compute the eigenvalues and eigenvectors of A.
b. Find a matrix P^{-1} such that PAP^{-1} is a diagonal matrix.
c. Compute A^3.

6. Let $L\colon V \to W$ be a linear transformation. Suppose $\dim(V) = n$ and $\dim(W) = m$. Show that it is possible to pick bases in V and W such that the matrix representation of L with respect to these bases is an $m \times n$ matrix $A = [a_{jk}]$ with $a_{jk} = 0$ if j is not equal to k and a_{kk} equals zero or one. Moreover, the number of ones that occur in A is exactly equal to $\dim(\mathrm{Rg}(L)) = \mathrm{rank}(A)$.

7. Let $A_p = [a_{pjk}]$ be a sequence of $n \times n$ matrices. We say that $\lim_{p\to\infty} A_p = A = [a_{jk}]$, if and only if $\lim_{p\to\infty} a_{pjk} = a_{jk}$ for each j and k. For each of the matrices in problem 2, determine whether or not $\lim_{n\to\infty} A^n$ exists, and then evaluate the limit if possible.
8. For the matrix A in problem 5, $\lim_{n\to\infty} A^n = ?$
9. Let $A = \begin{bmatrix} \lambda_1 & 0 \\ 0 & \lambda_2 \end{bmatrix}$. Let $g(\lambda) = \Sigma_{k=0}^{p} g_k \lambda^k$ be any polynomial in λ. Define $g(A)$ by $g(A) = \Sigma_{k=0}^{p} g_k A^k$. Show that
$$g(A) = \begin{bmatrix} g(\lambda_1) & 0 \\ 0 & g(\lambda_2) \end{bmatrix}$$
Generalize this formula to the case when A is an $n \times n$ diagonal matrix.
10. Suppose A and B are similar matrices. Thus, $A = PBP^{-1}$ for some nonsingular matrix P. Let $g(\lambda)$ be any polynomial in λ. Show that $g(A) = Pg(B)P^{-1}$.
11. If c is a nonnegative real number, we can compute its square root. In fact c^m is defined for any real number m if $c > 0$. Suppose A is a diagonalizable matrix with nonnegative eigenvalues. Thus $A = P^{-1}DP$, where D equals $[\lambda_j \delta_{jk}]$. Define $A^m = P^{-1}D^m P$, where $D^m = [\lambda_j^m \delta_{jk}]$.
    a. Show that $(D^{1/2})^2 = D$, assuming $\lambda_j \geq 0$.
    b. Show that $(A^{1/2})^2 = A$, with the same assumption as in a.
12. For each matrix A below compute $A^{1/2}$.

    a. $\begin{bmatrix} 8 & 2 \\ 1 & 7 \end{bmatrix}$ b. $\begin{bmatrix} 4 & 2 \\ 2 & 1 \end{bmatrix}$ c. $\begin{bmatrix} 4 & 1 & -1 \\ 1 & 4 & -1 \\ -1 & -1 & 4 \end{bmatrix}$
13. For each matrix in the preceding problem compute when possible $A^{-1/6}$ and $A^{2/3}$.
14. Compute $A^{1/2}$ and $A^{-1/2}$ where A is the matrix in problem 5. Verify that $A^{1/2}A^{-1/2} = I$.

5.3 APPLICATIONS

In this section, instead of presenting any new material we discuss and analyze a few problems by employing the techniques of the preceding sections. Several things should be observed. One is the recasting of the problem into the language of matrices, and the other is the method of analyzing matrices via their eigenvalues and eigenvectors.

The reader should also be aware that these problems are contrived, in the sense that they do not model (to the author's knowledge) any real phenomenon. They were made up so that the arithmetic would not be too tedious and yet the flavor of certain types of analyses would be allowed to seep through. Modeling real world problems mathematically usually leads to a lengthy analysis of the physical situation. This is done so that the mathematical model is believable; that is, we wish to analyze a mathematical structure, for us a matrix, and then be able, from this analysis, to infer something about the original problem. In most cases this just takes too much time—not the analysis, but making the model believable.

EXAMPLE 1 An individual has some money that is to be invested in three different accounts. The first, second, and third investments realize a profit of 8, 10, and 12 percent per year, respectively. Suppose our investor decides that at the end of each year, one-fourth of the money earned in the second investment and three-fourths of that earned in the third investment should be transferred into the first account, and that one-fourth of the third account's earnings will be transferred to the second account. Assuming that each account starts out with the same amount, how long will it take for the money in the first account to double?

Solution. The first step in analyzing this problem is to write down any equations relating the unknowns. Thus let a_{jk}, $j = 1, 2, 3$; $k = 1, 2, \ldots$ represent the amount invested in account j during the kth year. Thus, if a dollars were originally invested in each of the three accounts we have $a_{11} = a_{21} = a_{31} = a$. Moreover, we also have

$$\begin{aligned} a_{1(k+1)} &= 1.08a_{1k} + \tfrac{1}{4}(0.1a_{2k}) + \tfrac{3}{4}(0.12a_{3k}) \\ a_{2(k+1)} &= a_{2k} + \tfrac{3}{4}(0.1a_{2k}) + \tfrac{1}{4}(0.12a_{3k}) \\ a_{3(k+1)} &= a_{3k} \end{aligned} \tag{5.7}$$

Rewriting this as a matrix equation, where $U_k = [a_{1k}, a_{2k}, a_{3k}]^T$, we have

$$U_{k+1} = \begin{bmatrix} 1.08 & 0.025 & 0.09 \\ 0.0 & 1.075 & 0.03 \\ 0.0 & 0.0 & 1.0 \end{bmatrix} U_k \tag{5.8}$$

Setting A equal to the 3×3 matrix in (5.8) we have

$$U_k = AU_{k-1} = A^2U_{k-2} = A^{k-1}U_1 \tag{5.9}$$

Our problem is to determine k such that $a_{1k} = 2a$. Clearly the eigenvalues of A are 1.08, 1.075, and 1. Routine calculations give us the following eigenvectors:

$$\begin{aligned} \lambda_1 &= 1.08 \qquad & \mathbf{f}_1 &= (1,0,0) \\ \lambda_2 &= 1.075 \qquad & \mathbf{f}_2 &= (-5,1,0) \\ \lambda_3 &= 1 \qquad & \mathbf{f}_3 &= (-5,-2,5) \end{aligned}$$

Setting $P^{-1} = \begin{bmatrix} 1 & -5 & -5 \\ 0 & 1 & -2 \\ 0 & 0 & 5 \end{bmatrix}$, we have $P = \begin{bmatrix} 1 & 5 & 3 \\ 0 & 1 & \frac{2}{5} \\ 0 & 0 & \frac{1}{5} \end{bmatrix}$. Thus, $A = P^{-1}\begin{bmatrix} 1.08 & 0 & 0 \\ 0 & 1.075 & 0 \\ 0 & 0 & 1 \end{bmatrix} P$ and

$$A^k = \begin{bmatrix} (1.08)^k & 5[(1.08)^k - (1.075)^k] & 3(1.08)^k - 2(1.075)^k - 1 \\ 0 & (1.075)^k & \frac{2}{5}[(1.075)^k - 1] \\ 0 & 0 & 1 \end{bmatrix}$$

a_{1k}, which is the first component of the vector $A^{k-1}U_1$, must equal

$$a_{1k} = [9(1.08)^{k-1} - 7(1.075)^{k-1} - 1]a \tag{5.10}$$

where $U_1 = (a, a, a)^T$.

If a_{1k} equals $2a$, we then have

$$9(1.08)^{k-1} - 7(1.075)^{k-1} - 1 = 2 \tag{5.11}$$

Equations such as (5.11) are extremely difficult to solve; so we content ourselves with the following approximation:

$$\begin{aligned}(1.08)^{k-1} &= (1.075 + 0.005)^{k-1} \\ &= (1.075)^{k-1} + e\end{aligned}$$

If k is not too large, e will be approximately equal to $(0.005)(k-1)$, a relatively small number. In any case, since $e > 0$, we certainly have $(1.08)^{k-1} \geq (1.075)^{k-1}$. Therefore, if we find k such that

$$9(1.075)^{k-1} - 7(1.075)^{k-1} - 1 = 2 \tag{5.12}$$

then certainly (5.11) will hold with the equality sign replaced by $\geq$. Now (5.12) implies $(1.075)^{k-1} = \frac{3}{2}$. Thus $k - 1 = (\ln 3 - \ln 2)/\ln 1.075 \approx 5.607$ or $k \approx 6.607$ years. In other words the amount of money in the first account will certainly have doubled after 7 years. ■

This example also illustrates another aspect of mathematical anaylsis, approximations. Equation (5.11) is extremely difficult to solve, while (5.12) is much easier and more importantly supplies us with a usable solution.

EXAMPLE 2 Suppose we have two different species and their populations during the kth year are represented by x_k and y_k. Suppose that left alone the populations grow according to the following equations:

$$\begin{aligned}x_{k+1} &= 3x_k - y_k \\ y_{k+1} &= -x_k + 2y_k\end{aligned}$$

Our problem is the following. What percentage of each species can be removed (harvested) each year so that the populations remain constant?

Solution. Let r_1 and r_2 denote the fractions of x_k and y_k that are removed. We have $0 < r_1, r_2 < 1$, and at the end of each year we effectively have $(1 - r_1)x_k$ and $(1 - r_2)y_k$ members left in each species. These comments imply

$$\begin{bmatrix} x_{k+1} \\ y_{k+1} \end{bmatrix} = \begin{bmatrix} 3(1-r_1) & -(1-r_2) \\ -(1-r_1) & 2(1-r_2) \end{bmatrix} \begin{bmatrix} x_k \\ y_k \end{bmatrix}$$

Since we want $x_{k+1} = x_k$ and $y_{k+1} = y_k$, we need r_1 and r_2 such that

$$\begin{bmatrix} 3(1 - r_1) & -(1 - r_2) \\ -(1 - r_1) & 2(1 - r_2) \end{bmatrix} = \begin{bmatrix} 1 & 0 \\ 0 & 1 \end{bmatrix}$$

Clearly this is impossible. Hence, we cannot remove any percentage and leave the populations fixed. Well, maybe we asked for too much. Perhaps instead of wanting our removal scheme to work for any population, we should instead try to find a population for which there is some removal scheme. Thus, if A is the matrix

$$\begin{bmatrix} 3(1 - r_1) & -(1 - r_2) \\ -(1 - r_1) & 2(1 - r_2) \end{bmatrix}$$

we want to find numbers x and y such that $A[x\ y]^T = [x\ y]^T$. In other words, can we pick r_1 and r_2 so that 1 not only is an eigenvalue for A but also has an eigenvector with positive components. Let's compute the characteristic polynomial of A.

$$\begin{aligned} p(\lambda;r_1,r_2) &= \det \begin{bmatrix} 3(1 - r_1) - \lambda & -(1 - r_2) \\ -(1 - r_1) & 2(1 - r_2) - \lambda \end{bmatrix} \\ &= [\lambda - 3(1 - r_1)][\lambda - 2(1 - r_2)] - (1 - r_1)(1 - r_2) \\ &= \lambda^2 - (5 - 3r_1 - 2r_2)\,\lambda + 5(1 - r_1)(1 - r_2) \end{aligned}$$

Now we want $p(1;r_1,r_2) = 0$. Setting $\lambda = 1$, we have after some simplification

$$r_1(5r_2 - 2) = 3r_2 - 1 \tag{5.13}$$

where we want r_k, $k = 1$ or 2, to lie between 0 and 1. Solving (5.13) for r_1 and checking the various possibilities to ensure that both r_k's lie between 0 and 1, we have

$$r_1 = \frac{3r_2 - 1}{5r_2 - 2} \qquad 0 < r_2 < \tfrac{1}{3} \text{ or } \tfrac{1}{2} < r_3 < 1 \tag{5.14}$$

For r_1 and r_2 so related let's calculate the eigenvectors of the eigenvalue 1. The matrix

$$A - I_2 = \begin{bmatrix} 2 - 3r_1 & -(1 - r_2) \\ -(1 - r_1) & 1 - 2r_2 \end{bmatrix}$$

is row equivalent to the matrix

$$\begin{bmatrix} \dfrac{1}{5r_2 - 2} & 1 \\ 0 & 0 \end{bmatrix}$$

Thus, the eigenvector (x, y) corresponding to $\lambda = 1$ satisfies $x = (2 - 5r_2)y$. If both x and y are to be positive we clearly need $2 - 5r_2$ positive. We therefore have the following solution to our original problem. First $r_1 = (3r_2 - 1)/(5r_2 - 2)$, where r_2 satisfies the inequalities in (5.14). To ensure that x and y are positive we then restrict r_2 to satisfy $0 < r_2 < \frac{1}{3}$ only. In conclusion, if we harvest less than $\frac{1}{3}$ of the second species and $(3r_2 - 1)/(5r_2 - 2)$ of the first species where the two species are in the ratio $1/(2 - 5r_2)$, we will have a constant population from one year to the next. ■

EXAMPLE 3 This last example is interesting in that it is not at all clear how to express the problem in terms of matrices. Suppose we construct the following sequence of numbers: let $a_0 = 0$ and $a_1 = 1$, $a_2 = \frac{1}{2}$, $a_3 = (1 + \frac{1}{2})/2 = \frac{3}{4}$, $a_4 = (\frac{1}{2} + \frac{3}{4})/2 = \frac{5}{8}$, and in general $a_{n+2} = (a_n + a_{n+1})/2$; that is, a_{n+2} is the average of the two preceding numbers. Do these numbers approach some constant as n gets larger and larger?

Solution. Let $U_n = [a_{n-1}\ a_n]^T$ for $n = 1, 2, \ldots$. Recalling how the a_n are defined, we have

$$U_{n+1} = \begin{bmatrix} a_n \\ a_{n+1} \end{bmatrix} = \begin{bmatrix} a_n \\ \dfrac{a_{n-1}}{2} + \dfrac{a_n}{2} \end{bmatrix}$$

$$= \begin{bmatrix} 0 & 1 \\ \frac{1}{2} & \frac{1}{2} \end{bmatrix} \begin{bmatrix} a_{n-1} \\ a_n \end{bmatrix} = AU_n \qquad (5.15)$$

where A is the 2×2 matrix appearing in (5.15). As in similar examples, we have $U_n = A^{n-1}U_1$, where U_1 equals $[0\ 1]^T$. The characteristic polynomial of A is

$$p(\lambda) = \det(A - \lambda I) = \det \begin{bmatrix} -\lambda & 1 \\ \frac{1}{2} & \frac{1}{2} - \lambda \end{bmatrix} = \tfrac{1}{2}(2\lambda + 1)(\lambda - 1)$$

Thus, the eigenvalues of A are 1 and $-\frac{1}{2}$. We will see later that, if the a_n have a limiting value, it is necessary for the number 1 to be an eigenvalue.

$\lambda = 1$, $\mathbf{f}_1 = (1,1)$
$\lambda = -\frac{1}{2}$, $\mathbf{f}_2 = (-2,1)$

Setting P^{-1} equal to $\begin{bmatrix} 1 & -2 \\ 1 & 1 \end{bmatrix}$, we calculate $P = \frac{1}{3}\begin{bmatrix} 1 & 2 \\ -1 & 1 \end{bmatrix}$. Thus, we have

$$A^n = \begin{bmatrix} 1 & -2 \\ 1 & 1 \end{bmatrix} \begin{bmatrix} 1 & 0 \\ 0 & (-\frac{1}{2})^n \end{bmatrix} \begin{bmatrix} \frac{1}{3} & \frac{2}{3} \\ -\frac{1}{3} & \frac{1}{3} \end{bmatrix}$$

Clearly, the larger n gets the closer the middle matrix on the right-hand side gets to the matrix $\begin{bmatrix} 1 & 0 \\ 0 & 0 \end{bmatrix}$ which means that the vector $U_n = A^{n-1}U_1$ gets close to the vector U_∞, where

$$U_\infty = \begin{bmatrix} 1 & -2 \\ 1 & 1 \end{bmatrix} \begin{bmatrix} 1 & 0 \\ 0 & 0 \end{bmatrix} \begin{bmatrix} \frac{1}{3} & \frac{2}{3} \\ -\frac{1}{3} & \frac{1}{3} \end{bmatrix} \begin{bmatrix} 0 \\ 1 \end{bmatrix}$$

$$= \begin{bmatrix} \frac{1}{3} & \frac{2}{3} \\ \frac{1}{3} & \frac{2}{3} \end{bmatrix} \begin{bmatrix} 0 \\ 1 \end{bmatrix} = \begin{bmatrix} \frac{2}{3} \\ \frac{2}{3} \end{bmatrix}$$

Thus, $U_n = [a_{n-1}\ a_n]^T$ gets close to $[\frac{2}{3}\ \frac{2}{3}]^T$, which means that the numbers a_n get close to the number $\frac{2}{3}$. Notice too that U_∞ is an eigenvector of A corresponding to the eigenvalue 1. In fact, this was to be expected from the equation $U_{n+1} = AU_n$; for if the U_n converge to something nonzero, called U_∞, then U_∞ must satisfy the equation $U_\infty = AU_\infty$. That is, U_∞ must be an eigenvector of the matrix A corresponding to the eigenvalue 1. This is why 1 must be an eigenvalue. ■

PROBLEM SET 5.3

1. Suppose we have a single species that increases from one year to the next according to the rule

$$x_{k+1} = 2x_k$$

What percentage of the population can be harvested and have the population remain constant from one year to the next? If the initial population equals 10, assuming no harvesting, what will the population be in 20 years?

2. The following sequence of numbers is called the Fibonacci sequence:

$$a_0 = 1,\ a_1 = 1,\ a_2 = 2,\ a_3 = 3,\ \ldots,\ a_{n+1} = a_n + a_{n-1}$$

Find a general formula for a_n and determine its limit if one exists.

3. Define the following sequence of numbers:

$$a_0 = a \quad a_1 = b \quad a_2 = \tfrac{1}{3}a_0 + \tfrac{2}{3}a_1 \quad \text{and} \quad a_{n+2} = \tfrac{1}{3}a_n + \tfrac{2}{3}a_{n+1}$$

Find a general formula for a_n and determine its limiting behavior.

4. Define the following sequence of numbers:

$$a_0 = a \quad a_1 = b \quad a_2 = ca_0 + da_1 \quad a_{n+2} = ca_n + da_{n+1}$$

   a. If c and d are nonnegative and $c + d = 1$, determine a formula for a_n and the limiting behavior of this sequence.
   b. What happens if we just assume that $c + d = 1$?
   c. What happens if there are no restrictions on c and d?

5. Suppose there are two cities, the sum of whose populations remains constant. Assume that each year a certain fraction of one city's population moves to the second city and the same fraction of the second city's population moves to the first city. Let x_k and y_k denote each city's population in the kth year.
   a. Find formulas for x_{k+1} and y_{k+1} in terms of x_k and y_k. These formulas will of course depend on the fraction r of the populations that move.
   b. What is the limiting population of each city in terms of the original populations and r?

6. We again have the same two cities as in problem 5, only this time let's assume that r_1

represents the fraction of the first city's population that moves to the second city and r_2 the fraction of the second city's population that moves to the first city.

a. Find formulas for x_{k+1} and y_{k+1}.

b. What is the limiting population of each city?

7. Suppose that the populations of two species change from one year to the next according to the equations

$$x_{k+1} = 4x_k + 3y_k$$
$$y_{k+1} = 3x_k + 9y_k$$

Are there any initial populations and harvesting schemes that leave the populations constant from one year to the next?

8. Let $u_0 = a > 0$, $u_1 = b > 0$. Define $u_2 = u_0 u_1$, $u_{n+2} = u_n u_{n+1}$. What is lim u_n? Hint: log $u_n = ?$

SUPPLEMENTARY PROBLEMS

1. Define and give examples of the following:

a. Eigenvector and eigenvalue

b. Eigenspace

c. Characteristic matrix and polynomial

2. Find a linear transformation from $\mathbb{R}^3$ to $\mathbb{R}^3$ such that 4 is an eigenvalue of multiplicity 2 and 5 is an eigenvalue of multiplicity 1.

3. Let L be a linear transformation from V to V. Let λ be any eigenvalue of L, and let K_λ be the eigenspace of λ. That is, $K_\lambda = \{\mathbf{x}: L\mathbf{x} = \lambda\mathbf{x}\}$. Show that K_λ is a subspace of V, and that it is invariant under L, cf. number 13 in the Supplementary Problems for Chapter 3.

4. Let λ be an eigenvector of L. A nonzero vector $\mathbf{x}$ is said to be a generalized eigenvector of L, if there is a positive integer k such that

$$[L - \lambda I]^k \mathbf{x} = \mathbf{0}$$

Show that the set of all generalized eigenvectors of λ along with the zero vector is an invariant subspace.

5. Let

$$A = \begin{bmatrix} 2 & 1 & 0 \\ 0 & 2 & 1 \\ 0 & 0 & 2 \end{bmatrix}$$

a. Find all the eigenvalues of A.

b. Find the eigenspaces of A.

c. Find the generalized eigenspaces of A; cf. problem 4.

6. Let

$$A = \begin{bmatrix} 1 & 2 & -4 \\ 0 & -1 & 6 \\ 0 & -1 & 4 \end{bmatrix}$$

a. Determine the eigenvalues and eigenspaces of A.

b. Show that A is not similar to a diagonal matrix.
c. Find the generalized eigenspaces of A.

7. Let L be a linear transformation from $\mathbb{R}^2$ to $\mathbb{R}^2$. Let $\mathbf{x}$ be a vector in $\mathbb{R}^2$ for which $\mathbf{x}$ and $L\mathbf{x}$ are not zero, but for which $L^2\mathbf{x} = \mathbf{0}$.
a. Show that $F = \{L\mathbf{x},\mathbf{x}\}$ is linearly independent.
b. Show that the matrix representation A of L with respect to the basis F of part a equals $\begin{bmatrix} 0 & 1 \\ 0 & 0 \end{bmatrix}$.
c. Deduce that $L^2\mathbf{y} = 0$ for every vector $\mathbf{y}$ in $\mathbb{R}^2$.

8. Suppose that L is a linear transformation from $\mathbb{R}^3$ to $\mathbb{R}^3$, and there is a vector $\mathbf{x}$ for which $L^3\mathbf{x} = \mathbf{0}$, but $\mathbf{x}$, $L\mathbf{x}$, and $L^2\mathbf{x}$ are not equal to the zero vector. Set $F = \{L^2\mathbf{x}, L\mathbf{x},\mathbf{x}\}$.
a. Show that F is a basis of $\mathbb{R}^3$.
b. Find the matrix representation A of L with respect to the basis F.
c. Show that $L^3\mathbf{y} = \mathbf{0}$ for every vector $\mathbf{y}$ in $\mathbb{R}^3$.

9. Let

$$A = \begin{bmatrix} 1 & 0 & 0 & 2 \\ 0 & 1 & 3 & 0 \\ 0 & 0 & 2 & 0 \\ 0 & 0 & 0 & 2 \end{bmatrix}$$

Find a matrix P such that PAP^{-1} is a diagonal matrix. Compute A^{10}.

10. A mapping T from a vector space V into V is called affine if $T\mathbf{x} = L\mathbf{x} + \mathbf{a}$, where L is a linear transformation and $\mathbf{a}$ is a fixed vector in V.
a. Show that $T^n\mathbf{x} = L^n\mathbf{x} + \Sigma_{k=0}^{n-1} L^k\mathbf{a}$. Thus every power of T is also an affine transformation.
b. Let $A = \begin{bmatrix} 2 & 3 \\ 3 & 2 \end{bmatrix}$. Define $T\mathbf{x} = A\mathbf{x} + \begin{bmatrix} 1 \\ 3 \end{bmatrix}$. If $\mathbf{x} = (1,1)$, compute $T^{10}\mathbf{x}$. Hint: $1 + r + \cdots + r^n = [1 - r^{n+1}][1 - r]^{-1}$.

11. Let $V = \{\Sigma_{n=0}^{2} a_n \cos nt\colon a_n \text{ any real number}\}$.
a. Show that $\{1, \cos t, \cos 2t\}$ is a basis of V.
b. Define $L\colon V \to V$ by $L(\Sigma_{n=0}^{2} a_n \cos nt) = \Sigma_{n=0}^{2} n^2 a_n \cos nt$. Find the eigenvalues and eigenvectors of L. Note that $L[f] = -f''$.
c. Find $L^{1/2}$, i.e., find a linear transformation T such that $T^2 = L$.
d. For any positive integers p and q find $L^{p/q}$.

12. Consider the following set of directions. Start at the origin, facing toward the positive x_1 axis, turn 45 degrees counterclockwise, and walk 1 unit in this direction. Thus, you will be at the point $(1/\sqrt{2}, 1/\sqrt{2})$. Then turn 45 degrees counterclockwise and walk $\frac{1}{2}$ unit in this new direction; then turn 45 degrees counterclockwise and walk $\frac{1}{4}$ unit in this direction. If a cycle consists of a 45-degree counterclockwise rotation plus a walk that is half as long as the previous walk, where will you be after 20 cycles?

13. Let $\mathcal{M}$ be the set of 2×2 matrices A for which

$$A\begin{bmatrix} 1 \\ 0 \end{bmatrix} = \lambda \begin{bmatrix} 1 \\ 0 \end{bmatrix}$$

for some real number λ. That is, A is in $\mathcal{M}$ if $(1,0)$ is an eigenvector of A.

a. Show that $\mathcal{M}$ is a subspace of M_{22}.
b. Find a basis for $\mathcal{M}$.

14. Let $\mathbf{x}$ be a fixed nonzero vector in $\mathbb{R}^n$. Let $\mathcal{M}_{\mathbf{x}}$ be the set of $n \times n$ matrices for which $\mathbf{x}$ is an eigenvector.
a. Show that $\mathcal{M}_{\mathbf{x}}$ is a subspace of M_{nn}.
b. If $\mathbf{x} = \mathbf{e}_1$, find a basis for $\mathcal{M}_{\mathbf{x}}$.
c. Can you find a basis for $\mathcal{M}_{\mathbf{x}}$ if $\mathbf{x}$ is arbitrary? Hint: If B is an invertible $n \times n$ matrix for which $B\mathbf{x}_1 = \mathbf{x}_2$, then BAB^{-1} is in $\mathcal{M}_{\mathbf{x}_2}$ if A is in $\mathcal{M}_{\mathbf{x}_1}$.

6
GEOMETRY IN $\mathbb{R}^n$

In Chapter 2, where we first started our study of vector spaces, a vector was intuitively described as something possessing both magnitude and direction. This led us to the idea of expressing vectors, at least in $\mathbb{R}^2$, as pairs of real numbers. One of the topics we discuss in this chapter is the relation between these pairs of numbers and the length of a vector. We also show how the angle between two vectors is related to their ordered pair representation.

6.1 LENGTH AND DOT PRODUCT

We first define the length of a vector in $\mathbb{R}^2$ and then show how to compute the angle between two vectors. Let $\mathbf{a} = (a_1, a_2)$ be any vector in $\mathbb{R}^2$. By the length of $\mathbf{a}$ we mean the distance from the point with coordinates (a_1, a_2) to the origin (see Figure 6.1). The Pythagorean theorem tells us that this distance equals $(a_1^2 + a_2^2)^{1/2}$. We therefore define the length of any vector in $\mathbb{R}^2$ or $\mathbb{R}^n$ as follows:

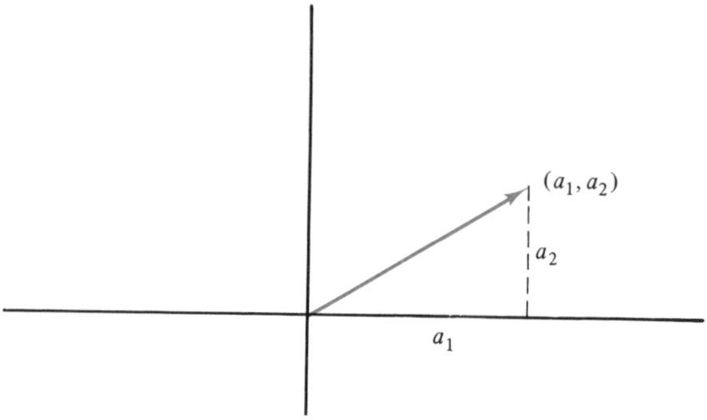

Figure 6.1

Definition 6.1. Let $\mathbf{x} = (x_1, x_2, \ldots, x_n)$ be any vector in $\mathbb{R}^n$. The length or norm of $\mathbf{x}$ denoted by $\|\mathbf{x}\|$, is

$$\|\mathbf{x}\| = (x_1^2 + \cdots + x_n^2)^{1/2} = \left(\sum_{j=1}^{n} x_j^2\right)^{1/2} \tag{6.1}$$

EXAMPLE 1 Compute the lengths of the following vectors:

a. $\|(2,-3)\| = (4 + 9)^{1/2} = (13)^{1/2}$
b. $\|(2,-1,3)\| = (4 + 1 + 9)^{1/2} = (14)^{1/2}$
c. $\|(7,-8,1,3,6)\| = (49 + 64 + 1 + 9 + 36)^{1/2} = (159)^{1/2}$ ■

In part b of Example 1, we computed the length of a vector in $\mathbb{R}^3$. Figure 6.2 shows that in this case we may also interpret the length of the vector (x_1, x_2, x_3) as the distance from the point with coordinates (x_1, x_2, x_3) to the origin. The following theorem lists a few useful properties of the norm or length of a vector.

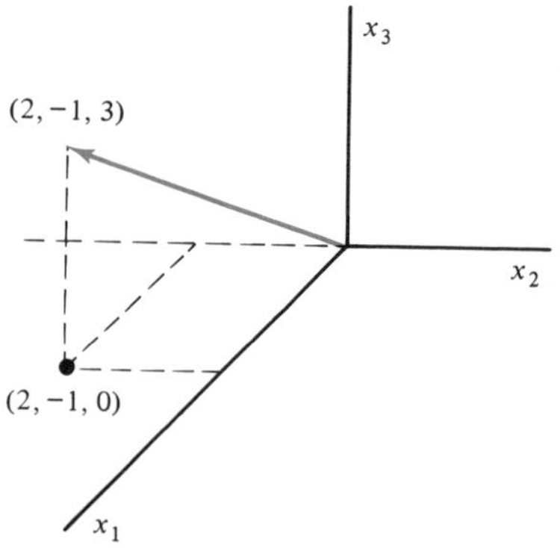

Figure 6.2

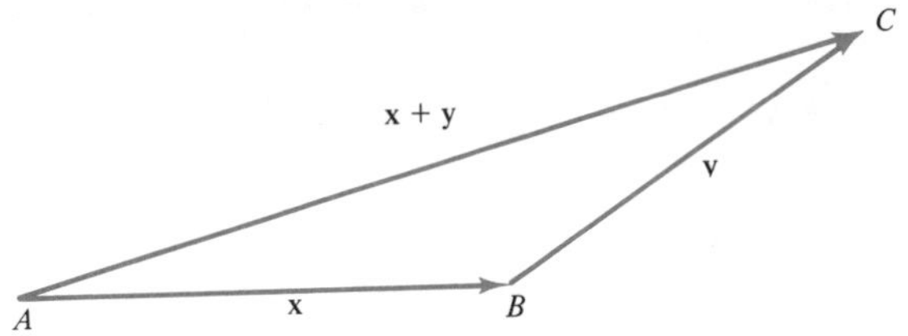

Figure 6.3

Theorem 6.1

1. Let $\mathbf{x}$ be any vector in $\mathbb{R}^n$. Then $\|\mathbf{x}\| \geq 0$, and $\|\mathbf{x}\| = 0$ if and only if $\mathbf{x} = \mathbf{0}$.
2. $\|c\mathbf{x}\| = |c|\,\|\mathbf{x}\|$ for any constant c and any vector $\mathbf{x}$.
3. For any two vectors $\mathbf{x}$ and $\mathbf{y}$, we have $\|\mathbf{x} + \mathbf{y}\| \leq \|\mathbf{x}\| + \|\mathbf{y}\|$.

The first two results are almost obvious, both geometrically and analytically, as we will see in a few lines. The third, which is called the triangle inequality, is not quite so obvious, but it does have a nice geometrical interpretation for $\mathbf{x}$ and $\mathbf{y}$ in $\mathbb{R}^3$.

Figure 6.3 shows the result of adding $\mathbf{y}$ to $\mathbf{x}$. The three points A, B, and C determine a triangle the lengths of whose sides are $\|\mathbf{x}\|$, $\|\mathbf{y}\|$, and $\|\mathbf{x} + \mathbf{y}\|$. Since the shortest distance between any two points is the straight line connecting them, we see that inequality 3 must indeed be true.

Proof (Theorem 6.1)

1. Let $\mathbf{x} = (x_1, \ldots, x_n)$. Then $\|\mathbf{x}\| = (x_1^2 + \cdots + x_n^2)^{1/2} \geq 0$. Moreover $\|\mathbf{x}\| = 0$ if and only if $x_j = 0$ for each j, i.e., $\|\mathbf{x}\| = 0$ if and only if $\mathbf{x} = \mathbf{0}$.
2. Let c be any scalar and $\mathbf{x}$ any vector in $\mathbb{R}^n$. Then $\|c\mathbf{x}\| = [\Sigma_{j=1}^n (cx_j)^2]^{1/2} = (c^2)^{1/2}[\Sigma_{j=1}^n x_j^2]^{1/2} = |c|\,\|\mathbf{x}\|$.
3. The reader is asked to prove this property in one of the problems at the end of this section.

Property 2 is used when we wish to construct a vector that has a given direction and length equal to 1. This is accomplished by taking any nonzero vector $\mathbf{x}$ that has the desired direction and then dividing it by its length, for if $\mathbf{u} = \mathbf{x}/\|\mathbf{x}\|$, then $\|\mathbf{u}\| = 1$.

EXAMPLE 2

a. $\|-2(1,3)\| = \|(-2,-6)\| = (4 + 36)^{1/2} = 2\|(1,3)\|$
b. Construct a unit vector that is parallel to the line going from the point $P = (-1,2)$ to the point $Q = (3,4)$. See Figure 6.4. By a unit vector we mean one whose length equals 1. A vector that has the desired direction can be found by subtracting the coordinates of the point P from those of Q. Thus

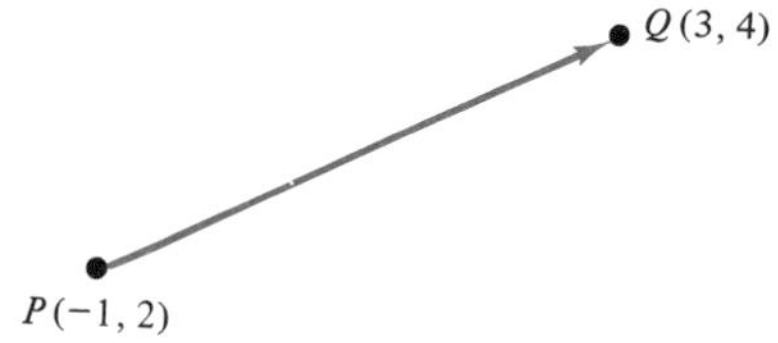

Figure 6.4

$\mathbf{x} = (3,4) - (-1,2) = (4,2)$ points in the desired direction, but it may not have length 1. In fact $\|\mathbf{x}\| = (16 + 4)^{1/2} = (20)^{1/2}$. Thus, the desired unit vector $\mathbf{u}$ equals $(4,2)/\sqrt{20}$. ■

We next wish to take two vectors $\mathbf{A} = (a_1, a_2)$ and $\mathbf{B} = (b_1, b_2)$ in $\mathbb{R}^2$ and derive a formula relating their coordinates and the cosine of the angle between them. See Figure 6.5. If we picture the triangle formed by these two vectors, the length of the side opposite the angle θ equals $\|\mathbf{A} - \mathbf{B}\| = \|(a_1 - b_1, a_2 - b_2)\|$. The law of cosines then implies

$$\|\mathbf{A} - \mathbf{B}\|^2 = \|\mathbf{A}\|^2 + \|\mathbf{B}\|^2 - 2\|\mathbf{A}\|\,\|\mathbf{B}\|\cos\theta \qquad (6.2)$$

Computing these lengths in terms of the coordinates, we have

$$(a_1 - b_1)^2 + (a_2 - b_2)^2 = a_1^2 + a_2^2 + b_1^2 + b_2^2 - 2\|\mathbf{A}\|\,\|\mathbf{B}\|\cos\theta$$

Squaring the terms in parentheses and then canceling gives us

$$-2a_1b_1 - 2a_2b_2 = -2\|\mathbf{A}\|\,\|\mathbf{B}\|\cos\theta$$

We finally arrive at the formula

$$\cos\theta = \frac{a_1b_1 + a_2b_2}{\|\mathbf{A}\|\,\|\mathbf{B}\|} \qquad (6.3)$$

where $\mathbf{A} = (a_1, a_2)$, $\mathbf{B} = (b_1, b_2)$, and θ is the smaller of the two angles determined by $\mathbf{A}$ and $\mathbf{B}$. We may draw a similar diagram for two vectors $\mathbf{A} =$

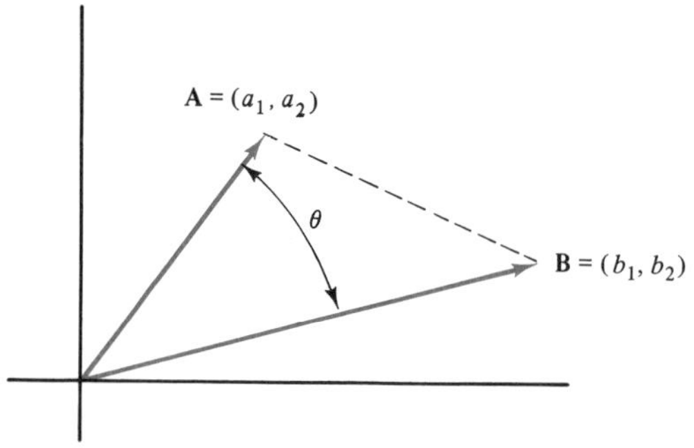

Figure 6.5

(a_1,a_2,a_3) and $\mathbf{B}=(b_1,b_2,b_3)$ in $\mathbb{R}^3$. If the same calculations are performed, we have

$$\cos\theta=\frac{a_1b_1+a_2b_2+a_3b_3}{\|\mathbf{A}\|\,\|\mathbf{B}\|} \tag{6.4}$$

where again, θ is the smaller of the two angles between **A** and **B**. The numerator in (6.3) and (6.4) appears so often in various formulas that it has been given a special name.

Definition 6.2. Let $\mathbf{x}=(x_1,\ldots,x_n)$ and $\mathbf{y}=(y_1,\ldots,y_n)$ be any two vectors in $\mathbb{R}^n$. The dot, or scalar, or inner product, of these two vectors is defined to be

$$\langle \mathbf{x},\mathbf{y}\rangle = x_1y_1+\cdots+x_ny_n=\sum_{j=1}^{n} x_jy_j \tag{6.5}$$

The phrase dot product arises because this operation is commonly denoted by a dot, i.e., $\mathbf{x}\cdot\mathbf{y}$. The term scalar is used because the operation produces a number and not a vector; while the term inner distinguishes this product from the outer product, $\mathbf{x}^T\mathbf{y}$, an $n\times n$ matrix whose j,k entry is x_jy_k.

EXAMPLE 3

a. $\langle(1,2),(6,4)\rangle=6+8=14$
b. $\langle(1,2),(-4,2)\rangle=-4+4=0$
c. $\langle(2,-3,4),(1,6,-2)\rangle=2-18-8=-24$
d. $\langle(1,0,4,6),(-2,3,2,8)\rangle=-2+0+8+48=54$ ■

We now rewrite formulas (6.3) and (6.4) as

$$\langle \mathbf{A},\mathbf{B}\rangle=\|\mathbf{A}\|\,\|\mathbf{B}\|\cos\theta \tag{6.6}$$

EXAMPLE 4 Compute the cosine of the angle between the following pairs of vectors:

a. $(1,-2)$ and $(4,3)$. From (6.6) we have

$$\cos\theta=\frac{\langle(1,-2),(4,3)\rangle}{\sqrt{5}\sqrt{25}}$$
$$=\frac{4-6}{5\sqrt{5}}=\frac{-2}{5\sqrt{5}}$$

b. $(-2,3,4)$ and $(3,-1,8)$

$$\cos\theta = \frac{\langle(-2,3,4),(3,-1,8)\rangle}{\sqrt{29}\sqrt{74}}$$

$$= \frac{23}{\sqrt{29}\sqrt{74}}$$

c. $(4,2,-1)$ and $(-3,4,-4)$

$$\cos\theta = \frac{\langle(4,2,-1),(-3,4,-4)\rangle}{\sqrt{21}\sqrt{41}}$$

$$= \frac{-12+8+4}{\sqrt{21}\sqrt{41}} = 0$$

Since $\cos\theta = 0$ if and only if θ equals 90 degrees, we see that the two vectors $(4,2,-1)$ and $(-3,4,-4)$ are perpendicular. ■

Two comments are in order. First, formula (6.6) is only valid (at this point) in $\mathbb{R}^2$ or $\mathbb{R}^3$. We later extend (6.6) to vectors in $\mathbb{R}^n$, by defining the cosine of the angle between two vectors to be that number which satisfies (6.6). The second comment, which we state as a theorem, has to do with when two vectors are perpendicular.

Theorem 6.2. Let $\mathbf{x}$ and $\mathbf{y}$ be any two nonzero vectors in $\mathbb{R}^2$ or $\mathbb{R}^3$. Then $\mathbf{x}$ and $\mathbf{y}$ are perpendicular if and only if $\langle\mathbf{x},\mathbf{y}\rangle = 0$.

Proof. Two nonzero vectors are perpendicular if and only if the angle between them equals 90 degrees; equivalently the cosine of that angle must equal zero. Using formula (6.6), we see that happens only when the inner product of the two vectors equals zero.

The inner product $\langle\mathbf{x},\mathbf{y}\rangle$ of two vectors satisfies many properties, some of which are listed in the next theorem.

Theorem 6.3. Let $\mathbf{x} = (x_1, \ldots, x_n)$ and $\mathbf{y} = (y_1, \ldots, y_n)$ be any two vectors in $\mathbb{R}^n$. Then

a. $\langle\mathbf{x},\mathbf{x}\rangle = \|\mathbf{x}\|^2$
b. $\langle\mathbf{x},\mathbf{y}\rangle = \langle\mathbf{y},\mathbf{x}\rangle$
c. Let a and b be any two scalars and let $\mathbf{z}$ be any vector in $\mathbb{R}^n$; then

$$\langle a\mathbf{x} + b\mathbf{y},\mathbf{z}\rangle = a\langle\mathbf{x},\mathbf{z}\rangle + b\langle\mathbf{y},\mathbf{z}\rangle$$

Proof

a. $\langle \mathbf{x},\mathbf{x}\rangle = x_1x_1 + x_2x_2 + \cdots + x_nx_n = \|\mathbf{x}\|^2$

b. $\langle \mathbf{x},\mathbf{y}\rangle = x_1y_1 + \cdots + x_ny_n$

$$= y_1x_1 + \cdots + y_nx_n$$

$$= \langle \mathbf{y},\mathbf{x}\rangle$$

c. $\langle a\mathbf{x} + b\mathbf{y},\mathbf{z}\rangle = \sum_{j=1}^{n} (ax_j + by_j)z_j$

$$= \sum_{j=1}^{n} ax_jz_j + \sum_{j=1}^{n} by_jz_j$$

$$= a\langle \mathbf{x},\mathbf{z}\rangle + b\langle \mathbf{y},\mathbf{z}\rangle$$

Property a is referred to by saying that the inner product is positive definite, i.e., $\langle \mathbf{x},\mathbf{x}\rangle \geq 0$, and it equals zero only if $\mathbf{x}$ is the zero vector. The second property is summarized by saying that the inner product is symmetric. The third property is the statement that the dot product is linear in its first argument. Symmetry immediately implies that

$$\langle \mathbf{z},a\mathbf{x} + b\mathbf{y}\rangle = a\langle \mathbf{z},\mathbf{x}\rangle + b\langle \mathbf{z},\mathbf{y}\rangle$$

We want to extend formula (6.6) to dimensions higher than three. Before we can do so, however, we need to know that the expression $\langle \mathbf{x},\mathbf{y}\rangle/\|\mathbf{x}\|\,\|\mathbf{y}\|$ can be interpreted as the cosine of some angle between 0 and 180 degrees. Figure 6.6 shows the graph of $\cos\theta$ for $0 \leq \theta \leq \pi$ radians. Notice that $\cos\theta$ is a number that always lies between -1 and 1. Thus, we would like the absolute value of $\langle \mathbf{x},\mathbf{y}\rangle/\|\mathbf{x}\|\,\|\mathbf{y}\|$ to be no greater than 1. This is the content of the next theorem.

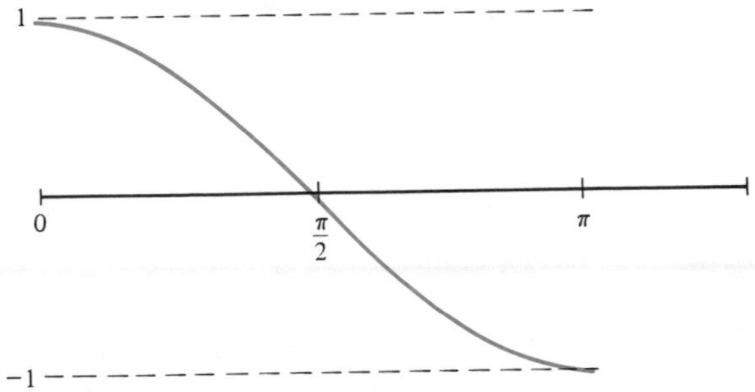

Figure 6.6

Theorem 6.4. (Cauchy-Schwarz inequality). Let $\mathbf{x}$ and $\mathbf{y}$ be any two vectors in $\mathbb{R}^n$. Then

$$|\langle \mathbf{x},\mathbf{y}\rangle| \leq \|\mathbf{x}\|\,\|\mathbf{y}\| \tag{6.7}$$

Proof. Define the following function of t by

$$f(t) = \|\mathbf{x} + t\mathbf{y}\|^2 = \langle \mathbf{x} + t\mathbf{y}, \mathbf{x} + t\mathbf{y} \rangle$$
$$= t^2\|\mathbf{y}\|^2 + 2t\langle \mathbf{x},\mathbf{y} \rangle + \|\mathbf{x}\|^2 \tag{6.8}$$

Thus, $f(t)$ is a quadratic function of t and is never negative. If $\mathbf{y} = 0$, then (6.7) is certainly true. Hence, we may assume that $\mathbf{y}$ is not the zero vector. Completing the square we rewrite (6.8) as

$$f(t) = \|\mathbf{y}\|^2 \left[t + \frac{\langle \mathbf{x},\mathbf{y} \rangle}{\|\mathbf{y}\|^2} \right]^2 + \|\mathbf{x}\|^2 - \frac{\langle \mathbf{x},\mathbf{y} \rangle^2}{\|\mathbf{y}\|^2} \tag{6.9}$$

Regardless of the value of t we must have $f(t) \geq 0$. We now pick t_0 in order to obtain the minimum value of $f(t)$. Setting $t_0 = -\langle \mathbf{x},\mathbf{y} \rangle / \|\mathbf{y}\|^2$, we have

$$0 \leq f(t_0) = \|\mathbf{x}\|^2 - \left[\frac{\langle \mathbf{x},\mathbf{y} \rangle}{\|\mathbf{y}\|} \right]^2$$

Thus, we see that $\langle \mathbf{x},\mathbf{y} \rangle^2 \leq [\|\mathbf{x}\|\,\|\mathbf{y}\|]^2$, from which (6.7) immediately follows.

EXAMPLE 5 Write the Cauchy-Schwarz inequality for any two vectors in $\mathbb{R}^2$ in terms of their coordinates.

Solution. If $\mathbf{x} = (x_1, x_2)$ and $\mathbf{y} = (y_1, y_2)$, we have

$$|\langle \mathbf{x},\mathbf{y} \rangle| = |x_1y_1 + x_2y_2| \leq (x_1^2 + x_2^2)^{1/2}(y_1^2 + y_2^2)^{1/2}$$ ■

Definition 6.3. Let $\mathbf{x}$ and $\mathbf{y}$ be two nonzero vectors in $\mathbb{R}^n$. We define the angle θ between $\mathbf{x}$ and $\mathbf{y}$ to be that angle which lies between 0 and 180 degrees and satisfies

$$\cos\theta = \frac{\langle \mathbf{x},\mathbf{y} \rangle}{\|\mathbf{x}\|\,\|\mathbf{y}\|}$$

Definition 6.4. We say that two nonzero vectors $\mathbf{x}$ and $\mathbf{y}$ are perpendicular if $\langle \mathbf{x},\mathbf{y} \rangle$ equals zero.

Definitions 6.3 and 6.4 extend our usual notions of the angle between two vectors, and the concept of perpendicularity, to $\mathbb{R}^n$, for n greater than 3, in a consistent manner.

EXAMPLE 6 Determine which of the following pairs of vectors are perpendicular.

a. (1,0) and (0,1) are perpendicular since $\langle (1,0),(0,1) \rangle = 0 + 0 = 0$.
b. (a,b), and $(-b,a)$ are perpendicular since $\langle (a,b), (-b,a) \rangle = -ab + ba = 0$.

c. (1,6,3) and (0,1,2) are not perpendicular since their inner product, which equals 12, is not zero.
d. (1,2,−6,1) and (0,1,4,3) are not perpendicular since their inner product equals −19.
e. (−2,8,3,4) and (6,4,4,−8) are perpendicular since their dot product equals zero. ■

EXAMPLE 7 Show that the diagonals of a rhombus must be perpendicular. A rhombus is a four-sided polygon with all four sides having the same length.

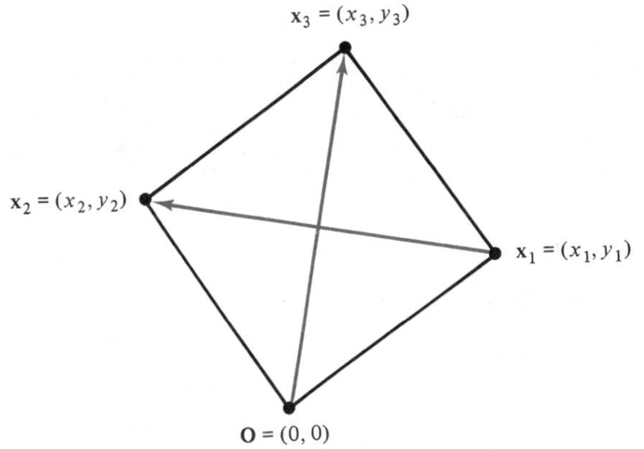

Solution. The accompanying figure has all four sides the same length. Thus,

$$\|\mathbf{x}_2\| = \|\mathbf{x}_1\| = \|\mathbf{x}_3 - \mathbf{x}_2\| = \|\mathbf{x}_3 - \mathbf{x}_1\|$$

Hence, we have by Theorem 6.3,

$$\begin{aligned}0 &= \|\mathbf{x}_3 - \mathbf{x}_1\|^2 - \|\mathbf{x}_3 - \mathbf{x}_2\|^2 \\ &= \langle \mathbf{x}_3 - \mathbf{x}_1, \mathbf{x}_3 - \mathbf{x}_1 \rangle - \langle \mathbf{x}_3 - \mathbf{x}_2, \mathbf{x}_3 - \mathbf{x}_2 \rangle \\ &= \langle \mathbf{x}_3, \mathbf{x}_3 \rangle - 2\langle \mathbf{x}_3, \mathbf{x}_1 \rangle + \langle \mathbf{x}_1, \mathbf{x}_1 \rangle \\ &\quad - \{\langle \mathbf{x}_3, \mathbf{x}_3 \rangle - 2\langle \mathbf{x}_3, \mathbf{x}_2 \rangle + \langle \mathbf{x}_2, \mathbf{x}_2 \rangle\} \\ &= 2\langle \mathbf{x}_3, \mathbf{x}_2 - \mathbf{x}_1 \rangle + \|\mathbf{x}_1\|^2 - \|\mathbf{x}_2\|^2 \\ &= 2\langle \mathbf{x}_3, \mathbf{x}_2 - \mathbf{x}_1 \rangle\end{aligned}$$

Thus, the vectors $\mathbf{x}_3$ and $\mathbf{x}_2 - \mathbf{x}_1$ are perpendicular. Since these vectors are parallel to the diagonals of the rhombus, the diagonals must be perpendicular to each other. ■

PROBLEM SET 6.1

1. Calculate the lengths of the following vectors:
a. (1,2) **b.** (−1,3,6) **c.** (1,1,2,8)

2. Find all unit vectors that are parallel to the vector (1,2,−4).

3. Compute the dot product of each of the following pairs of vectors:
   a. (1,0), (0,1) **b.** (a,b), (b,a) **c.** (1,2,1), (3,−6,2)
4. Sketch each of the following pairs of vectors. Compute their inner product and determine the cosine of the angle between them.
   a. (1,0), (1,0) **b.** (1,0), (1,1) **c.** (1,0), (0,1) **d.** (1,0), (−1,1)
   e. (1,0), (−1,0) **f.** (1,0), (−1,−1)
5. Find the cosine of the angle between each of the following pairs of vectors:
   a. (1,2), (3,−1) **b.** (1,0,−4), (6,1,2) **c.** (−2,3,0,1), (1,2,8,−2)
6. Show that if $\mathbf{x}$ and $\mathbf{y}$ are any two vectors in $\mathbb{R}^n$ for which $|\langle \mathbf{x},\mathbf{y}\rangle| = \|\mathbf{x}\|\,\|\mathbf{y}\|$, then one of them must be a scalar multiple of the other. [Hint: Let $f(t)$ be the function defined in (6.8). Show that the above equality holds if and only if $f(t) = 0$ for some value of t.]
7. Use the Cauchy-Schwarz inequality to prove the triangle inequality for any two vectors. (Hint: Compute $\|\mathbf{x} + \mathbf{y}\|^2$.)
8. Let V be the vector space $C[0,1]$ which consists of all real-valued continuous functions defined on the interval [0,1]. For any two functions $\mathbf{f}(t)$ and $\mathbf{g}(t)$ in V define

$$\langle \mathbf{f},\mathbf{g}\rangle = \int_0^1 \mathbf{f}(t)\mathbf{g}(t)\,dt$$

   a. Let $\mathbf{f}(t) = t^2$ and $\mathbf{g}(t) = 1 - t$. Compute $\langle \mathbf{f},\mathbf{g}\rangle$.
   b. Define the length of a vector $\mathbf{f}$ in V by

$$\|\mathbf{f}\|^2 = \langle \mathbf{f},\mathbf{f}\rangle = \int_0^1 \mathbf{f}(t)\mathbf{f}(t)\,dt$$

   If $\mathbf{f}(t) = \sin n\pi t$, compute its length.
   c. Show that properties 1 and 2 of Theorem 6.1 are valid.
   d. Show that $\langle\ ,\ \rangle$ is an inner product in the sense that Theorem 6.3 is valid.
   e. Prove the Cauchy-Schwarz inequality for this inner product. (Hint: Repeat the proof of Theorem 6.4.)
9. Let $A = [a_{jk}]$ be any real $m \times n$ matrix. Define $\|A\|^2 = \Sigma_{j=1}^m \Sigma_{k=1}^n a_{jk}^2$.
   a. For any vector $\mathbf{x}$ in $\mathbb{R}^n$ show that $\|A\mathbf{x}\| \le \|\mathbf{A}\|\,\|\mathbf{x}\|$.
   b. Let $A = \begin{bmatrix} 1 & 2 \\ 3 & -1 \end{bmatrix}$. Compute $\|A\|$ and verify directly that $\|A\mathbf{x}\| \le \|A\|\,\|\mathbf{x}\|$ for any $\mathbf{x}$ in $\mathbb{R}^2$.
10. The distance between a vector $\mathbf{x}$ and a subspace W is defined to be the smallest value of $\|\mathbf{x} - \mathbf{w}\|$ as $\mathbf{w}$ varies throughout W. Let V equal $\mathbb{R}^2$ and $W = S[(1,1)]$, the span of the vector (1,1). For each of the following vectors $\mathbf{x}$ compute the distance between $\mathbf{x}$ and W.
    a. (2,2) **b.** (−1,1) **c.** (1,2)
    In each case determine that vector $\mathbf{w}_0$ in W such that $\|\mathbf{x} - \mathbf{w}_0\|$ equals the distance between $\mathbf{x}$ and W. Then show that $\mathbf{x} - \mathbf{w}_0$ is perpendicular to (1,1) and hence to every vector in $S[(1,1)] = W$.
11. Let $\mathbf{x} = (x_1, \ldots, x_n)$ be any vector in $\mathbb{R}^n$. Show that $|x_j| \le \|\mathbf{x}\|$ for any j, and $\|\mathbf{x}\| \le |x_1| + |x_2| + \cdots + |x_n|$.
12. Let $\mathbf{x}_1, \ldots, \mathbf{x}_n, \ldots$ be a sequence of vectors in $\mathbb{R}^m$. We say that the sequence $\mathbf{x}_n$ converges to $\mathbf{x}$, $\lim_{n\to\infty} \mathbf{x}_n = \mathbf{x}$, if and only if $\lim_{n\to\infty} (\|\mathbf{x}_n - \mathbf{x}\|) = 0$.

a. Let $\mathbf{x}_n = (1 + (1/n),0)$. Show that $\mathbf{x}_n$ converges to $(1,0)$.
b. Let $\mathbf{x}_n$ be a sequence of vectors in $\mathbb{R}^2$. Show that $\lim_{n\to\infty} \mathbf{x}_n = \mathbf{x}$ if and only if the components of $\mathbf{x}_n$ converge to the corresponding components of $\mathbf{x}$; cf. problem 11.

13. Let $\mathbf{x}_n = (1 + (1/n),2 - (2/n),3)$. Show that $\mathbf{x}_n$ converges to $(1,2,3)$.

14. Show that the result of problem 12b is also true in $\mathbb{R}^n$.

15. Let $A = [a_{jk}]$, $A_p = [a_{pjk}]$, $p = 1, 2, \ldots$. Define the norm of $m \times n$ matrices as in problem 9. Show that $\lim_{p\to\infty} \|A - A_p\| = 0$ if and only if $\lim_{p\to\infty} |a_{jk} - a_{pjk}| = 0$, for each j and k. Note, this is problem 14 where we identify M_{mn} with $\mathbb{R}^{mn}$.

16. Suppose that $\mathbf{x}$ is perpendicular to every vector in some set A. Show that $\mathbf{x}$ must then be perpendicular to every vector in $S[A]$.

17. Let A be any $m \times n$ matrix. Show that $A\mathbf{x} = \mathbf{0}$ if and only if $\mathbf{x}$ is perpendicular to every row of A. Thus, $\mathbf{x}$ is in $\ker(A)$ if and only if $\mathbf{x}$ is perpendicular to every vector in the row space of A.

18. Let $V = \mathbb{R}^2$. For any $\mathbf{y}$ in V the mapping $L[\mathbf{x}] = \langle \mathbf{x},\mathbf{y} \rangle$ maps V to $\mathbb{R}$.
a. Show that L is a linear transformation.
b. Using the standard bases in V and $\mathbb{R}$, what is the matrix representation of L?
c. Repeat parts a and b where V is now $\mathbb{R}^n$.

19. Let 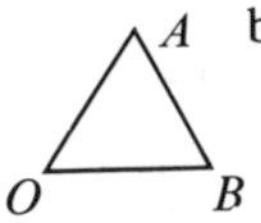

be an isosceles triangle with equal angles at O and B. Show that the line drawn from the vertex A to the midpoint of OB is perpendicular to OB.

20. Show that the diagonals of a square bisect not only each other but also each vertex angle.

6.2 PROJECTIONS AND BASES

For various reasons we sometimes wish to compute the component or projection of a vector in some particular direction. Geometrically it's easy to see how to do this. Suppose $\mathbf{x}$ is some vector in $\mathbb{R}^2$ and we wish to compute its perpendicular projection onto the direction indicated by the dashed line in Figure 6.7a.

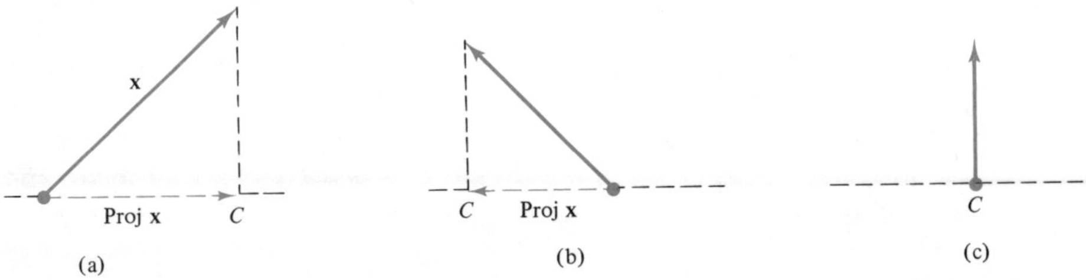

Figure 6.7

We locate a point C on this line so that the line joining the tip of **x** to C is perpendicular to the original line. The projection of **x** in the direction C is then given by the vector starting where **x** starts and ending at the point C. Figure 6.7*b* and c show two other configurations. Notice that in Fig. 6.7c, **x** is perpendicular to the dashed line. Since this forces the point C to coincide with the origin of **x**, the projection of **x** in this case is the zero vector.

In deriving a formula for the projection, we first start by assuming that the direction is given by a unit vector **u**, that is, $\|\mathbf{u}\| = 1$. Let θ denote the angle between the vector **x** and **u**. Let d denote the length of the projection; cf. Figure 6.8. Then $\cos\theta = d/\|\mathbf{x}\|$ and

$$d = \|\mathbf{x}\|\cos\theta = \|\mathbf{x}\|\frac{\langle\mathbf{x},\mathbf{u}\rangle}{\|\mathbf{x}\|\,\|\mathbf{u}\|} = \langle\mathbf{x},\mathbf{u}\rangle \tag{6.10}$$

Thus, d is merely the inner product of **x** with **u**. To get the projection of **x** onto **u**, $\mathrm{Proj}_u\mathbf{x}$, we just multiply **u** by d. We note that if **u** were not a unit vector we would have $d = \langle\mathbf{x},\mathbf{u}\rangle/\|\mathbf{u}\|$. In order to avoid having to carry along the factor $\|\mathbf{u}\|^{-1}$ we insist that **u** be a vector of length 1.

Definition 6.5. Let **x** be any vector in $\mathbb{R}^n$, and let **u** be any unit vector in $\mathbb{R}^n$. The projection of **x** onto **u**, $\mathrm{Proj}_u\mathbf{x}$, is defined to be

$$\mathrm{Proj}_u\mathbf{x} = \langle\mathbf{x},\mathbf{u}\rangle\mathbf{u} \tag{6.11}$$

EXAMPLE 1 Let $\mathbf{x} = (2,-3)$. Compute $\mathrm{Proj}_u\mathbf{x}$ for each of the following unit vectors:

a. $\mathbf{u} = (1,0)$: $\mathrm{Proj}_u(2,-3) = \langle(2,-3),(1,0)\rangle(1,0) = 2(1,0) = (2,0)$

b. $\mathbf{u} = (1/\sqrt{2},1/\sqrt{2})$: $\mathrm{Proj}_u(2,-3) = \left\langle(2,-3),\left(\frac{1}{\sqrt{2}},\frac{1}{\sqrt{2}}\right)\right\rangle\left(\frac{1}{\sqrt{2}},\frac{1}{\sqrt{2}}\right) = \frac{2-3}{\sqrt{2}}\left(\frac{1}{\sqrt{2}},\frac{1}{\sqrt{2}}\right) = -\left(\frac{1}{2},\frac{1}{2}\right)$

c. $\mathbf{u} = (0,1)$: $\mathrm{Proj}_u(2,-3) = \langle(2,-3),(0,1)\rangle(0,1) = -3(0,1) = (0,-3)$ ■

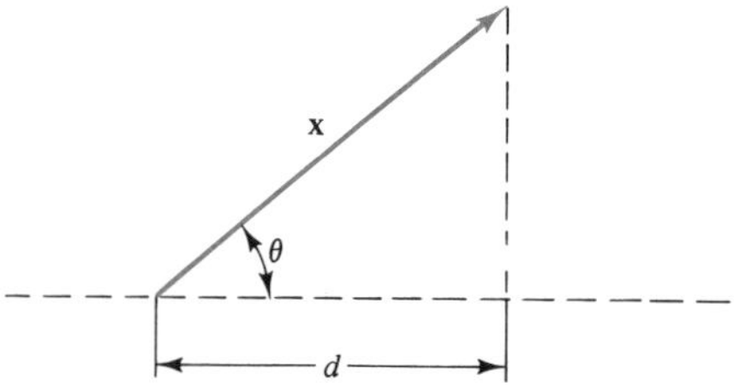

Figure 6.8

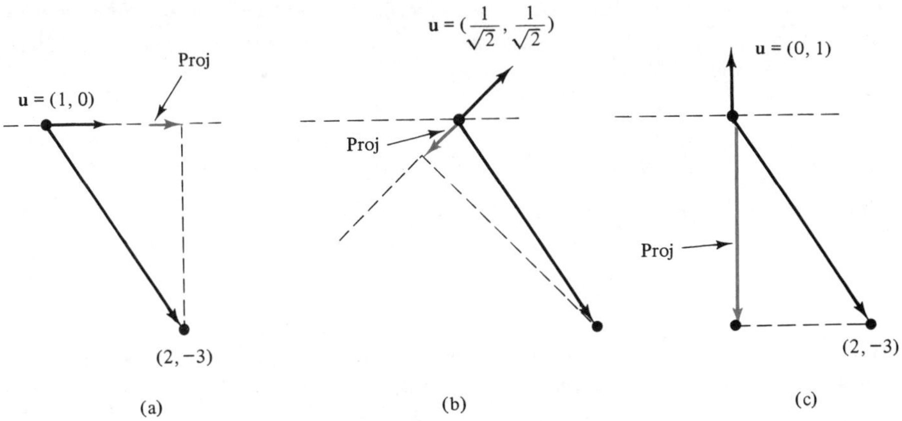

Figure 6.9

Figure 6.9 illustrates this example. It is clear, geometrically, that our construction gives us the "perpendicular component" of **x** in the direction specified by **u**. The following lemma shows that this is indeed true.

Lemma 6.1. Let **x** be any vector in $\mathbb{R}^n$, and let **u** be a unit vector. Then $\mathbf{x} - \text{Proj}_u\mathbf{x}$ is either the zero vector or it is perpendicular to **u**.

Proof

$$\begin{aligned}\langle \mathbf{x} - \text{Proj}_u\mathbf{x},\mathbf{u}\rangle &= \langle \mathbf{x},\mathbf{u}\rangle - \langle \text{Proj}_u\mathbf{x},\mathbf{u}\rangle \\ &= \langle \mathbf{x},\mathbf{u}\rangle - \langle \langle \mathbf{x},\mathbf{u}\rangle\mathbf{u},\mathbf{u}\rangle \\ &= \langle \mathbf{x},\mathbf{u}\rangle - \langle \mathbf{x},\mathbf{u}\rangle\langle \mathbf{u},\mathbf{u}\rangle \\ &= \langle \mathbf{x},\mathbf{u}\rangle - \langle \mathbf{x},\mathbf{u}\rangle = 0\end{aligned}$$

Remember that **u** is assumed to be a unit vector and therefore $\langle \mathbf{u},\mathbf{u}\rangle = 1$. Thus, if $\mathbf{x} - \text{Proj}_u\mathbf{x}$ is not the zero vector it must be perpendicular to **u**. See Figure 6.10.

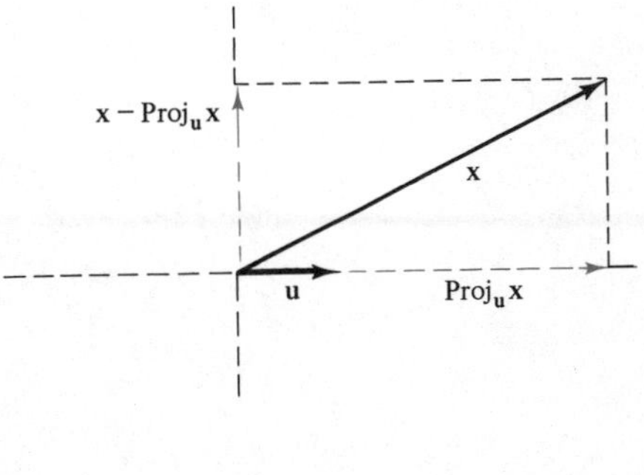

Figure 6.10

Lemma 6.1 tells us that, given any vector **x** and any unit vector **u**, we can write **x** as the sum of two vectors, one parallel to **u** and the other perpendicular to **u**.

$$\mathbf{x} = \text{Proj}_u\mathbf{x} + [\mathbf{x} - \text{Proj}_u\mathbf{x}] \tag{6.12}$$

Why is $\text{Proj}_u\mathbf{x}$ parallel to **u**?

EXAMPLE 2 Write $\mathbf{x} = (4,2)$ as the sum of two vectors, one of which is parallel to the line joining the two points $P = (-6,4)$ and $Q = (3,-5)$, while the second vector is perpendicular to this line. See Figure 6.11. Let $\mathbf{y} = (-6 - 3, 4 - (-5)) = (-9,9)$. Since we found **y** by subtracting the coordinates of Q from those of P, **y** is a vector parallel to the line through the two points P and Q. Thus a unit vector parallel to this line is $\mathbf{y}/\|\mathbf{y}\| = (1/\sqrt{2})(-1,1) = \mathbf{u}$.

A vector parallel to **u** is $\text{Proj}_u(\mathbf{x}) = \langle(4,2),(-1,1)/\sqrt{2}\rangle\mathbf{u} = (1,-1)$. A vector perpendicular to the line is $\mathbf{x} - \text{Proj}_u\mathbf{x} = (4,2) - (1,-1) = (3,3)$. Thus, $\mathbf{x} = (4,2) = (1,-1) + (3,3)$, where $(1,-1)$ is parallel to the line and $(3,3)$ is perpendicular to the line. ■

We next want to relate these ideas to those involving the coordinates of a vector. Referring to Example 1a and c, notice that the vectors $\mathbf{u}_1 = (1,0)$ and $\mathbf{u}_2 = (0,1)$ are our standard basis. For $\mathbf{x} = (2,-3)$ we had $\text{Proj}_{\mathbf{u}_1}\mathbf{x} = 2\mathbf{u}_1$ and $\text{Proj}_{u_2}\mathbf{x} = -3\mathbf{u}_2$. In other words, the coordinates of **x** with respect to the standard basis can be found by taking the dot product of **x** with each of the basis vectors. That may not happen for an arbitrary basis, as Example 3 shows.

EXAMPLE 3 The pair $\mathbf{u}_1 = (1,1)/\sqrt{2}$ and $\mathbf{u}_2 = (1,2)/\sqrt{5}$ form a basis of $\mathbb{R}^2$. Moreover each of them has length 1. Find the coordinates of $\mathbf{x} = (2,-3)$ with respect to this basis and also compute the inner product of **x** with $\mathbf{u}_1$ and $\mathbf{u}_2$. An

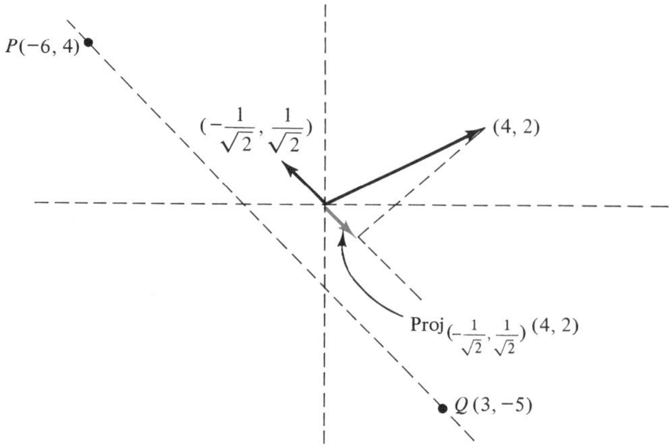

Figure 6.11

easy calculation shows that $\mathbf{x} = 7(\sqrt{2})\mathbf{u}_1 - 5(\sqrt{5})\mathbf{u}_2$. Notice though that $\langle \mathbf{x},\mathbf{u}_1\rangle = -1/\sqrt{2}$, and $\langle \mathbf{x},\mathbf{u}_2\rangle = -4/\sqrt{5}$. Thus, the inner products do not equal the coordinates of **x** with respect to this basis. Actually, we shouldn't have expected any such relationship because the two vectors $\mathbf{u}_1$ and $\mathbf{u}_2$ are not perpendicular and the inner product of **x** with $\mathbf{u}_1$ gives us the size of the perpendicular projection of **x** onto $\mathbf{u}_1$. ■

With the above example in mind we might expect that if $U = \{\mathbf{u}_1,\mathbf{u}_2\}$ consists of two perpendicular unit vectors in $\mathbb{R}^2$, then $[\mathbf{x}]_u = [\langle \mathbf{x},\mathbf{u}_1\rangle, \langle \mathbf{x},\mathbf{u}_2\rangle]$. Indeed, we will prove that

$$\mathbf{x} = \langle \mathbf{x},\mathbf{u}_1\rangle \mathbf{u}_1 + \langle \mathbf{x},\mathbf{u}_2\rangle \mathbf{u}_2 \tag{6.13}$$

Thus, suppose that $\mathbf{U} = \{\mathbf{u}_1,\mathbf{u}_2\}$ is a basis of $\mathbb{R}^2$ that consists of two perpendicular unit vectors. Then $\mathbf{x} = c_1\mathbf{u}_1 + c_2\mathbf{u}_2$. Let's now take the inner product of **x** with $\mathbf{u}_1$ and then with $\mathbf{u}_2$.

$$\langle \mathbf{x},\mathbf{u}_1\rangle = \langle c_1\mathbf{u}_1 + c_2\mathbf{u}_2,\mathbf{u}_1\rangle = c_1\langle \mathbf{u}_1,\mathbf{u}_1\rangle + c_2\langle \mathbf{u}_2,\mathbf{u}_1\rangle = c_1$$

since $\langle \mathbf{u}_1,\mathbf{u}_1\rangle = 1$ and $\langle \mathbf{u}_1,\mathbf{u}_2\rangle = 0$. Similarly we have $\langle \mathbf{x},\mathbf{u}_2\rangle = c_2$. Thus, (6.13) is valid, and as we shall see in a short while, it is also valid in $\mathbb{R}^n$.

Definition 6.6. A set of nonzero vectors $\{\mathbf{u}_j : j = 1, \ldots, p\}$ in $\mathbb{R}^n$ is said to be orthogonal if they are mutually perpendicular, i.e., $\langle \mathbf{u}_j,\mathbf{u}_k\rangle = 0$ if $j \neq k$.

EXAMPLE 4 The set $\{(1,1,0,0), (0,0,1,1), (1,-1,1,-1)\}$ is orthogonal since $\langle (1,1,0,0), (0,0,1,1)\rangle = 0$, $\langle (1,1,0,0), (1,-1,1,-1)\rangle = 0$, and $\langle (0,0,1,1), (1,-1,1,-1)\rangle = 0$. However, the set $\{(1,1,1), (1,-1,0), (1,1,2)\}$ is not orthogonal since $\langle (1,1,1), (1,1,2)\rangle$ equals 4, not zero. ■

Lemma 6.2. Any set of orthogonal vectors must be linearly independent.

Proof. Let $\{\mathbf{u}_k : k = 1, \ldots, p\}$ be an orthogonal set of vectors and suppose that we have constants c_j such that

$$\mathbf{0} = c_1\mathbf{u}_1 + \cdots + c_p\mathbf{u}_p \tag{6.14}$$

Taking the inner product of (6.14) with $\mathbf{u}_1$ we have

$$\begin{aligned} 0 = \langle \mathbf{0},\mathbf{u}_1\rangle &= \left\langle \sum_{j=1}^{p} c_j\mathbf{u}_j,\mathbf{u}_1 \right\rangle \\ &= \sum_{j=1}^{p} c_j\langle \mathbf{u}_j,\mathbf{u}_1\rangle \\ &= c_1\langle \mathbf{u}_1,\mathbf{u}_1\rangle \end{aligned}$$

Since $\mathbf{u}_1$ is not the zero vector, we know that $\langle \mathbf{u}_1,\mathbf{u}_1 \rangle \neq 0$. Thus $c_1 = 0$. By taking the inner product of (6.14) with any one of the $\mathbf{u}_k$'s, we similarly see that $c_k = 0$ for each k. Thus, our orthogonal set of vectors is linearly independent.

Definition 6.7. A set of vectors $U = \{\mathbf{u}_1, \ldots, \mathbf{u}_n\}$ is said to be an orthonormal basis of $\mathbb{R}^n$ if it is a basis consisting of orthogonal unit vectors. Thus, we have $\langle \mathbf{u}_j,\mathbf{u}_k \rangle = \delta_{jk}$.

The following is probably the most useful idea in this section.

Theorem 6.5. Let $U = \{\mathbf{u}_1, \ldots, \mathbf{u}_n\}$ be an orthonormal basis of $\mathbb{R}^n$. Then the coordinates of any vector $\mathbf{x}$ can be found by taking the inner product of $\mathbf{x}$ with each of the basis vectors $\mathbf{u}_k$. That is,

$$\mathbf{x} = \langle \mathbf{x},\mathbf{u}_1 \rangle \mathbf{u}_1 + \cdots + \langle \mathbf{x},\mathbf{u}_n \rangle \mathbf{u}_n \tag{6.15}$$

Proof. Since U is a basis we know there are unique constants c_k, $1 \leq k \leq n$, such that

$$\mathbf{x} = c_1\mathbf{u}_1 + \cdots + c_n\mathbf{u}_n \tag{6.16}$$

Taking the dot product of both sides of (6.16) with the kth basis vector $\mathbf{u}_k$, we have

$$\begin{aligned} \langle \mathbf{x},\mathbf{u}_k \rangle &= \left\langle \sum_{j=1}^{n} c_j\mathbf{u}_j,\mathbf{u}_k \right\rangle \\ &= \sum_{j=1}^{n} c_j \langle \mathbf{u}_j,\mathbf{u}_k \rangle = \sum_{j=1}^{n} c_j \delta_{jk} \\ &= c_k \end{aligned}$$

Thus, each coordinate of $\mathbf{x}$ with respect to U is the inner product of $\mathbf{x}$ with the corresponding basis vector in U. Another interpretation of (6.15) is that a vector equals the sum of its projections onto the vectors of an orthonormal basis.

EXAMPLE 5 Verify that each of the following is an orthonormal basis of $\mathbb{R}^3$, and then compute the coordinates of the vector $\mathbf{x} = (6,-2,1)$ with respect to these bases.

a. $U = \{(1,0,0), (0,1,0), (0,0,1)\}$. Clearly each of these vectors has length 1, and they are mutually perpendicular. Thus, (6.15) is applicable. Computing the inner product of x with each basis vector we have

$$\begin{aligned} \langle (6,-2,1), (1,0,0) \rangle &= 6 \\ \langle (6,-2,1), (0,1,0) \rangle &= -2 \\ \langle (6,-2,1), (0,0,1) \rangle &= 1 \end{aligned}$$

Thus, $\mathbf{x}$ is equal to

$$\begin{aligned}\mathbf{x} &= (6,-2,1) = 6\mathbf{u}_1 + (-2)\mathbf{u}_2 + 1\mathbf{u}_3 \\ &= 6(1,0,0) - 2(0,1,0) + (0,0,1)\end{aligned}$$

b. $U = \{(1,1,1)/\sqrt{3},\ (2,-1,-1)/\sqrt{6},\ (0,1,-1)/\sqrt{2}\}$. We first verify that U is an orthonormal set.

$$\begin{aligned}\|\mathbf{u}_1\|^2 &= \tfrac{1}{3} + \tfrac{1}{3} + \tfrac{1}{3} = 1 \qquad & \langle \mathbf{u}_1,\mathbf{u}_2\rangle &= (2-1-1)/\sqrt{18} = 0 \\ \|\mathbf{u}_2\|^2 &= \tfrac{4}{6} + \tfrac{1}{6} + \tfrac{1}{6} = 1 & \langle \mathbf{u}_1,\mathbf{u}_3\rangle &= (1-1)/\sqrt{6} = 0 \\ \|\mathbf{u}_3\|^2 &= \tfrac{1}{2} + \tfrac{1}{2} = 1 & \langle \mathbf{u}_2,\mathbf{u}_3\rangle &= (-1+1)/\sqrt{12} = 0\end{aligned}$$

Thus, U consists of mutually orthogonal unit vectors. Lemma 6.2 tells us that U is linearly independent. Since $\dim(\mathbb{R}^3)$ equals 3, U must be a basis. Computing the inner products of $\mathbf{x}$ with each of the basis vectors we have

$$\langle (6,-2,1),\ (1,1,1)/\sqrt{3}\ \rangle = \frac{6-2+1}{\sqrt{3}} = \frac{5}{\sqrt{3}}$$

$$\langle (6,-2,1),\ (2,-1,-1)/\sqrt{6}\rangle = \frac{12+2-1}{\sqrt{6}} = \frac{13}{\sqrt{6}}$$

$$\langle (6,-2,1),\ (0,1,-1)/\sqrt{2}\rangle = \frac{-2-1}{\sqrt{2}} = \frac{-3}{\sqrt{2}}$$

Thus, $\mathbf{x} = (5/\sqrt{3})\mathbf{u}_1 + (13/\sqrt{6})\mathbf{u}_2 - (3/\sqrt{2})\mathbf{u}_3$. ■

In order to appreciate the convenience of an orthonormal basis, the reader should compute the coordinates of $\mathbf{x}$ with respect to U without using Theorem 6.5.

Another fact, which is sometimes useful, is the relationship between the inner product of two vectors and their coordinates with respect to an orthonormal basis.

Theorem 6.6. Let $U = \{\mathbf{u}_j : j = 1, \ldots, n\}$ be any orthonormal basis of $\mathbb{R}^n$. Let $\mathbf{x}$ and $\mathbf{y}$ be any two vectors. Their coordinates with respect to the orthonormal basis U are $[x_1, \ldots, x_n]_U$ and $[y_1, \ldots, y_n]_U$, respectively. Then

$$\langle \mathbf{x},\mathbf{y}\rangle = x_1y_1 + \cdots + x_ny_n = \sum_{j=1}^{n} x_jy_j \tag{6.17a}$$

$$\|\mathbf{x}\|^2 = \sum_{j=1}^{n} x_j^2 \tag{6.17b}$$

Proof. By hypothesis $\mathbf{x} = \Sigma_{j=1}^{n} x_j\mathbf{u}_j$ and $\mathbf{y} = \Sigma_{j=1}^{n} y_j\mathbf{u}_j$. Thus

$$\begin{aligned}\langle \mathbf{x},\mathbf{y}\rangle &= \left\langle \sum_{j=1}^{n} x_j\mathbf{u}_j, \sum_{k=1}^{n} y_k\mathbf{u}_k \right\rangle \\ &= \sum_{j=1}^{n}\sum_{k=1}^{n} x_jy_k\langle \mathbf{u}_j,\mathbf{u}_k\rangle \\ &= \sum_{j=1}^{n}\sum_{k=1}^{n} x_jy_k\delta_{jk} = \sum_{j=1}^{n} x_jy_j\end{aligned}$$

To verify (6.17b) we only need to remember that

$$\|\mathbf{x}\|^2 = \langle \mathbf{x},\mathbf{x}\rangle = \sum_{j=1}^{n} x_j^2$$

EXAMPLE 6 $U = \{(1,1)/\sqrt{2}, (-1,1)/\sqrt{2}\}$ is an orthonormal basis of $\mathbb{R}^2$. Verify (6.17) for the vectors (1,1) and (2,6).

Solution. The inner product of these two vectors is

$$\langle (1,1), (2,6)\rangle = 2 + 6 = 8$$

Using Theorem 6.5 to compute the coordinates of these vectors with respect to U, we have

$$\begin{aligned}[(1,1)]_u &= [\langle(1,1), (1,1)/\sqrt{2}\rangle, \langle(1,1), (-1,1)/\sqrt{2}\rangle] = [\sqrt{2},0] \\ [(2,6)]_u &= [\langle(2,6), (1,1)/\sqrt{2}\rangle, \langle(2,6), (-1,1)/\sqrt{2}\rangle] \\ &= [8/\sqrt{2}, 4/\sqrt{2}]\end{aligned}$$

Thus, $x_1 = \sqrt{2}$, $x_2 = 0$, $y_1 = 8/\sqrt{2}$, and $y_2 = 4/\sqrt{2}$, and we have

$$x_1y_1 + x_2y_2 = 8 + 0 = \langle \mathbf{x},\mathbf{y}\rangle$$

Using the coordinates of **x** and **y** with respect to U to compute their lengths, we have

$$\|\mathbf{x}\|^2 = (\sqrt{2})^2 + 0 = 2 \qquad \|\mathbf{y}\|^2 = \tfrac{64}{2} + \tfrac{16}{2} = 40$$

■

PROBLEM SET 6.2

1. Compute $\text{Proj}_u\mathbf{x}$, where $\mathbf{x} = (7,-8)$ for each of the following unit vectors:
   a. $(1,-2)/5^{1/2}$ **b.** $(2,3)/(13)^{1/2}$ **c.** $(1,0)$
2. Compute the projection of $\mathbf{x} = (-2,3)$ in a direction parallel to the straight line joining the points (1,7) and $(-3,8)$. There are two possible choices for a direction; take the one that points from (1,7) to $(-3,8)$.
3. Let $\mathbf{x} = (7,-5)$. Let $U = \{(1,5)/\sqrt{26}, (-5,1)/\sqrt{26}\}$.
   a. Show that U is an orthonormal basis of $\mathbb{R}^2$.
   b. Find $\text{Proj}_{u_j}\mathbf{x}$, where $\mathbf{u}_j$ is the jth unit vector in U.
   c. Compute the coordinates of **x** with respect to U.

4. Let $\mathbf{x} = (1,-2)$. Show that any vector orthogonal to $\mathbf{x}$ is a scalar multiple of $(2,1)$.

5. Let $A = \{(1,2,-1), (-2,1,0)\}$. Show that A is an orthogonal set of vectors, and if $\mathbf{x}$ is any vector orthogonal to both vectors in A, then $\mathbf{x}$ must be a scalar multiple of $(1,2,5)$.

6. Let $V = \{(2,-3,1),(2,3,5),(-9,-4,6)\}$.
   a. Show that V is an orthogonal set of vectors.
   b. Let $\mathbf{x} = (7,-3,4)$. Compute the projection of $\mathbf{x}$ onto the direction given by $\mathbf{v}_j$, where $\mathbf{v}_j$ is the jth vector in the set V.
   c. Compute the coordinates of $(7,-3,4) = \mathbf{x}$ with respect to the basis V. (Hint: It's easy to construct an orthonormal basis from V.)

7. Find the angle between the following pairs of vectors:
   a. $(1,1)$, $(0,1)$ **b.** $(1,1,1)$, $(0,1,0)$
   c. $(1,1,1,1)$, $(0,1,0,0)$ **d.** $(6,7,-2,3)$, $(-1,-2,1,1)$

8. Let $U = \{(1,-1)/\sqrt{2}, (1,1)/\sqrt{2}\}$. Use the fact that U is an orthonormal basis to compute the coordinates of the following vectors with respect to U:
   a. $(9,-2)$ **b.** $(6,4)$ **c.** $(1,-1)$ **d.** $(1,0)$

9. Let $U = \{(2,-3,1)/\sqrt{14}, (2,3,5)/\sqrt{38}, (-9,4,6)/\sqrt{133}\}$. Show that U is an orthonormal basis (cf. problem 6) and then compute $[\mathbf{x}]_u$ for the following vectors:
   a. $(1,0,0)$ **b.** $(-1,6,4)$ **c.** $(18,2,-4)$

10. Let $\mathbf{u}$ be an arbitrary unit vector in $\mathbb{R}^n$.
    a. If $\mathbf{x}$ is the zero vector, show that $\text{Proj}_u\mathbf{x} = \mathbf{0}$.
    b. If $\mathbf{x}$ and $\mathbf{u}$ are perpendicular, show that $\text{Proj}_u\mathbf{x} = \mathbf{0}$.
    c. Show that $\text{Proj}_u\mathbf{x}$ is a linear transformation from $\mathbb{R}^n$ to $\mathbb{R}^n$.
    d. What is the dimension of the kernel of this linear transformation?

11. Let $V = P_3$. Let $\mathbf{f}$ and $\mathbf{g}$ be any two polynomials in V. Define $\langle \mathbf{f},\mathbf{g}\rangle = \int_0^1 \mathbf{f}(t)\mathbf{g}(t)\,dt$; cf. problem 8 in Section 6.1.
    a. Find a unit vector $\mathbf{u}$ that points in the same direction as $\mathbf{f}(t) = t$.
    b. Find the projection of $\mathbf{t}^2$ onto the vector $\mathbf{u}$ of part **a**.
    c. Find the cosine of the angle between the vectors $\mathbf{t}^2$ and $\mathbf{t}$.

12. Let $V = P_1$. Define the inner product of two vectors as we did in problem 11. Show that $\{\mathbf{1},\mathbf{t} - \frac{1}{2}\}$ is an orthogonal set of vectors. Find an orthonormal basis for V.

13. Let $V = P_2$. Define the inner product as we did in problems 11 and 12. Let $\mathbf{f}'$ denote the derivative of $\mathbf{f}$.
    a. Find all polynomials in P_2 that are perpendicular to their derivatives.
    b. For any two polynomials f and g in V, compute $\langle \mathbf{f},\mathbf{g}'\rangle - \langle \mathbf{f}',\mathbf{g}\rangle$.

14. Let $V = M_{22}$, the vector space of 2×2 matrices. For $A = [a_{jk}]$ and $B = [b_{jk}]$ in M_{22} define $\langle A,B\rangle = \Sigma_{j=1}^2\Sigma_{k=1}^2 a_{jk}b_{jk}$.
    a. Compute the norms of the matrices $\begin{bmatrix} 1 & 0 \\ 0 & 1 \end{bmatrix}$ and $\begin{bmatrix} a & b \\ c & d \end{bmatrix}$
    b. Let E_{jk}, $1 \leq j, k \leq 2$, be the standard basis of V. Do these four matrices form an orthonormal basis?

15. In problem 10 we saw that for a fixed unit vector $\mathbf{u}$ in $\mathbb{R}^n$, $L[\mathbf{x}] = \text{Proj}_u\mathbf{x}$ is a linear transformation from $\mathbb{R}^n$ to $\mathbb{R}^n$. Let

$$\mathbf{u} = \frac{1}{\sqrt{2}}(\mathbf{e}_1 - \mathbf{e}_3) \qquad n \geq 3$$

a. What is the matrix representation of L with respect to the standard basis?
b. Find an orthonormal basis for Rg(L).
c. Find an orthonormal basis for ker(L).
d. Show that the union of the two orthonormal sets in parts **b** and **c** is an orthonormal basis of $\mathbb{R}^n$.
e. What is the matrix representation of L with respect to the basis of part d? (List the vectors from b and then the vectors from c.)

6.3 CONSTRUCTION OF ORTHONORMAL BASES

We indicated in Chapter 2 that every vector space has a basis. A natural question now is whether or not every vector space has an orthonormal basis. This of course makes sense only if the vector space has an inner product. Clearly, the answer is yes for $\mathbb{R}^n$ since the standard basis $\{\mathbf{e}_1, \ldots, \mathbf{e}_n\}$ is orthonormal. What about subspaces of $\mathbb{R}^n$? The answer is again yes. In fact there is a technique for constructing an orthonormal basis from any given basis. This technique goes by the name of Gram-Schmidt. We illustrate it with an example before going into the details of the algorithm.

EXAMPLE 1 Let $\mathbf{f}_1 = (0,1,1)$ and $\mathbf{f}_2 = (0,2,0)$. Let $W = S[\mathbf{f}_1, \mathbf{f}_2]$. Construct an orthonormal basis for W.

Solution. Geometrically W is a plane (two-dimensional subspace of $\mathbb{R}^3$). In fact W is the plane $x_1 = 0$. Clearly $\mathbf{e}_2$ and $\mathbf{e}_3$ form an orthonormal basis for W. What we wish to do, though, is to use the given basis for W in constructing our orthonormal basis. We first set $\mathbf{u}_1 = \mathbf{f}_1/\|\mathbf{f}_1\| = (0,1,1)/\sqrt{2}$. The unit vector $\mathbf{u}_1$ will be the first vector in our basis. We now want a unit vector $\mathbf{u}_2$ that is perpendicular to $\mathbf{u}_1$ and also lies in W. Such a vector is easy to construct by using the fact that $\mathbf{f}_2 - \text{Proj}_{u_1}\mathbf{f}_2$ must be perpendicular to $\mathbf{u}_1$. Moreover, since this vector is a linear combination of $\mathbf{f}_1$ and $\mathbf{f}_2$ ($\mathbf{u}_1$ is a scalar multiple of $\mathbf{f}_1$), it will lie in W. Thus, set

$$\mathbf{v}_2 = \mathbf{f}_2 - \text{Proj}_{\mathbf{u}_1}\mathbf{f}_2$$
$$k = 2, \cdots, n$$
$$= (0,2,0) - \langle (0,2,0), (0,1,1)/\sqrt{2}\rangle \frac{(0,1,1)}{\sqrt{2}}$$
$$= (0,2,0) - (0,1,1) = (0,1,-1)$$
$$\mathbf{u}_2 = \frac{\mathbf{v}_2}{\|\mathbf{v}_2\|} = \frac{(0,1,-1)}{\sqrt{2}}$$

A quick computation shows that $\langle \mathbf{u}_1, \mathbf{u}_2\rangle = 0$.

The only difficulty here is that the vector $\mathbf{v}_2$ might equal zero. But this cannot happen, since the vectors $\mathbf{f}_1$ and $\mathbf{f}_2$ are linearly independent. ■

Let's assume now that $\{\mathbf{f}_1, \ldots, \mathbf{f}_n\}$ is a set of linearly independent vectors. The Gram-Schmidt procedure given below provides us with an orthonormal set of vectors $\{\mathbf{u}_1, \ldots, \mathbf{u}_n\}$, such that $S[\mathbf{u}_1, \ldots, \mathbf{u}_p] = S[\mathbf{f}_1, \ldots, \mathbf{f}_p]$, for $p = 1, 2, \ldots, n$. Define the unit vectors $\mathbf{u}_k$ inductively by

$$\begin{aligned}
\mathbf{u}_1 &= \frac{\mathbf{f}_1}{\|\mathbf{f}_1\|} \\
\mathbf{v}_2 &= \mathbf{f}_2 - \langle \mathbf{f}_2, \mathbf{u}_1 \rangle\, \mathbf{u}_1 \\
\mathbf{u}_2 &= \frac{\mathbf{v}_2}{\|\mathbf{v}_2\|} \\
\mathbf{v}_k &= \mathbf{f}_k - [\langle \mathbf{f}_k, \mathbf{u}_1 \rangle \mathbf{u}_1 + \langle \mathbf{f}_k, \mathbf{u}_2 \rangle \mathbf{u}_2 + \cdots + \langle \mathbf{f}_k, \mathbf{u}_{k-1} \rangle \mathbf{u}_{k-1}] \\
\mathbf{u}_k &= \frac{\mathbf{v}_k}{\|\mathbf{v}_k\|} \qquad k = 2, \cdots, n
\end{aligned} \tag{6.18}$$

Theorem 6.7. Let $\{\mathbf{f}_k : k = 1, \cdots, n\}$ be a linearly independent set of vectors. Define $\mathbf{u}_k$ by (6.18). Then $\{\mathbf{u}_k : k = 1, \ldots, n\}$ is an orthonormal set or vectors and $S[\mathbf{u}_1, \ldots, \mathbf{u}_p] = S[\mathbf{f}_1, \ldots, \mathbf{f}_p]$ for $p = 1, \ldots, n$.

Proof. We prove this theorem by induction. For $p = 1$, we have $\mathbf{u}_1 = \mathbf{f}_1/\|\mathbf{f}_1\|$. Clearly $\{\mathbf{u}_1\}$ is an orthonormal set and $S[\mathbf{u}_1] = S[\mathbf{f}_1]$. Again we note that $\mathbf{f}_1 \neq 0$ since the $\mathbf{f}_j$'s are linearly independent. We now assume that the theorem is true for p and deduce its truth for $p + 1$. From (6.18) we have

$$\mathbf{v}_{p+1} = \mathbf{f}_{p+1} - [\mathrm{Proj}_{\mathbf{u}_1}(\mathbf{f}_{p+1}) + \cdots + \mathrm{Proj}_{\mathbf{u}_p}(\mathbf{f}_{p+1})] \tag{6.19a}$$
$$\mathbf{u}_{p+1} = \mathbf{v}_{p+1}\|\mathbf{v}_{p+1}\| \tag{6.19b}$$

We have by assumption that $S[\mathbf{u}_1, \ldots, \mathbf{u}_p] = S[\mathbf{f}_1, \ldots, \mathbf{f}_p]$. Thus, the vector $\mathbf{v}_{p+1}$ cannot be the zero vector since $\mathbf{f}_{p+1}$ is not in $S[\mathbf{f}_1, \ldots, \mathbf{f}_p] = S[\mathbf{u}_1, \ldots, \mathbf{u}_p]$. Hence we may divide $\mathbf{v}_{p+1}$ by its length to get $\mathbf{u}_{p+1}$, a unit vector. Since (6.19a) can be solved for $\mathbf{f}_{p+1}$, the reader can easily show that $S[\mathbf{u}_1, \ldots, \mathbf{u}_{p+1}] = S[\mathbf{f}_1, \ldots, \mathbf{f}_{p+1}]$. It remains to show that $\mathbf{u}_{p+1}$, equivalently $\mathbf{v}_{p+1}$, is orthogonal to each of the preceding $\mathbf{u}_k$.

$$\begin{aligned}
\langle \mathbf{v}_{p+1}, \mathbf{u}_k \rangle &= \langle \mathbf{f}_{p+1} - \mathrm{Proj}_{u_1}\mathbf{f}_{p+1} - \mathrm{Proj}_{u_2}\mathbf{f}_{p+1} - \cdots - \mathrm{Proj}_{u_p}\mathbf{f}_{p+1}, \mathbf{u}_k \rangle \\
&= \left\langle \mathbf{f}_{p+1} - \sum_{j=1}^{p} \langle \mathbf{f}_{p+1}, \mathbf{u}_j \rangle \mathbf{u}_j, \mathbf{u}_k \right\rangle \\
&= \langle \mathbf{f}_{p+1}, \mathbf{u}_k \rangle - \sum_{j=1}^{p} \langle \mathbf{f}_{p+1}, \mathbf{u}_j \rangle \langle \mathbf{u}_j, \mathbf{u}_k \rangle
\end{aligned}$$

By assumption the set $\{\mathbf{u}_k: k = 1, \cdots, p\}$ is orthonormal. Thus $\langle \mathbf{u}_j, \mathbf{u}_k \rangle = \delta_{jk}$ and we have

$$\langle \mathbf{v}_{p+1}, \mathbf{u}_k \rangle = \langle \mathbf{f}_{p+1}, \mathbf{u}_k \rangle - \langle \mathbf{f}_{p+1}, \mathbf{u}_k \rangle = 0$$

EXAMPLE 2 Construct an orthonormal basis for $\mathbb{R}^3$ from the vectors {(1,0,1),(2,1,0),(1,1,1) by using the Gram-Schmidt algorithm

$$\mathbf{u}_1 = \frac{(1,0,1)}{\sqrt{2}}$$

$$\begin{aligned}\mathbf{v}_2 &= (2,1,0) - \langle(2,1,0), (1,0,1)/\sqrt{2}\rangle((1,0,1)/\sqrt{2}) \\ &= (2,1,0) - (1,0,1) = (1,1,-1)\end{aligned}$$

$$\mathbf{u}_2 = \frac{(1,1,-1)}{\sqrt{3}}$$

$$\begin{aligned}\mathbf{v}_3 &= (1,1,1) - \langle(1,1,1), (1,0,1)/\sqrt{2}\rangle((1,0,1)/\sqrt{2}) \\ &\quad - \langle(1,1,1),(1,1,-1)/\sqrt{3}\rangle((1,1,-1)/\sqrt{3}) \\ &= (1,1,1) - (1,0,1) - \frac{(1,1,-1)}{3} = \frac{(-1,2,1)}{3}\end{aligned}$$

$$\mathbf{u}_3 = \frac{(-1,2,1)}{\sqrt{6}}$$

■

In the preceding section we defined and showed how to calculate the projection of a vector onto a unit vector. We now wish to define the projection of a vector onto a subspace, and we do so in terms of the distance between the vector and the subspace.

Definition 6.8. Let $\mathbf{x}$ be any vector in $\mathbb{R}^n$ and let W be any subspace of $\mathbb{R}^n$. The projection of $\mathbf{x}$ onto W, $\text{Proj}_w(\mathbf{x})$, is defined to be that vector $\mathbf{y}$ in W which minimizes $\|\mathbf{x} - \mathbf{w}\|$ for $\mathbf{w}$ any vector in W.

It is not at all clear that there is such a vector; or perhaps there might be more than one, and which should we pick?

Theorem 6.8. Let W be any m-dimensional subspace of $\mathbb{R}^n$. Let $U = \{\mathbf{u}_1, \ldots, \mathbf{u}_m\}$ be any orthonormal basis of W. Then for any vector $\mathbf{x}$ in $\mathbb{R}^n$ there is a unique $\mathbf{y}$ that minimizes $\|\mathbf{x} - \mathbf{w}\|$ for $\mathbf{w}$ in W. Moreover,

$$\mathbf{y} = \text{Proj}_w(\mathbf{x}) = \sum_{k=1}^{m} \langle\mathbf{x},\mathbf{u}_k\rangle\mathbf{u}_k \tag{6.20}$$

Proof. Let $U = \{\mathbf{u}_1, \ldots, \mathbf{u}_m, \mathbf{v}_1, \ldots, \mathbf{v}_{n-m}\}$ be any orthonormal basis of $\mathbb{R}^n$ whose first m vectors are the given orthonormal basis of W. Theorem 6.7 tells us how to construct such a basis, and we also have

$$\mathbf{x} = \sum_{k=1}^{m} \langle\mathbf{x},\mathbf{u}_k\rangle\mathbf{u}_k + \sum_{k=1}^{n-m} \langle\mathbf{x},\mathbf{v}_k\rangle\mathbf{v}_k$$

Let $\mathbf{w}$ be any vector in W. Then $\mathbf{w} = \Sigma_{k=1}^{m}\langle \mathbf{w},\mathbf{u}_k\rangle \mathbf{u}_k$. From (6.17) we have

$$\|\mathbf{x}-\mathbf{w}\|^2 = \sum_{k=1}^{m} [\langle \mathbf{x}-\mathbf{w},\mathbf{u}_k\rangle]^2 + \sum_{k=1}^{n-m} [\langle \mathbf{x},\mathbf{v}_k\rangle]^2$$

Clearly, the second sum is constant regardless of the choice of $\mathbf{w}$. However, the first sum equals zero if and only if $\langle \mathbf{w},\mathbf{u}_k\rangle = \langle \mathbf{x},\mathbf{u}_k\rangle$ for each k. In other words, the minimum value of $\|\mathbf{x}-\mathbf{w}\|$ occurs only when $\mathbf{w} = \Sigma_{k=1}^{m} \langle \mathbf{x},\mathbf{u}_k\rangle \mathbf{u}_k$. Moreover, it is clear from the above equation that this minimum length equals

$$\|\mathbf{x} - \text{Proj}_{\mathbf{w}}\mathbf{x}\| = \left[\sum_{k=1}^{n-m} \langle \mathbf{x},\mathbf{v}_k\rangle^2\right]^{1/2}$$

and that $\mathbf{x} - \text{Proj}_w(\mathbf{x})$ is perpendicular to every vector in W.

Formula (6.20) also says that the projection of any vector $\mathbf{x}$ onto a subspace W equals the sum of its projections onto the vectors of an orthonormal basis of W. If $\mathbf{x}$ is an arbitrary vector in $\mathbb{R}^n$, what is the projection of $\mathbf{x}$ onto $\mathbb{R}^n$?

Definition 6.9. We define the distance between a vector $\mathbf{x}$ and a subspace W to equal $\|\mathbf{x} - \text{Proj}_{\mathbf{w}}\mathbf{x}\|$.

EXAMPLE 3 Let $\mathbf{x} = (1,2,3)$. Compute the projection of $\mathbf{x}$ onto each of the following subspaces.

a. W is the x_1, x_2 plane in $\mathbb{R}^3$. An orthonormal basis for W is the set $\{(1,0,0), (0,1,0)\}$. Thus,

$$\begin{aligned}\text{Proj}_w(1,2,3) &= \langle(1,2,3), (1,0,0)\rangle(1,0,0) + \langle(1,2,3), (0,1,0),\rangle(0,1,0)\\ &= (1,0,0) + 2(0,1,0) = (1,2,0)\end{aligned}$$

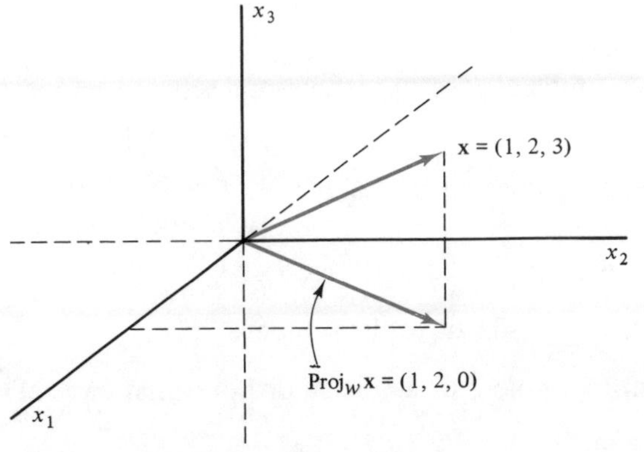

b. $W = S[(1,1,0), (0,1,-1)]$. Since the vectors $(1,1,0)$ and $(0,1,-1)$ are linearly independent, we may use them to construct an orthonormal basis for W. This basis is $\{(1,1,0)/\sqrt{2}, (-1,1,-2)/\sqrt{6}\}$ and we have

$$\begin{aligned}\text{Proj}_w(1,2,3) &= \langle(1,2,3),\mathbf{u}_1\rangle\,\mathbf{u}_1 + \langle(1,2,3),\mathbf{u}_2\rangle\,\mathbf{u}_2 \\ &= (\tfrac{1}{2})(1,1,0) + (\tfrac{5}{6})(-1,1,-2) \\ &= (-\tfrac{1}{3},\tfrac{4}{3},-\tfrac{5}{3})\end{aligned}$$

■

We conclude this section with a discussion of the properties of a change of basis matrix P relating two orthonormal bases. Let $U = \{\mathbf{u}_k\colon k = 1, \ldots, n\}$ and $V = \{\mathbf{v}_k\colon k = 1, \ldots, n\}$ be two orthonormal bases of $\mathbb{R}^n$. Let $P = [p_{jk}]$ be the matrix that gives the vectors $\mathbf{u}_k$ as linear combinations of the $\mathbf{v}_k$. That is,

$$\mathbf{u}_k = \sum_{j=1}^{n} p_{jk}\mathbf{v}_j \tag{6.21a}$$

and if $P^{-1} = Q = [q_{jk}]$, then

$$\mathbf{v}_k = \sum_{j=1}^{n} q_{jk}\mathbf{u}_j \tag{6.21b}$$

We remind the reader that the kth column of P consists of the coordinates of the vector $\mathbf{u}_k$ with respect to the basis V. A similar comment of course applies to the columns of $P^{-1} = Q$. However, these two bases are orthonormal. Thus, Theorem 6.5 may be used to compute the coordinates

$$p_{jk} = \langle\mathbf{u}_k,\mathbf{v}_j\rangle \qquad q_{jk} = \langle\mathbf{v}_k,\mathbf{u}_j\rangle \tag{6.22}$$

Since our inner product is symmetric, we have

$$p_{jk} = \langle\mathbf{u}_k,\mathbf{v}_j\rangle = \langle\mathbf{v}_j,\mathbf{u}_k\rangle = q_{kj}$$

In other words the matrix $Q = P^{-1}$ is the transpose of the matrix P, or $P^{-1} = P^T$, an extremely useful fact. We also note that the formulas

$$PP^T = P^TP = I$$

imply that both the rows and columns of P form orthonomal sets of vectors.

Definition 6.10. A matrix P is said to be orthogonal if $P^T = P^{-1}$.

EXAMPLE 4 Let $U = \{(1,0,1)/\sqrt{2},(1,1,-1)/\sqrt{3},(-1,2,1)/\sqrt{6}\}$. Find the change of basis matrices P and P^{-1} relating this basis to the standard basis.

Solution. We know from Example 2 that U is an orthonormal basis of $\mathbb{R}^3$. Thus $P^{-1} = P^T$.

$$P^{-1}=\begin{bmatrix}\frac{1}{\sqrt{2}} & \frac{1}{\sqrt{3}} & -\frac{1}{\sqrt{6}}\\ 0 & \frac{1}{\sqrt{3}} & \frac{2}{\sqrt{6}}\\ \frac{1}{\sqrt{2}} & -\frac{1}{\sqrt{3}} & \frac{1}{\sqrt{6}}\end{bmatrix} \qquad P=\begin{bmatrix}\frac{1}{\sqrt{2}} & 0 & \frac{1}{\sqrt{2}}\\ \frac{1}{\sqrt{3}} & \frac{1}{\sqrt{3}} & -\frac{1}{\sqrt{3}}\\ -\frac{1}{\sqrt{6}} & \frac{2}{\sqrt{6}} & \frac{1}{\sqrt{6}}\end{bmatrix} \quad \blacksquare$$

Clearly, when we deal with orthonormal bases, the amount of computational work is considerably lessened. There is of course the initial labor involved in constructing such a basis, but it is usually well worth the effort.

PROBLEM SET 6.3

1. Use the Gram-Schmidt procedure to construct an orthonormal basis for each of the following subspaces of $\mathbb{R}^3$:
   a. $W=\{(x_1,x_2,x_3): x_1-x_2=0\}$
   b. $W=S[(1,-1,2),(6,1,1)]$
2. Construct an orthonormal basis for $\mathbb{R}^3$ from the following basis, $\{(0,5,1),(0,1,-5),(1,-2,3)\}$.
3. Let W be the subspace of $\mathbb{R}^4$ spanned by the vectors $\mathbf{f}_1=(1,1,0,1)$ and $\mathbf{f}_2=(3,1,4,1)$. Compute the projection of $\mathbf{x}=(3,0,3,3)$ onto W.
4. Find the distance from the point $(1,-2,3)$ to the plane $2x_1-3x_2+6x_3=0$.
5. Find the distance from the point $(1,-2,3)$ to the plane $2x_1-3x_2+6x_3=2$.
6. Show that $U=\left\{\left(\frac{1}{\sqrt{2}},\frac{1}{\sqrt{2}}\right),\left(\frac{-1}{\sqrt{2}},\frac{1}{\sqrt{2}}\right)\right\}$ and $V=\{(1,0),(0,1)\}$ are both orthonormal bases of $\mathbb{R}^2$. Find a change of basis matrix P relating U and V and verify that it is orthogonal.
7. Construct an orthonormal basis for $\mathbb{R}^4$ from the basis $\{(0,1,1,1),(1,0,1,1),(1,1,0,1),(1,1,1,0)\}$.
8. Show that $\{(x_1,x_2),(y_1,y_2)\}$ is an orthonormal basis for $\mathbb{R}^2$ if and only if the matrix $\begin{bmatrix}x_1 & y_1\\ x_2 & y_2\end{bmatrix}$ is orthogonal.
9. Find an orthonormal basis for the kernels of each of the following matrices:
   a. $\begin{bmatrix}1 & 2\\ 3 & 6\end{bmatrix}$ **b.** $\begin{bmatrix}1 & -1 & 2\\ 4 & 6 & 3\end{bmatrix}$ **c.** $\begin{bmatrix}1 & 0 & -1 & 3\\ -3 & 1 & 0 & 1\end{bmatrix}$
10. Find an orthonormal basis for the ranges of each of the matrices in problem 9.
11. We've seen that if $U=\{\mathbf{u}_j: j=1,\ldots,n\}$ and $V=\{\mathbf{v}_j: j=1,\ldots,n\}$ are two orthonormal bases of $\mathbb{R}^n$, then the matrix $P=[p_{jk}]$ relating them is orthogonal.

Conversely, show that if P is orthogonal and U is an orthonormal basis then $V = \{\mathbf{v}_j\}$, where the vectors in V are defined by

$$\mathbf{v}_j = \sum_{k=1}^{n} p_{kj}\mathbf{u}_k$$

is also an orthonormal basis.

12. Show that if W is any subspace of $\mathbb{R}^n$ and $\mathbf{x}$ is any vector in $\mathbb{R}^n$, then there is a unique unit vector $\mathbf{w}_0$ in W such that the angle between $\mathbf{x}$ and $\mathbf{w}_0$ is minimized. That is, the angle between $\mathbf{x}$ and $\mathbf{w}$ for any vector $\mathbf{w}$ in W is no smaller than that between $\mathbf{x}$ and $\mathbf{w}_0$.

13. Let $V = P_2$. Define $\langle \mathbf{f},\mathbf{g} \rangle = \int_0^1 \mathbf{f}(t)\mathbf{g}(t)dt$. The set $B = \{1,t,t^2\}$ is a basis for V. Construct an orthonormal basis for V from B by using the Gram-Schmidt procedure.

14. Let $V = P_2$. Define $\langle \mathbf{f},\mathbf{g} \rangle = f_0g_0 + f_1g_1 + f_2g_2$, where $\mathbf{f}(t) = f_0 + f_1t + f_2t^2$ and $\mathbf{g}(t) = g_0 + g_1t + g_2t^2$. Show that $\{1,t,t^2\}$ is an orthonormal basis if we use this inner product, but not if we use the inner product of problem 13.

15. Let $V = C[0,1]$. Define $\langle \mathbf{f},\mathbf{g} \rangle$ as we did in problem 13.

a. Compute the length of the vector $\sin \pi t$.

b. Show that the set $\{1, \sin \pi t, \cos \pi t, \cdots, \sin n\pi t, \cos n\pi t, \cdots\}$ is orthogonal.

16. Let $V = S[(1,0,1),(1,1,1)]$. Show that $(-1,0,1)$ is perpendicular to every vector in V.

17. Let V and W be subspaces of $\mathbb{R}^n$. We say that V and W are perpendicular if $\langle \mathbf{x},\mathbf{y} \rangle = 0$ for every $\mathbf{x}$ in V and $\mathbf{y}$ in W.

a. Suppose $\mathbf{f}_1$ and $\mathbf{f}_2$ are two perpendicular vectors. Show $S[\mathbf{f}_1]$ and $S[\mathbf{f}_2]$ are perpendicular.

b. Let $\{\mathbf{v}_1, \ldots, \mathbf{v}_p\}$ be an orthogonal set of vectors. Show that $S[\mathbf{v}_1, \ldots, \mathbf{v}_k]$ and $S[\mathbf{v}_{k+1}, \ldots, \mathbf{v}_p]$ are perpendicular.

18. Let V be any substance of $\mathbb{R}^n$. Define $V^\perp$ (V perp) by $V^\perp = \{\mathbf{x}: \langle \mathbf{x},\mathbf{y} \rangle = 0 \text{ for every } \mathbf{y} \text{ in } V\}$.

a. Show that $V^\perp$ is a subspace of $\mathbb{R}^n$.

b. Show that V and $V^\perp$ are perpendicular in the sense of problem 17.

c. $\{\mathbf{0}\}^\perp = \mathbb{R}^n$, $(\mathbb{R}^n)^\perp = \{\mathbf{0}\}$.

d. $(V^\perp)^\perp = V$. Hint: What are the dimensions of the two spaces?

19. Let V be any subspace of $\mathbb{R}^2$. Show that any vector $\mathbf{x}$ in $\mathbb{R}^2$ can be written uniquely in the form $\mathbf{x} = \mathbf{v} + \mathbf{w}$, for some $\mathbf{v}$ in V and $\mathbf{w}$ in $V^\perp$. Consider the two special cases $V = \{\mathbf{0}\}$ and $V = \mathbb{R}^2$ first. Then consider the case $V = S[\mathbf{v}]$ for some fixed vector $\mathbf{v}$.

20. Let V be any subspace of $\mathbb{R}^3$. Show that $\dim(V) + \dim(V^\perp) = 3$. Generalize this to $\mathbb{R}^n$.

21. Use the result of problem 20 to show that for any subspace V of $\mathbb{R}^n$ the following is true. Given any $\mathbf{x}$ in $\mathbb{R}^n$, we can write $\mathbf{x}$ uniquely in form $\mathbf{x} = \mathbf{v} + \mathbf{w}$ for some $\mathbf{v}$ in V and $\mathbf{w}$ in $V^\perp$.

22. Given any unit vector $\mathbf{u}$, $\text{Proj}_u\mathbf{x}$ is a linear transformation from $\mathbb{R}^n$ to $\mathbb{R}^n$; cf. problem 10 in Section 6.2.

a. Find a "nice" matrix representation for $\text{Proj}_u\mathbf{x}$.

b. Describe geometrically the two subspaces ker $(\text{Proj}_\mathbf{u})$ and $\text{Rg}(\text{Proj}_\mathbf{u})$.

c. What are the dimensions of the kernel and range of $\text{Proj}_\mathbf{u}$?

23. Let A be a matrix representation of a linear transformation $L{:}\mathbb{R}^2 \to \mathbb{R}^2$, where L rotates the plane through some angle θ. Show that A is an orthogonal matrix.

24. Let W be any subspace of $\mathbb{R}^n$. Define $L(\mathbf{x}) = \text{Proj}_w\mathbf{x}$.
  a. Show L is a linear transformation.
  b. Find the range and kernel of L.
  c. Find a "nice" matrix representation for L.

6.4 SYMMETRIC MATRICES

Given a matrix A we defined the transpose of A seemingly for no special reason. There is, however, an important relationship between A and A^T that is not apparent until we have an inner product. To demonstrate this relationship we take the inner product of $A\mathbf{x}$ with $\mathbf{y}$. Thus suppose $A = [a_{jk}]$ is an $m \times n$ matrix, $\mathbf{x}$ a vector in $\mathbb{R}^n$, and $\mathbf{y}$ a vector in $\mathbb{R}^m$. Then $A^T = [a^T_{jk}]$ is an $n \times m$ matrix, $A\mathbf{x}$ is in $\mathbb{R}^m$ and $A^T\mathbf{y}$ is in $\mathbb{R}^n$.

$$\begin{aligned}\langle A\mathbf{x},\mathbf{y}\rangle &= \left\langle\left(\sum_{k=1}^{n} a_{1k}x_k, \ldots, \sum_{k=1}^{n} a_{mk}x_k\right),\mathbf{y}\right\rangle \\ &= \sum_{j=1}^{m}\left(\sum_{k=1}^{n} a_{jk}x_k\right)y_j = \sum_{k=1}^{n} x_k\left(\sum_{j=1}^{m} a_{jk}y_j\right) \\ &= \sum_{k=1}^{n} x_k\left(\sum_{j=1}^{m} a^T_{kj}y_j\right) \\ &= \langle \mathbf{x}, A^T\mathbf{y}\rangle\end{aligned}$$

This is such a useful formula that we write it again

$$\langle A\mathbf{x},\mathbf{y}\rangle = \langle \mathbf{x}, A^T\mathbf{y}\rangle \tag{6.23}$$

Notice that if $A = A^T$, then (6.23) becomes $\langle A\mathbf{x},\mathbf{y}\rangle = \langle \mathbf{x}, A\mathbf{y}\rangle$.

In Chapter 5 we stated that every symmetric matrix was similar to a diagonal matrix. Put another way, we know that given any symmetric matrix there is a basis of $\mathbb{R}^n$, which consists of eigenvectors of A. It turns out that it is also possible to construct this basis in such a manner that it is an orthonormal basis. This useful feature of symmetric matrices is a consequence of the next lemma.

Lemma 6.3. Let A be a symmetric matrix. Let λ_1 and λ_2 be two distinct eigenvalues of A. Then any pair of eigenvectors $\mathbf{f}_1$ and $\mathbf{f}_2$ corresponding to λ_1 and λ_2, respectively, must be perpendicular.

Proof

$$\begin{aligned}\lambda_1\langle \mathbf{f}_1,\mathbf{f}_2\rangle &= \langle \lambda_1\mathbf{f}_1,\mathbf{f}_2\rangle \\ &= \langle A\mathbf{f}_1,\mathbf{f}_2\rangle = \langle \mathbf{f}_1,A\mathbf{f}_2\rangle \\ &= \langle \mathbf{f}_1,\lambda_2\mathbf{f}_2\rangle = \lambda_2\langle \mathbf{f}_1,\mathbf{f}_2\rangle\end{aligned}$$

Thus, we have $(\lambda_1 - \lambda_2)\langle \mathbf{f}_1,\mathbf{f}_2\rangle = 0$. Since $\lambda_1 - \lambda_2 \neq 0$, we must have $\langle \mathbf{f}_1,\mathbf{f}_2\rangle = 0$, i.e., the eigenvectors are perpendicular.

EXAMPLE 1 Let $A = \begin{bmatrix} 2 & 3 \\ 3 & 4 \end{bmatrix}$. Find the eigenvectors of A and verify Lemma 6.3 for this symmetric matrix.

Solution. A quick calculation shows that the characteristic polynomial of A is $p(\lambda) = \lambda^2 - 6\lambda - 1$. The eigenvalues are $3 \pm \sqrt{10}$ and their corresponding eigenvectors are

$$\lambda_1 = 3 + \sqrt{10} \qquad \mathbf{f}_1 = (-1 - \sqrt{10}, 3)$$
$$\lambda_2 = 3 - \sqrt{10} \qquad \mathbf{f}_2 = (-1 + \sqrt{10}, 3)$$

Computing the inner product of the eigenvectors we have

$$\begin{aligned}\langle \mathbf{f}_1, \mathbf{f}_2 \rangle &= \langle (-1 - \sqrt{10}, 3), (-1 + \sqrt{10}, 3) \rangle \\ &= (-1 - \sqrt{10})(-1 + \sqrt{10}) + 9 = 0\end{aligned}$$

■

The procedure for constructing an orthonormal basis from the eigenvectors of a symmetric matrix is now relatively easy. We first find the eigenvalues of A from its characteristic polynomial

$$p(\lambda) = (\lambda - \lambda_1)^{m_1}(\lambda - \lambda_2)^{m_2} \cdots (\lambda - \lambda_p)^{m_p}$$

We next find a basis for each of the eigenspaces $\ker(A - \lambda_j I)$. An orthonormal basis for each of these eigenspaces is constructed by using the Gram-Schmidt procedure. These orthonormal bases are then adjoined to form a basis of $\mathbb{R}^n$. It is Lemma 6.3 which guarantees that combining these individually orthogonal sets will produce an orthogonal set. Since each vector has length 1 to begin with, they will remain unit vectors. In other words, suppose that $\{\mathbf{f}_1, \ldots, \mathbf{f}_{m_1}\}$ and $\{\mathbf{g}_1, \ldots, \mathbf{g}_{m_2}\}$ are orthonormal bases of $\ker(A - \lambda_1 \mathrm{I})$ and $\ker(A - d_2 \mathrm{I})$, respectively. Then since $\langle \mathbf{f}_j, \mathbf{g}_k \rangle = 0$ for every j and k, we conclude that $\{\mathbf{f}_1, \ldots, \mathbf{g}_{m_2}\}$ is also an orthonormal set.

EXAMPLE 2 Find an orthogonal matrix P such that PAP^T is a diagonal matrix, where A is the matrix

$$\begin{bmatrix} 7 & -2 & -1 \\ -2 & 10 & 2 \\ -1 & 2 & 7 \end{bmatrix}$$

Solution. Since A is symmetric, we know that there is an orthonormal basis of $\mathbb{R}^3$ consisting of eigenvectors. If P^{-1} is the matrix whose columns are these eigenvectors then $P^{-1} = P^T$ and we have $PAP^{-1} = PAP^T$, a diagonal matrix. Computing the characteristic polynomial of A we have

$$p(\lambda) = \det(A - \lambda I) = (6 - \lambda)^2(12 - \lambda)$$

■

The eigenvalues of A are 6 with multiplicity 2 and 12 with multiplicity 1. We list the corresponding eigenvectors.

$$\lambda_1 = 6 \qquad \mathbf{f}_1 = (1,0,1) \qquad \mathbf{f}_2 = (2,1,0)$$
$$\lambda_2 = 12 \qquad \mathbf{f}_3 = (-1,2,1)$$

Since the eigenvectors $\mathbf{f}_1$ and $\mathbf{f}_2$ correspond to the eigenvalue 6, they are automatically perpendicular to the eigenvector $\mathbf{f}_3$. To construct our orthonormal basis we use the Gram-Schmidt procedure for the first two eigenvectors and then divide $\mathbf{f}_3$ by its length.

$$\mathbf{u}_1 = \frac{(1,0,1)}{\sqrt{2}}$$

$$\mathbf{v}_2 = \mathbf{f}_2 - \langle \mathbf{f}_2,\mathbf{u}_1\rangle \mathbf{u}_1 = (1,1,-1)$$

$$\mathbf{u}_2 = \frac{\mathbf{v}_2}{\|\mathbf{v}_2\|} = \frac{(1,1,-1)}{\sqrt{3}}$$

$$\mathbf{u}_3 = \frac{\mathbf{f}_3}{\|\mathbf{f}_3\|} = \frac{(-1,2,1)}{\sqrt{6}}$$

Thus,

$$P^{-1} = \begin{bmatrix} \frac{1}{\sqrt{2}} & \frac{1}{\sqrt{3}} & -\frac{1}{\sqrt{6}} \\ 0 & \frac{1}{\sqrt{3}} & \frac{2}{\sqrt{6}} \\ \frac{1}{\sqrt{2}} & -\frac{1}{\sqrt{3}} & \frac{1}{\sqrt{6}} \end{bmatrix} = P^T$$

and

$$PAP^T = \begin{bmatrix} 6 & 0 & 0 \\ 0 & 6 & 0 \\ 0 & 0 & 12 \end{bmatrix}$$

We summarize this discussion in the following theorem.

Theorem 6.9. Let A be an $n \times n$ symmetric matrix ($A = A^T$). Then $\mathbb{R}^n$ has an orthonormal basis of eigenvectors of A, and there is an orthogonal matrix P such that $PAP^T = D$ is a diagonal matrix. The diagonal elements of D are the eigenvalues of A and the columns of P^T are eigenvectors of A.

Another consequence of formula (6.23), which we use in the next section, is the fact that if A is any $m \times n$ matrix then $A^TA\mathbf{x} = \mathbf{0}$ if and only if $A\mathbf{x} = \mathbf{0}$.

Lemma 6.4. Let A be an $m \times n$ matrix. Then the following statements are true:

a. $A^T A\mathbf{x} = \mathbf{0}$ if and only if $A\mathbf{x} = \mathbf{0}$
b. Rank$(A^T A) = \text{rank}(A) = \text{rank}(A^T)$

Proof. Clearly, if $A\mathbf{x} = \mathbf{0}$, then $A^T A\mathbf{x} = \mathbf{0}$. Thus, to verify a it suffices to show that if $A^T A\mathbf{x} = \mathbf{0}$, then $A\mathbf{x} = \mathbf{0}$. Assuming $A^T A\mathbf{x} = \mathbf{0}$ we have

$$0 = \langle A^T A\mathbf{x}, \mathbf{x}\rangle = \langle A\mathbf{x}, A\mathbf{x}\rangle = \|A\mathbf{x}\|^2$$

Thus we also have $A\mathbf{x} = \mathbf{0}$. To see that rank$(A^T A)$ equals rank(A) we note that part **a** has shown that ker$(A^T A)$ equals ker(A). Thus

$$\begin{aligned}\text{Rank}(A^T A) &= n - \dim(\ker A^T A) = n - \dim(\ker A)\\ &= \text{rank}(A)\end{aligned}$$

That rank$(A) = \text{rank}(A^T)$ was proved in Chapter 3; cf. Theorem 3.5.

PROBLEM SET 6.4

1. For each of the following matrices verify formula (6.23):

a. $\begin{bmatrix} 1 & 2 \\ 3 & 4 \end{bmatrix}$ **b.** $\begin{bmatrix} 3 & 6 & 8 \\ 1 & 2 & 4 \end{bmatrix}$ **c.** $\begin{bmatrix} 2 & 4 \\ 1 & -1 \\ 5 & 6 \end{bmatrix}$

2. Verify (6.23) for each of the following matrices:

a. $\begin{bmatrix} 1 & 3 \\ 3 & 1 \end{bmatrix}$ **b.** $\begin{bmatrix} 0 & 1 & 0 \\ 1 & 0 & 2 \\ 0 & 2 & 0 \end{bmatrix}$ **c.** $\begin{bmatrix} 6 & 0 & 0 \\ 0 & 3 & 0 \\ 0 & 0 & 2 \end{bmatrix}$

3. An $n \times n$ symmetric matrix is said to be positive definite if $\langle A\mathbf{x}, \mathbf{x}\rangle$ is positive for each nonzero vector $\mathbf{x}$ in $\mathbb{R}^n$.

a. Show that the matrix $\begin{bmatrix} 3 & 1 \\ 1 & 2 \end{bmatrix}$ is positive definite.

b. Show that the matrix $\begin{bmatrix} a & b \\ b & d \end{bmatrix}$ is positive definite if and only if $a > 0$ and $ad - b^2 > 0$.

c. Find a similar criterion for 3×3 symmetric matrices.

4. Show that a symmetric matrix is positive definite if and only if each of its eigenvalues is positive. (Hint: If A is positive definite so is PAP^T when P is a nonsingular matrix.)

5. Find an orthonormal basis of eigenvectors for each of the following matrices:

a. $\begin{bmatrix} 1 & 2 \\ 2 & -3 \end{bmatrix}$ **b.** $\begin{bmatrix} 6 & 0 \\ 0 & 4 \end{bmatrix}$ **c.** $\begin{bmatrix} 3 & -1 \\ -1 & 2 \end{bmatrix}$

6. Find an orthonormal basis of eigenvectors for each of the following matrices:

a. $\begin{bmatrix} 8 & -1 & 1 \\ -1 & 8 & 1 \\ 1 & 1 & 8 \end{bmatrix}$ **b.** $\begin{bmatrix} -2 & 3 & 0 \\ 3 & 4 & 0 \\ 0 & 0 & 2 \end{bmatrix}$

7. Find an orthonormal basis of eigenvectors for each matrix in problem 2.

8. For each of the matrices A of problem 5 find P and D such that $PAP^T = D$, where D is a diagonal matrix.

9. For each of the matrices A of problem 6 find P and D such that $PAP^T = D$, where D is a diagonal matrix.

10. Let A be any symmetric matrix. Show that PAP^T is also a symmetric matrix.

11. Let A be any $n \times n$ matrix. Show that A is symmetric if and only if there is an orthonormal basis of $\mathbb{R}^n$ consisting of eigenvectors of A.

12. Let A be an $m \times n$ matrix. Show that if $A\mathbf{x} = \mathbf{b}$ has a solution, then $\mathbf{b}$ must be perpendicular to $\ker(A^T)$. [Hint: If $A\mathbf{x} = \mathbf{b}$ and $\mathbf{y}$ is in $\ker(A^T)$, we have $\langle \mathbf{b},\mathbf{y} \rangle = \langle A\mathbf{x},\mathbf{y} \rangle = ?$]

13. Show that the converse of problem 12 is also true. That is, show that if $\mathbf{b}$ is perpendicular to $\ker(A^T)$, then the equation $A\mathbf{x} = \mathbf{b}$ has a solution.

14. Let V be a vector space with an inner product $\langle\ ,\ \rangle$. A linear transformation L: $V \to V$ is said to be symmetric if $\langle L\mathbf{x},\mathbf{y} \rangle = \langle \mathbf{x},L\mathbf{y} \rangle$ for every pair of vectors $\mathbf{x}$ and $\mathbf{y}$ in V. Let V be P_1 and for $\mathbf{f}$, $\mathbf{g}$ in V define $\langle \mathbf{f},\mathbf{g} \rangle = \int_0^1 \mathbf{f}(t)\mathbf{g}(t)\, dt$. Decide which, if any, of the following linear transformations is symmetric.
   a. $L[\mathbf{f}] = \mathbf{f}'$ b. $L[\mathbf{f}] = \mathbf{f}''$ c. $L[\mathbf{f}] = t\mathbf{f}'$

15. Let L be a linear transformation from $\mathbb{R}^2$ to $\mathbb{R}^2$.
   a. Show that for each $\mathbf{x}$ in $\mathbb{R}^2$ there is a unique $\mathbf{y}$ in $\mathbb{R}^2$ such that $\langle \mathbf{x},L[\mathbf{z}] \rangle = \langle \mathbf{y},\mathbf{z} \rangle$ for every $\mathbf{z}$ in $\mathbb{R}^2$. This vector, $\mathbf{y}$, will be denoted as $L^T[\mathbf{x}]$. Thus, we have the formula $\langle \mathbf{x},L[\mathbf{z}] \rangle = \langle L^T[\mathbf{x}],\mathbf{z} \rangle$.
   b. Show that if A is the matrix representation of L with respect to the standard basis of $\mathbb{R}^2$, then A^T is the matrix representation of L^T.

   Note, a linear transformation is said to be symmetric if $L = L^T$. Thus, we have shown that L is symmetric if and only if its matrix representation, with respect to the standard basis, is a symmetric matrix.

16. Generalize problem 15 to $\mathbb{R}^n$.

17. For each of the linear transformations in problem 14, find L^T.

18. Let $V = C[0,1]$. Define $\langle \mathbf{f},\mathbf{g} \rangle = \int_0^1 \mathbf{f}(t)\mathbf{g}(t)\, dt$. Let L: $V \to V$ be defined as $L[\mathbf{f}] = \int_0^t \mathbf{f}(s)\, ds$. Define the transpose of L as in problem 15, and find a formula for it.

6.5 LEAST SQUARES

The first problem we considered at the start of this text was that of solving a system of linear equations. At that time, we noticed that there are systems which have no solution. In the language of matrices and linear transformations, this translates to the statement that $A\mathbf{x} = \mathbf{b}$ does not have a solution unless $\mathbf{b}$ is in the range of A. What we can do at this time is to find a "best" possible solution. That is, we find that vector $\mathbf{y}$ in $\mathrm{Rg}(A)$ which is closest to $\mathbf{b}$. Thus, if we cannot solve $A\mathbf{x} = \mathbf{b}$, we solve the equation

$$A\mathbf{x} = \mathrm{Proj}_{R_g A}(\mathbf{b}) \tag{6.24}$$

Since (6.24) will in general have many solutions, we restrict our discussion to the case where A is one-to-one or equivalently ker(A) consists of just the zero vector.

With the preceding in mind let $A = [a_{jk}]$ be an $m \times n$ matrix with rank(A) $= n \leq m$. Note that if $m < n$, then A cannot be one-to-one. In terms of a system of linear equations the restriction $n \leq m$ says that there are at least as many equations as unknowns.

Theorem 6.10. Let A be an $m \times n$ matrix with $n \leq m$ and rank(A) $= n$. Let $\mathbf{b}$ be any vector in $\mathbb{R}^m$. Then the solution to $A\mathbf{x} = \mathrm{Proj}_{R_gA}(\mathbf{b})$ is given by

$$\mathbf{x} = (A^TA)^{-1}(A^T\mathbf{b}) \tag{6.25}$$

Proof. The trick in the proof is to pick a nice basis for $\mathbb{R}^n$. We first note that A^TA is a symmetric $n \times n$ matrix. Moreover, by Lemma 6.4 we know that rank $(A^TA) =$ rank $(A) = n$. Theorem 6.9 guarantees an orthonormal basis $U = \{\mathbf{u}_1, \ldots, \mathbf{u}_n\}$ of $\mathbb{R}^n$ such that $A^TA\mathbf{u}_k = d_k\mathbf{u}_k$. Moreover,

$$\begin{aligned}\langle A\mathbf{u}_j, A\mathbf{u}_k\rangle &= \langle A^TA\mathbf{u}_j, \mathbf{u}_k\rangle \\ &= d_j\langle \mathbf{u}_j, \mathbf{u}_k\rangle = d_j\delta_{jk}\end{aligned}$$

Thus, the vectors $A\mathbf{u}_j$ are mutually perpendicular, and since ker(A) is just the zero vector, we have $\|A\mathbf{u}_j\|^2 = d_j > 0$. We therefore conclude that $\{A\mathbf{u}_j/\sqrt{d_j}: j = 1, \ldots, n\}$ is an orthonormal basis of Rg(A). By (6.20) we have

$$\begin{aligned}\mathrm{Proj}_{R_gA}(\mathbf{b}) &= \sum_{j=1}^{n} \langle \mathbf{b}, A\mathbf{u}_j/\sqrt{d_j}\rangle \frac{A\mathbf{u}_j}{\sqrt{d_j}} \\ &= \sum_{j=1}^{n} \frac{1}{d_j} \langle A^T\mathbf{b}, \mathbf{u}_j\rangle A\mathbf{u}_j \\ &= A\left[\sum_{j=1}^{n} \frac{1}{d_j} \langle A^T\mathbf{b}, \mathbf{u}_j\rangle \mathbf{u}_j\right]\end{aligned}$$

Setting

$$\mathbf{x} = \sum_{j=1}^{n} \frac{1}{d_j} \langle A^Tb, \mathbf{u}_j\rangle \mathbf{u}_j$$

we see, since A is $1-1$, that $\mathbf{x}$ is the unique solution to $A\mathbf{x} = \mathrm{Proj}_{R_gA}(\mathbf{b})$. Moreover since $(A^TA)\mathbf{u}_j = d_j\mathbf{u}_j$ we also have $(A^TA)^{-1}\mathbf{u}_j = (1/d_j)\mathbf{u}_j$. Thus,

$$\begin{aligned}\mathbf{x} &= \sum_{j=1}^{n} \langle A^T\mathbf{b}, \mathbf{u}_j\rangle (A^TA)^{-1}\mathbf{u}_j \\ &= (A^TA)^{-1}\left[\sum_{j=1}^{n} \langle A^T\mathbf{b}, \mathbf{u}_j\rangle \mathbf{u}_j\right]\end{aligned}$$

But the set $\{\mathbf{u}_j\}$ is an orthonormal basis of $\mathbb{R}^n$; hence the term in brackets is equal to $A^T\mathbf{b}$. Thus, $\mathbf{x} = (A^TA)^{-1}A^T\mathbf{b}$ is that unique vector in $\mathbb{R}^n$ such that $A\mathbf{x}$ is closest to $\mathbf{b}$.

This formula has an immediate application to curve fitting. Suppose we have a set of data points (x_j, y_j), $1 \leq j \leq n$, and we wish to find a straight line passing through these points. If there are more than two data points, such a line is usually nonexistent. See Figure 6.12. What is normally done in this situation is to find that straight line $y = mx + b$, such that the sum $\Sigma_{j=1}^n e_j^2 = \Sigma_{j=1}^n [y_j - (mx_j + b)]^2$ is minimized. The numbers e_j equal the error in approximating y_j by $mx_j + b$. Thus, in a certain sense, picking m and b in order to minimize the above sum gives us the best straight-line approximation to our data. This line is often referred to as the least squares fit.

At this time the reader might find it advisable to review the discussion immediately preceding Example 4 in Section 3.4.

Let A be the $n \times 2$ matrix

$$\begin{bmatrix} 1 & x_1 \\ \cdot & \cdot \\ \cdot & \cdot \\ \cdot & \cdot \\ 1 & x_n \end{bmatrix}$$

Think of $\mathbb{R}^n$ as pairs of numbers of the form $(b,m)^T$, and $A: \mathbb{R}^2 \to \mathbb{R}^n$. We wish to find a solution to the equation

$$A\begin{bmatrix} b \\ m \end{bmatrix} = \begin{bmatrix} y_1 \\ \cdot \\ \cdot \\ \cdot \\ y_n \end{bmatrix}$$

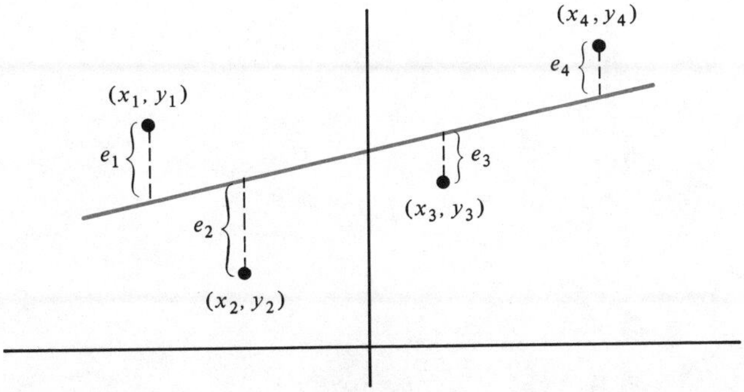

Figure 6.12

The pair (b,m) is a solution to this equation if and only if the line $y = mx + b$ passes through each of the data points (x_j,y_j). Realizing that this is unlikely we look for that pair (b,m) such that $A(b,m)^T$ is closest to $(y_1, \ldots, y_n)^T$. Since rank(A) is two (assuming at least two different x_j's), we may apply Theorem 6.10. Thus, our approximate solution is

$$\begin{bmatrix} b \\ m \end{bmatrix} = (A^TA)^{-1}A^T \begin{bmatrix} y_1 \\ y_2 \\ \cdot \\ \cdot \\ \cdot \\ y_n \end{bmatrix}$$

One easily calculates that

$$A^TA = \begin{bmatrix} n & \sum_{j=1}^n x_j \\ \sum_{j=1}^n x_j & \sum_{j=1}^n x_j^2 \end{bmatrix} \quad \text{and that} \quad A^T\mathbf{y}^{\mathrm{T}} = \begin{bmatrix} \sum_{j=1}^n y_j \\ \sum_{j=1}^n x_jy_j \end{bmatrix}$$

where $\mathbf{y} = (y_1, \ldots, y_n)$. Using Cramer's rule we have

$$b = \frac{\det \begin{bmatrix} \sum_{j=1}^n y_j & \sum_{j=1}^n x_j \\ \sum_{j=1}^n x_jy_j & \sum_{j=1}^n x_j^2 \end{bmatrix}}{\det \begin{bmatrix} n & \sum_{j=1}^n x_j \\ \sum_{j=1}^n x_j & \sum_{j=1}^n x_j^2 \end{bmatrix}} \qquad m = \frac{\det \begin{bmatrix} n & \sum_{j=1}^n y_j \\ \sum_{j=1}^n x_j & \sum_{j=1}^n x_jy_j \end{bmatrix}}{\det \begin{bmatrix} n & \sum_{j=1}^n x_j \\ \sum_{j=1}^n x_j & \sum_{j=1}^n x_j^2 \end{bmatrix}} \tag{6.26}$$

EXAMPLE 1 Find the least squares fit to the following data:

$(1,-1), (2,3), (3,4), (7,5)$

Solution. We first construct the following table:

x_j	1	2	3	7	$\sum_{j=1}^{4} x_j = 13$
x_j^2	1	4	9	49	$\sum_{j=1}^{4} x_j^2 = 63$
y_j	-1	3	4	5	$\sum_{j=1}^{4} y_j = 11$
$x_j y_j$	-1	6	12	35	$\sum_{j=1}^{4} x_j y_j = 52$

A is a 4×2 matrix and we have

$$A = \begin{bmatrix} 1 & 1 \\ 1 & 2 \\ 1 & 3 \\ 1 & 7 \end{bmatrix} \qquad A^T A = \begin{bmatrix} 4 & 13 \\ 13 & 63 \end{bmatrix} \qquad A^T y = \begin{bmatrix} 11 \\ 52 \end{bmatrix}$$

$$b = \frac{\det \begin{bmatrix} 11 & 13 \\ 52 & 63 \end{bmatrix}}{\det \begin{bmatrix} 4 & 13 \\ 13 & 63 \end{bmatrix}} = \frac{17}{83} \qquad m = \frac{\det \begin{bmatrix} 4 & 11 \\ 13 & 52 \end{bmatrix}}{\det \begin{bmatrix} 4 & 13 \\ 13 & 63 \end{bmatrix}} = \frac{65}{83}$$

Thus, $y = \frac{65}{83}x + \frac{17}{83}$ is the least squares straight-line approximation to our data. ■

There is no a priori reason why one should always insist upon fitting a straight line to data. For example, we might wish to fit a parabola to the data. That is, find a_0, a_1, and a_2 such that $y = a_0 + a_1 x + a_2 x^2$ is the quadratic least squares fit, cf. problem 5.

PROBLEM SET 6.5

1. Let $A = \begin{bmatrix} 2 & 1 \\ 0 & 2 \\ 3 & 4 \end{bmatrix}$

- **a.** Determine the range of A, and show that (1,1,0) is not in the range, i.e., the equation $A\mathbf{x} = (1,1,0)^T$ does not have a solution.
- **b.** Compute $A^T A$ and show that it is one to one.
- **c.** Solve the equation $A^T A\mathbf{x} = A^T \mathbf{b}$, where $\mathbf{b} = (1,1,0)$.
- **d.** If $\mathbf{x}$ is your solution from part **c**, show that $\|A\mathbf{x} - \mathbf{b}\|$ is smaller than $\|\mathbf{w} - \mathbf{b}\|$ for any vector $\mathbf{w}$ in the range of A.

2. Determine the straight-line least squares fit for the following data: (1,1), (2,−3), (4,0), (5,1), (10,3).

3. Determine the straight-line least squares fit for the following data:
   a. (0,6), (3,0), (4,−2)
   b. (−2,4), (3,9), (4,7)
4. Consider the system of equations:

$$\begin{aligned} 3x_1 + 4x_2 + 8x_3 &= 0 \\ x_1 \qquad\quad - x_3 &= 1 \\ 2x_1 + x_2 + 4x_3 &= 0 \\ x_1 + x_2 + x_3 &= 0 \end{aligned}$$

   a. This system is overdetermined (more equations than unknowns) and may not have a solution. Show that if there is a solution, it is unique.
   b. Show that this system does not have a solution, and then find **x** in $\mathbb{R}^3$ such that $A\mathbf{x}$ is that vector in the range of A closest to (0,1,0,0).
5. Determine the least squares quadratic fit for the following data; i.e., find $p(x) = a_0 + a_1x + a_2x^2$, such that $\Sigma_{j=1}^{n} [p(x_j) - y_j]^2$ is minimized, where (x_j,y_j) are the given data. Remember, you will need to solve an equation of the form $A^TA\mathbf{x} = A^T\mathbf{b}$.
   a. (−2,4), (3,9), (4,7)
   b. (1,1), (2,−3), (4,0), (5,1), (10,3)

SUPPLEMENTARY PROBLEMS

1. Define each of the following and give an example of each:
   a. Length of a vector
   b. Angle between two vectors
   c. Orthonormal basis
   d. Projection onto a subspace
   e. Orthogonal matrix
2. Let **x** be a vector in $\mathbb{R}^n$.
   a. Suppose $\langle \mathbf{x},\mathbf{y} \rangle = 0$ for every **y** in $\mathbb{R}^n$, show that **x** must be the zero vector.
   b. Suppose $\langle \mathbf{x},\mathbf{y} \rangle = 0$ for every vector **y** in some spanning set F of $\mathbb{R}^n$. Show that **x** must be the zero vector.
3. Compute the inner product and the cosine of the angle between each of the following pairs of vectors:
   a. (−4,5), (1,2) b. (−2,3,7), (2,−4,5) c. (−1,−2,3,5), (1,1,0,8)
4. Let $\mathbf{x}_0 = (1,-2,6)$. Show that the vector (1,−2,0) is that vector in the subspace $x_3 = 0$ which is closest to $\mathbf{x}_0$, by showing that

$$f(x,y) = \|(1,-2,6) - x(1,0,0) - y(0,1,0)\|^2$$

   obtains its minimum when $x = 1$ and $y = -2$. Repeat this for $\mathbf{x}_0 = (a,b,c)$, an arbitrary vector in $\mathbb{R}^3$.
5. Let $V = M_{22}$. Define the inner product of two 2×2 matrices by

$$\langle A,B \rangle = \sum_{j=1}^{2} \sum_{k=1}^{2} a_{jk}b_{jk}$$

   a. Show that this inner product satisfies properties a through c of Theorem 6.3.
   b. Show that Theorem 6.4 is valid.

c. Let $A = \begin{bmatrix} c & 6 \\ -3 & 2 \end{bmatrix}$, where c is a fixed constant. Let $E_1 = \begin{bmatrix} 1 & 0 \\ 0 & 0 \end{bmatrix}$. Let $f(t) = \|A - tE_1\|^2$. Find that value of t which minimizes $f(t)$.

6. Given two vectors $\mathbf{x} = (x_1, x_2, x_3)$ and $\mathbf{y} = (y_1, y_2, y_3)$ in $\mathbb{R}^3$, define their cross product as (cf. problem 8 in Supplementary Problems to Chapter 4).

$$\mathbf{x} \times \mathbf{y} = (x_2y_3 - x_3y_2, x_3y_1 - x_1y_3, x_1y_2 - x_2y_1)$$

a. Show that the cross product of $\mathbf{x}$ and $\mathbf{y}$ is perpendicular to both $\mathbf{x}$ and $\mathbf{y}$.
b. Show that $\mathbf{x} \times \mathbf{y} = \mathbf{0}$ if and only if $\mathbf{x}$ and $\mathbf{y}$ are linearly dependent.
c. Show that $\mathbf{i} \times \mathbf{j} = \mathbf{k}$, $\mathbf{j} \times \mathbf{k} = \mathbf{i}$, and $\mathbf{k} \times \mathbf{i} = \mathbf{j}$.
d. Verify that $\mathbf{x} \times \mathbf{y} = -\mathbf{y} \times \mathbf{x}$.
e. Show that $\mathbf{x} \times (\mathbf{y} + \mathbf{z}) = (\mathbf{x} \times \mathbf{y}) + (\mathbf{x} \times \mathbf{z})$.
f. Find three vectors $\mathbf{x}$, $\mathbf{y}$, and $\mathbf{z}$ for which $\mathbf{x} \times (\mathbf{y} \times \mathbf{z}) \neq (\mathbf{x} \times \mathbf{y}) \times \mathbf{z}$. Hint: Use parts b and c.

Thus, the cross product of two vectors is a noncommutative, nonassociative operation that produces a vector perpendicular to both of the original vectors.

7. Define $\mathbf{x} \times \mathbf{y}$ as in problem 6. Show that

$$\|\mathbf{x} \times \mathbf{y}\|^2 + \langle \mathbf{x},\mathbf{y} \rangle^2 = \|\mathbf{x}\|^2 \|\mathbf{y}\|^2$$

a. Deduce from the above formula that $\|\mathbf{x} \times \mathbf{y}\| = \|\mathbf{x}\| \|\mathbf{y}\| \sin \theta$, where θ is the angle between the vectors $\mathbf{x}$ and $\mathbf{y}$.
b. If P is the parallelogram determined by $\mathbf{x}$ and $\mathbf{y}$, show that area$(P) = \|\mathbf{x} \times \mathbf{y}\|$.

8. If $\mathbf{x} \times \mathbf{y}$ is defined as in problem 6, show that

$$(\mathbf{x} \times \mathbf{y})' = (\mathbf{x}' \times \mathbf{y}) + (\mathbf{x} \times \mathbf{y}')$$

where we assume that both $\mathbf{x}$ and $\mathbf{y}$ are vector-valued functions of a real variable and the $'$ denotes differentiation.

9. Let P be an orthogonal $n \times n$ matrix. Let $\mathbf{x}$ and $\mathbf{y}$ be any two vectors in $\mathbb{R}^n$.
a. Show that $\langle \mathbf{x},\mathbf{y} \rangle = \langle P\mathbf{x},P\mathbf{y} \rangle$. Deduce from this that the linear transformation L given by $L\mathbf{x} = P\mathbf{x}$ preserves the lengths of vectors and the angles between them.
b. Conversely, show that if P is an $n \times n$ matrix for which $\langle \mathbf{x},\mathbf{y} \rangle = \langle P\mathbf{x},P\mathbf{y} \rangle$ for every pair of vectors in $\mathbb{R}^n$, then P is an orthogonal matrix.

10. A mapping T from $\mathbb{R}^n$ to $\mathbb{R}^n$ is said to be affine if $T\mathbf{x} = A\mathbf{x} + \mathbf{a}$, where A is an $n \times n$ matrix and $\mathbf{a}$ is a fixed vector in $\mathbb{R}^n$. Clearly, if A is an orthogonal matrix, then T is distance preserving, i.e., $\|T\mathbf{x} - T\mathbf{y}\| = \|\mathbf{x} - \mathbf{y}\|$. Show that the converse of this is also true. That is, if T is any mapping that preserves distance then T is affine and the matrix A is orthogonal.

11. Let $\mathbf{x}(t)$ and $\mathbf{y}(t)$ be two vector-valued functions from $\mathbb{R}$ to $\mathbb{R}^2$. If $\mathbf{x}(t) = (x_1(t), x_2(t))$, define $x'(t) = (x_1', x_2')$.
a. Let $c(t)$ be a real-valued differentiable function. Show that $[c\mathbf{x}]' = c'\mathbf{x} + c\mathbf{x}'$.
b. Show that $\langle \mathbf{x},\mathbf{y} \rangle' = \langle \mathbf{x}',\mathbf{y} \rangle + \langle \mathbf{x},\mathbf{y}' \rangle$.
c. If $\mathbf{x}(t)$ is a vector-valued function with constant nonzero length, show that $\mathbf{x}$ and $\mathbf{x}'$ are perpendicular if $\mathbf{x}'$ is not the zero vector.

12. A linear transformation L from $\mathbb{R}^n$ to $\mathbb{R}^n$ is positive definite if $\langle L\mathbf{x},\mathbf{x} \rangle \geq 0$ for all vectors $\mathbf{x}$, and whenever the inner product of $\mathbf{x}$ and $L\mathbf{x}$ equals zero, then $\mathbf{x}$ equals $\mathbf{0}$. Let $A = [a_{jk}]$ be the matrix representation of L with respect to the standard basis. Assume that A is a symmetric matrix.

a. If $n = 2$, show that L is positive definite if and only if $a_{11} > 0$, and $\det(A) > 0$.
b. If $n = 3$, show that L is positive definite if and only if $a_{11} > 0$, $\det(M_{33}) > 0$, and $\det(A) > 0$, where M_{33} is the 2×2 matrix in the upper left-hand corner of A.

13. Let P be a linear transformation from $\mathbb{R}^n$ to $\mathbb{R}^n$. Suppose $P^2 = P$. That is, $P(P(\mathbf{x})) = P\mathbf{x}$ for all vectors in $\mathbb{R}^n$. Such mappings are called projections.
a. Show that $(I - P)^2 = I - P$. Thus, if P is a projection so too is $I - P$.
b. Show that the following are equivalent:
(1) $\mathbf{x}$ is in the range of P.
(2) $P\mathbf{x} = \mathbf{x}$.
(3) $(I - P)\mathbf{x} = \mathbf{0}$.
c. Show that $\ker(P) = \text{Rg}(I - P)$.
d. A projection is said to be orthogonal if $\ker(P)$ is orthogonal to $\text{Rg}(P)$. Show that P is an orthogonal projection if and only if $P = P^T$.
e. Show that the projections defined in Section 6.2 are orthogonal projections in the sense of part **d**.

14. Let $T{:}P_2 \to P_3$ be a linear transformation given by

$$T[\mathbf{p}](t) = \int_0^t \mathbf{p}(s)\, ds$$

Define an inner product on P_2 by $\langle \mathbf{p},\mathbf{q} \rangle = p_0q_0 + p_1q_1 + p_2q_2$, where $\mathbf{p}(t) = p_0 + p_1t + p_2t^2$. Define an inner product on P_3 in a similar manner.
a. Show that $\mathbf{p}(t) \equiv 1$ is not in the range of T.
b. Show that T is one-to-one.
c. Find the least squares solution to $T(\mathbf{p}) = 1$. That is, find $\mathbf{p}$ in P_2 such that $T(\mathbf{p})$ is that vector in the range of T closest to the polynomial identically 1.

15. Let $V = M_{22}$. Define the inner product of two matrices $A = [a_{jk}]$ and $B = [b_{jk}]$ by

$$\langle A,B \rangle = \sum_{j=1}^{2} \sum_{k=1}^{2} a_{jk}b_{jk}$$

Let F be the set consisting of the three matrices below:

$$\left\{ \begin{bmatrix} 1 & 2 \\ -3 & 1 \end{bmatrix} \begin{bmatrix} 0 & 1 \\ -2 & 0 \end{bmatrix} \begin{bmatrix} 1 & -1 \\ -1 & 1 \end{bmatrix} \right\}$$

a. Show that F is a linearly independent set.
b. Construct orthonormal bases for V and for $S[F]$.

16. Define $T{:}P_2 \to P_3$ by $T[\mathbf{p}](t) = \int_0^t \mathbf{p}(s)ds - (t/2)\mathbf{p}(t)$.
a. Show that T is a linear transformation and determine its range and kernel.
b. If inner products on P_2 and P_3 are defined as in problem 14, construct orthonormal bases for the range and kernel of T.

7
NUMERICAL METHODS

In the preceding chapters we discussed solving systems of equations via Gaussian elimination, and how to analyze a matrix by determining its eigenvalues and eigenvectors. When discussing these techniques, we completely ignored the computational problems. We also never considered the possibility that our equations might not be known precisely. That is, we might wish to solve the equations $A\mathbf{x} = \mathbf{b}$, but, because of errors in data measurement, we know the coefficients a_{jk} and b_j only approximately. Thus, we solve the equation $A_1\mathbf{x} = \mathbf{b}_1$, where A_1 is close to A and $\mathbf{b}_1$ is close to $\mathbf{b}$. We would then hope that the solution to this system is close to the solution of the first system.

The first section in this chapter discusses roundoff and truncation error, and demonstrates how arithmetic operations can drastically decrease the precision of our answers. Section 7.2 describes the standard pivoting method employed with Gaussian elimination. Section 7.3 is an elementary discussion of how to calculate eigenvalues and eigenvectors by an iterative procedure. Since this is only one chapter in a beginning text on linear algebra, the reader should not expect the techniques discussed to cover all possibilities but should consider this a brief introduction. For those interested in further study, the references listed in the bibliography are an excellent place to start.

7.1 TROUBLES WITH ARITHMETIC

Whether calculations are performed with pencil and paper or by computer, error is involved. In fact, most numbers are not even represented exactly. Some common examples are $\sqrt{2} = 1.4142$, $\pi = 3.1415$, and $\frac{1}{3} = 0.3333$. These illustrate a common cause of error, truncation. One can improve the accuracy by carrying more digits, e.g., $\sqrt{2} = 1.414213562$, but the number of significant digits allowed, especially when time and machine constraints enter the picture, must be relatively small. For purposes of illustration, we assume that we can carry only six significant digits. How much error is possible can depend on the size of the number. For example, $\sqrt{2} \approx 1.41421$, and $|\sqrt{2} - 1.41421| \leq 10^{-5}$, but 1,012,689 would be written as 1.01268×10^5. For the latter number we have $|(1.012689 - 1.01268) \times 10^6| = 9$, an error considerably larger than 10^{-5}. One way to avoid this sort of scaling problem is to talk about the relative error. Thus, if N is a number and n is its approximation, the relative error is defined to be $|(N - n)/N|$. For both of the preceding examples the relative error is on the order of 10^{-5}. It would seem, with modern computers routinely carrying 16 significant digits (this is not quite true since most computers use base 2 arithmetic rather than base 10), that we would have nothing to worry about. Unfortunately the operations of addition and multiplication can rapidly increase not only the actual error but also the relative error. For example, suppose that $N = 1.111111$ and $M = -1.111100$, and that these are exact. Our machine will approximate N with $n = 1.11111$ and M with $m = -1.11110$. The relative error for both is less than 10^{-6}. The exact value of $N + M$ is 0.000011. However, our computer will have $n + m = 0.00001$, and the relative error is now $|(0.000011 - 0.00001)/0.000011| = |1/1.1| \approx 1$. This is a drastic change in the relative error, and only one operation was performed. This behavior is due to the fact that N and M are of opposite sign but have almost the same magnitude. Errors may also occur when two numbers of substantially different magnitude are added together. For example, if $N = 1.0 \times 10^5$ and $M = 0.12$, then $N + M = 100{,}000.12$. Since we keep only six significant digits, our machine stores $N + M$ as 1.0×10^5, i.e., $N + M = N$, a decided loss of information. By itself, this might not be too disastrous. However, when one thinks of the thousands of times additions and multiplications are performed in many problems, appropriate feelings of terror engulf one's mind and body.

The preceding discussion was intended as a warning to the reader that whenever an algorithm is proposed or implemented, a fair amount of effort must be expended to avoid loss of information and errors so large that the computed answers are meaningless.

PROBLEM SET 7.1

1. Calculate the error and relative error when each of the following numbers is approximated by the first three digits in its decimal expansion:
   a. 1.0101 **b.** 10.101 **c.** 101.01 **d.** 1010.1
2. Let $M = 1$ and $N = 2000$. Compute $M + N$. Assuming we carry first three then four significant digits, compute both the error and the relative error.
3. Let A be an $n \times m$ matrix and let B be an $m \times p$ matrix. Show that mnp multiplications are required to produce the product matrix AB. How many additions are needed?
4. If A is an $n \times n$ matrix, determine the number of multiplications needed to transform A into a matrix B, which is in row echelon form.

7.2 GAUSSIAN ELIMINATION AND PIVOTING

Given an equation $A\mathbf{x} = \mathbf{b}$, there are two types of matrices for which this equation is relatively easy to solve: upper triangular and lower triangular. We illustrate the relative ease with which these two types of systems can be solved, and also the form that a computer program, written to solve such a system, might have.

EXAMPLE 1 Lower triangular system.

$$\begin{aligned} l_{11}y_1 \qquad\qquad\qquad &= b_1 \\ l_{21}y_1 + l_{22}y_2 \qquad\quad &= b_2 \\ l_{31}y_1 + l_{32}y_2 + l_{33}y_3 &= b_3 \end{aligned}$$

Solution. Clearly the solution to this system is

$$y_1 = \frac{b_1}{l_{11}} \qquad y_2 = \frac{b_2 - l_{21}y_1}{l_{22}} \qquad y_3 = \frac{b_3 - l_{31}y_1 - l_{32}y_2}{l_{33}}$$

We have of course assumed that the diagonal terms l_{jj} are not zero.

The following lines of BASIC will compute the solution of an $n \times n$ system that is nonsingular and in lower triangular form.

$y_1 = b_1/l_{11}$

For $j = 2$ to n

$y_j = b_j$

For $k = 1$ to $j - 1$

$y_j = y_j - l_{jk} * y_k$

Next k

$$y_j = y_j / l_{jj}$$

Next j

We assumed that the lower triangular matrix $L = [l_{jk}]$ was nonsingular, i.e., $l_{jj} \neq 0$, for each j. ■

EXAMPLE 2 Upper triangular system.

$$\begin{aligned} u_{11}x_1 + u_{12}x_2 + u_{13}x_3 &= y_1 \\ u_{22}x_2 + u_{23}x_3 &= y_2 \\ u_{33}x_3 &= y_3 \end{aligned}$$

The solution, found by backward substitution, is

$$x_3 = \frac{y_3}{u_{33}} \qquad x_2 = \frac{y_2 - u_{23}x_3}{u_{22}} \qquad x_1 = \frac{y_1 - u_{12}x_2 - u_{13}x_3}{u_{11}}$$

The lines of BASIC to solve such a system would be

$$x_n = y_n / u_{nn}$$

For $j = n - 1$ to 1, step -1

$$x_j = y_j$$

For $k = j + 1$ to n

$$x_j = x_j - u_{jk} * x_k$$

Next k

$$x_j = x_j / u_{jj}$$

Next j

We assumed in this case also that $U = [u_{jk}]$ was nonsingular, i.e., $u_{jj} \neq 0$, for each j. ■

Let us quickly review how one solves a system of linear equations with Gaussian elimination. Given the system $A\mathbf{x} = \mathbf{b}$, we perform elementary row operations on the augmented matrix $[A{:}\mathbf{b}]$, transforming it to an equivalent row echelon form. We've seen that this is equivalent to multiplying $[A{:}\mathbf{b}]$ by a sequence of elementary row matrices $E_1, \ldots, E_p$. Moreover, if we assume that none of these matrices performs a row interchange, then each of them will be a lower triangular matrix. For example, if we had the system

$$\begin{aligned} 3x_1 + x_2 - x_3 &= 2 \\ x_1 + x_2 \phantom{{}+2x_3} &= 1 \\ -x_1 - x_2 + 2x_3 &= 1 \end{aligned}$$

then the coefficient matrix A equals

$$A = \begin{bmatrix} 3 & 1 & -1 \\ 1 & 1 & 0 \\ -1 & -1 & 2 \end{bmatrix}$$

The two matrices E_1 and E_2:

$$E_1 = \begin{bmatrix} 1 & 0 & 0 \\ -\frac{1}{3} & 1 & 0 \\ \frac{1}{3} & 0 & 1 \end{bmatrix} \qquad E_2 = \begin{bmatrix} 1 & 0 & 0 \\ 0 & 1 & 0 \\ 0 & 1 & 1 \end{bmatrix}$$

perform the following elementary row operations:

E_1 : $-\frac{1}{3}$ row 1 to row 2, and $\frac{1}{3}$ row 1 to row 3
E_2 : row 2 to row 3

When the matrix product $U = E_2 E_1 A$ is calculated, we see that it equals

$$U = \begin{bmatrix} 3 & 1 & -1 \\ 0 & \frac{2}{3} & \frac{1}{3} \\ 0 & 0 & 2 \end{bmatrix}$$

an upper triangular matrix. Moreover, since the matrices E_1 and E_2 are lower triangular (no row interchanges), $E_2 E_1$ is also lower triangular. Thus,

$$A = (E_1^{-1} E_2^{-1})U = LU$$

where $L = (E_2 E_1)^{-1}$ is lower triangular. Hence, to solve $A\mathbf{x} = \mathbf{b}$, it suffices to solve the two systems

$$L\mathbf{y} = \mathbf{b} \qquad U\mathbf{x} = \mathbf{y}$$

The first system is lower triangular, while the second is upper triangular. Since the solution to $L\mathbf{y} = \mathbf{b}$ is $\mathbf{y} = E_2 E_1 \mathbf{b}$, we have $\mathbf{y} = [2, \frac{1}{3}, 2]^T$. Thus, $\mathbf{x} = [1,0,1]$. This factorization of the matrix A into the product of a lower triangular and upper triangular matric is called the LU decomposition of A.

Before discussing this solution scheme in more detail, we need to look at an example of what can happen when roundoff errors occur.

EXAMPLE 3 Solve the following system of equations twice. Once under the assumption that only three significant digits can be carried, and then exactly.

Exact arithmetic	Truncated arithmetic
$0.001x_1 - x_2 = 1$ $x_1 + x_2 = -4$	$0.001x_1 - x_2 = 1$ $x_1 + x_2 = -4$
$0.001x_1 - x_2 = -1$ $1001x_2 = -1004$	$0.001x_1 - x_2 = 1$ $1000x_2 = -1000$
$x_2 \approx -1.002997$ $x_1 \approx -2.997002$	$x_2 = -1$ $x_1 = 0$

Clearly, if we blindly implemented Gaussian elimination with a computer, we would have a very unreliable algorithm. ■

The immediate cause of our problem is that the coefficient of x_1 in the first equation is considerably smaller than the corresponding coefficient in the second equation. What happens if we interchange the order of the equations so that we have

$$\begin{aligned} x_1 + x_2 &= -4 \\ 0.001x_1 - x_2 &= 1 \end{aligned}$$

Solving this system and only carrying three significant digits, we have

$$\begin{aligned} x_1 + x_2 &= -4 \\ -x_2 &= 1 \\ x_2 = -1 \qquad & x_1 = -3 \end{aligned}$$

Notice that both components of our computed solution are reasonable approximations of the actual solution.

To minimize the type of error encountered when we first solved this system, we implement the following procedure. Before using the entry in the first row and first column to place zeros in the $j,1$ positions, for $j \geq 2$, we search the first column to find that entry with the largest absolute value. The row in which this entry occurs is then interchanged with the first row. We then perform row operations to produce zeros in the $(2,1)$, $(3,1)$, . . . , and $(n,1)$ positions of our matrix. We illustrate this with the following system:

$$\begin{aligned} x_1 + 6x_2 + 3x_3 &= 1 \\ 2x_1 + x_2 + x_3 &= 0 \\ -3x_1 + 3x_2 + 2x_3 &= 1 \end{aligned}$$

The coefficient matrix A of this system is

$$\begin{bmatrix} 1 & 6 & 3 \\ 2 & 1 & 1 \\ -3 & 3 & 2 \end{bmatrix}$$

Since $-3 = a_{31}$ is such that $|a_{31}| \geq |a_{j1}|$, for $j = 1,2,3$, we interchange the first and third rows. This produces the matrix A_1,

$$A_1 = \begin{bmatrix} -3 & 3 & 2 \\ 2 & 1 & 1 \\ 1 & 6 & 3 \end{bmatrix}$$

We need to keep track of this, so set $r_1 = 3$. That is, row 1 was interchanged with row 3. Our next step is to zero out all the entries in the first column below the diagonal entry. Set E_1 equal to

$$\begin{bmatrix} 1 & 0 & 0 \\ \frac{2}{3} & 1 & 0 \\ \frac{1}{3} & 0 & 1 \end{bmatrix}$$

Notice that all the entries in E_1 have absolute value no greater than 1. This will always happen when a_{11} has the largest absolute value of all the entries in the first column. We now calculate E_1A_1,

$$E_1A_1 = \begin{bmatrix} -3 & 3 & 2 \\ 0 & 3 & \frac{7}{3} \\ 0 & 7 & \frac{11}{3} \end{bmatrix}$$

The next step is to find that entry in the second column in rows 2 or 3 with the largest absolute value. For this matrix, the entry in the third row is the one we seek. We now interchange rows 2 and 3. Set $r_2 = 3$; i.e., row 2 and row 3 are switched. Notice that the subscript tells us not only which rows are interchanged, but also that it is the second interchange. This operation produces the matrix A_2,

$$A_2 = \begin{bmatrix} -3 & 3 & 2 \\ 0 & 7 & \frac{11}{3} \\ 0 & 3 & \frac{7}{3} \end{bmatrix}$$

To zero out all entries in the second column below the diagonal entry, we use the matrix

$$E_2 = \begin{bmatrix} 1 & 0 & 0 \\ 0 & 1 & 0 \\ 0 & -\frac{3}{7} & 1 \end{bmatrix}$$

Thus, we finally have

$$U = E_2A_2 = \begin{bmatrix} -3 & 3 & 2 \\ 0 & 7 & \frac{11}{3} \\ 0 & 0 & \frac{16}{21} \end{bmatrix}$$

Let $R_j(k)$ be the elementary row matrix that interchanges rows j and k. We clearly have the following equation relating A and U:

$$E_2R_2(3)E_1R_1(3)A = U$$

or

$$A = R_1(3)E_1^{-1}R_2(3)E_2^{-1}U$$

Thus, to solve the equation $A\mathbf{x} = \mathbf{b}$, we need to solve the equation

$$U\mathbf{x} = E_2R_2(3)E_1R_1(3)\mathbf{b} = [1 \quad \tfrac{4}{3} \quad \tfrac{2}{21}]^T$$

The solution to this system is $x_1 = -\frac{1}{8}$, $x_2 = \frac{1}{8}$, and $x_3 = \frac{1}{8}$.

This technique, which puts the largest entry of the first column in the 1,1 position, etc., is called partial pivoting. It seems like a lot of work, with many matrices to remember, but in reality the matrices E_j and $R_j(k)$ are easy to store. In fact, once we have determined E_j, there is no longer any need to keep track of the entries of A that lie below the main diagonal in its first j columns. (They are all zeros.) In practice it is common for only the nonobvious entries of E_j to be stored in the jth column of an updated version of A. Thus, for the matrix E_1 above, the numbers $\frac{2}{3}$ and $\frac{1}{3}$ would be stored in the 2,1 and 3,1 positions of A, respectively. At the same time that the entries of E_j are being stored, the top rows of A would be replaced by the corresponding rows of the matrix U. This overwriting procedure considerably reduces the needed amount of storage space.

In the following example we go through the details of this precedure and at each step write out the various matrices.

EXAMPLE 4 Solve the system:

$$\begin{aligned} x_1 + x_2 + x_3 &= 1 \\ 6x_1 \qquad + 3x_3 &= 2 \\ x_1 - 2x_2 + x_3 &= 4 \end{aligned}$$

Solution. The coefficient matrix A equals

$$\begin{bmatrix} 1 & 1 & 1 \\ 6 & 0 & 3 \\ 1 & -2 & 1 \end{bmatrix}$$

$$r_1 = 2 \qquad E_1 = \begin{bmatrix} 1 & 0 & 0 \\ -\frac{1}{6} & 1 & 0 \\ -\frac{1}{6} & 0 & 1 \end{bmatrix}$$

We now update the matrix A; that is, we compute $E_1R_1(2)A$, and in the first column of this matrix we store the two crucial entries of E_1. Thus,

$$A_1 = \begin{bmatrix} 6 & 0 & 3 \\ -\frac{1}{6} & 1 & \frac{1}{2} \\ -\frac{1}{6} & -2 & \frac{1}{2} \end{bmatrix}$$

$$r_2 = 3 \qquad E_2 = \begin{bmatrix} 1 & 0 & 0 \\ 0 & 1 & 0 \\ 0 & \frac{1}{2} & 1 \end{bmatrix} \qquad A_2 = \begin{bmatrix} 6 & 0 & 3 \\ -\frac{1}{6} & -2 & \frac{1}{2} \\ -\frac{1}{6} & \frac{1}{2} & \frac{3}{4} \end{bmatrix}$$

Thus, U is the following matrix:

$$\begin{bmatrix} 6 & 0 & 3 \\ 0 & -2 & \frac{1}{2} \\ 0 & 0 & \frac{3}{4} \end{bmatrix}$$

Hence, we need to solve the following system:

$$U \begin{bmatrix} x_1 \\ x_2 \\ x_3 \end{bmatrix} = E_2 R_2(3) E_1 R_1(2) \begin{bmatrix} 1 \\ 2 \\ 4 \end{bmatrix} = \begin{bmatrix} 2 \\ \frac{11}{3} \\ \frac{15}{6} \end{bmatrix}$$

The solution to this system is $x_3 = \frac{10}{3}$, $x_2 = -1$, and $x_1 = -\frac{4}{3}$. ■

We have assumed in the above discussion that the coefficient matrix A is nonsingular. The effect of this assumption is that we will always find an entry in the column we are examining that is nonzero. For example, if the first column of A has only zeros, then clearly A is not invertible. Moreover, if after the first column is zeroed out, all the entries in the second column below the first row are zero, it is easy to see that $\det(A) = 0$. Hence, A would be singular. One can argue in a similar fashion when A is an $n \times n$ matrix for $n > 3$. This means that the upper triangular matrix U will have nonzero entries on its main diagonal and we can therefore solve the system of equations $U\mathbf{x} = \mathbf{y}$.

For some problems, in order to avoid roundoff errors caused by dividing by small quantities, a procedure called complete pivoting is implemented. This technique maximizes the entries in the main diagonal of the matrix U. This is done by first searching for that entry in A that has largest absolute value. This term is then placed in the 1,1 position by a row and column interchange. Thus, partial pivoting involves only column searches, while complete pivoting searches through all the entries of A. Complete pivoting is usually not implemented since it involves many more comparisons than partial pivoting and will in practice not improve the accuracy obtained with partial pivoting.

The reader is warned that there are systems of equations for which partial or complete pivoting do not prevent large relative errors from occurring. A simple example of such a system arises when the coefficient matrix is the

Hilbert matrix. The $n \times n$ Hilbert matrix has as entry in the j,k position the number $(j+k-1)^{-1}$. Thus, the 3×3 Hilbert matrix equals

$$\begin{bmatrix} 1 & \frac{1}{2} & \frac{1}{3} \\ \frac{1}{2} & \frac{1}{3} & \frac{1}{4} \\ \frac{1}{3} & \frac{1}{4} & \frac{1}{5} \end{bmatrix}$$

For small values of n ($n = 6$) partial (or complete) pivoting will not produce very accurate solutions unless double precision arithmetic is used. For larger values of n ($n \geq 20$) even double precision arithmetic will in general not produce accurate results.

If very accurate solutions are desired, Gaussian elimination with partial pivoting is a first step. This can then be followed by an iteration scheme, which, in most cases, gives a better approximation to the true solution.

PROBLEM SET 7.2

1. Solve the following systems of equations using partial pivoting:

a. $$\begin{aligned} 2x_1 + 3x_2 &= 5 \\ 8x_1 + x_2 &= 1 \end{aligned}$$

b. $$\begin{aligned} 5x_2 - x_3 &= 1 \\ x_1 + x_2 + x_3 &= 5 \\ 2x_1 \qquad + x_3 &= -1 \end{aligned}$$

2. Solve the following system of equations for each of the given nonhomogeneous terms. Use partial pivoting and updating as was done in Example 3.

$$\begin{aligned} x_1 + x_2 + x_3 &= 6, \; -1 \\ 2x_1 \qquad - x_3 &= 0, \quad 1 \\ x_1 + 4x_2 \qquad &= 1, \quad 2 \end{aligned}$$

3. Solve the following system of equations:

$$H[x_1,x_2,x_3]^T = [1,0,0]^T$$

where H is the 3×3 Hilbert matrix. Use partial pivoting with Gaussian elimination, and assume that only four significant digits are carried. Then compute the actual solution and compare your two answers.

7.3 COMPUTING EIGENVALUES

In this section a computational method for calculating the eigenvalues and eigenvectors of a matrix is discussed. Theoretically, we can compute these quantities for any matrix A. We merely calculate $\det(A - \lambda) = p(\lambda)$ and then find those values λ_j for which $p(\lambda_j) = 0$. The kernel of the matrix $A - \lambda_j I$ then contains all the eigenvectors corresponding to the eigenvalue λ_j. At first glance, this seems to be a reasonable way to proceed. This route, though, is strewn with several intractable obstacles. First, if A is a fairly large matrix (100×100 would not be uncommon), the calculation of $\det(A - \lambda I)$ is quite formidable and time-consuming, except in some special cases, e.g., if A is upper triangular. Even if one is able to calculate $p(\lambda)$, the determination of its roots is an interesting problem in its own right.

Bearing in mind these difficulties, we now sketch an iterative technique that often gives good approximations for both eigenvalue and eigenvector.

Suppose first that A is a 2×2 matrix with two real and distinct eigenvalues λ_1 and λ_2 such that $|\lambda_1| < |\lambda_2|$. Let $\mathbf{f}_1$ and $\mathbf{f}_2$ be eigenvectors corresponding to λ_1 and λ_2, respectively. We wish to examine what happens when we take some fixed vector $\mathbf{x}_0$ and repeatedly multiply it by A. Thus, let $\mathbf{x}_0 = c_1\mathbf{f}_1 + c_2\mathbf{f}_2$ be an arbitrary vector in $\mathbb{R}^2$. Define $\mathbf{y}_1, \mathbf{y}_2, \ldots, \mathbf{y}_n, \ldots$, by

$$\begin{aligned}
\mathbf{y}_1 &= A\mathbf{x}_0 = A(c_1\mathbf{f}_1 + c_2\mathbf{f}_2) \\
&= c_1 A\mathbf{f}_1 + c_2 A\mathbf{f}_2 \\
&= c_1\lambda_1\mathbf{f}_1 + c_2\lambda_2\mathbf{f}_2 \\
\mathbf{y}_2 &= A\mathbf{y}_1 = A^2\mathbf{x}_0 \\
&= c_1\lambda_1^2\mathbf{f}_1 + c_2\lambda_2^2\mathbf{f}_2 \\
\mathbf{y}_n &= A^n\mathbf{x}_0 = c_1\lambda_1^n\mathbf{f}_1 + c_2\lambda_2^n\mathbf{f}_2
\end{aligned} \tag{7.1}$$

Since the terms λ_j^n approach 0 if $|\lambda_j| < 1$ and become unbounded if $|\lambda_j| > 1$, we need to introduce a scaling factor. Thus, let $\mathbf{x}_n$ equal

$$\begin{aligned}
\mathbf{x}_n &= \frac{\mathbf{y}_n}{\lambda_2^n} \\
&= \left(\frac{\lambda_1}{\lambda_2}\right)^n c_1\mathbf{f}_1 + c_2\mathbf{f}_2
\end{aligned} \tag{7.2}$$

Since $|\lambda_1| < |\lambda_2|$, the factor $(\lambda_1/\lambda_2)^n$ converges to zero as n approaches infinity. Thus, the sequence of vectors $\mathbf{x}_n$ converges to $c_2\mathbf{f}_2$, an eigenvector corresponding to the eigenvalue of A with largest absolute value. Thus, if n is large enough, $\mathbf{x}_n = A^n\mathbf{x}_0/\lambda_2^n$ will be approximately equal to an eigenvector of A, at least if $c_2 \neq 0$. To get an approximation to λ_2, we only need to compute $\|\mathbf{y}_{n+1}\|/\|x_n\|$, and then decide if λ_2 is positive or negative. This can be done by observing how the signs of the coordinates of $\mathbf{x}_n$ compare with the signs of the coordinates of $\mathbf{x}_{n-1}$. In practice, since λ_2 would not be known, instead of dividing by λ_2, one divides $\mathbf{y}_n$ by the magnitude of its largest coordinate or some other factor that has the same order of magnitude as λ_2, to get $\mathbf{x}_n$. Thus if $\mathbf{y}_n = (10, -15)$, then $\mathbf{x}_n = (\frac{2}{3}, -1)$.

Table 7.1 lists the results of the above iteration scheme for the matrix

$$A = \begin{bmatrix} 2 & 3 \\ 3 & -4 \end{bmatrix}$$

and $\mathbf{x}_0 = (1,1)$.

TABLE 7.1

n	$\mathbf{x}_n$	λ
0	(1, 1)	
1	(0.9999, −0.1999)	5
5	(0.6957, −1.000)	−4.1593
10	(−0.3912, 0.9999)	−5.3565
20	(−0.4140, 1.000)	−5.2435

The actual eigenvalues for this matrix are $\lambda_1 = -1 + 3\sqrt{2} \approx 3.242640686$, and $\lambda_2 = -1 - 3\sqrt{2} \approx -5.242640686$. We note that our approximation is not very good, being accurate only through the hundredths digit. We will see how this may be improved in a few minutes. Eigenvectors corresponding to λ_1 and λ_2 are $\mathbf{f}_1 = (1, \sqrt{2} - 1)$ and $\mathbf{f}_2 = (1 - \sqrt{2}, 1)$ respectively.

We have to be somewhat clever to use this technique in computing the remaining eigenvalue of our matrix. We remind the reader that if A is an $n \times n$ matrix with $\{\lambda_1, \lambda_2, \ldots, \lambda_n\}$ its eigenvalues, then the eigenvalues of the matrix $A - cI$ are precisely $\{\lambda_1 - c, \lambda_2 - c, \ldots, \lambda_n - c\}$; cf. problems 17, 18, and 19 in Section 5.1. In particular, if we set c equal to the value -5.2435, and repeat this procedure for the matrix

$$B = A - (-5.2435)I = \begin{bmatrix} 7.2435 & 3 \\ 3 & 1.2435 \end{bmatrix}$$

we get the data in Table 7.2. Thus, (1.000, 0.4142) is an approximate eigenvector of $A - (-5.2435)I$, and hence an approximate eigenvector of A. Moreover, $8.4861 + (-5.2435) = 3.2426$ will be an approximate eigenvalue of A. Comparing Tables 7.1 and 7.2, we note that even after 20 iterations the approximate eigenvalue -5.2435 is correct only to the hundredths digit, while the approximate eigenvalue 3.2426 is correct through the ten-thousandth digit after only three iterations. The reason for the more rapid convergence in the second case is readily apparent if we examine (7.2). For the matrix A, the ratio of its eigenvalues λ_1/λ_2 is approximately $\frac{3}{5}$. This ratio, for the matrix B, is on the order of

TABLE 7.2

n	$\mathbf{x}_n$	λ	$\lambda - 5.2435$
0	(1, 1)		
1	(0.9999, 0.4142)	10.2435	5.0000
2	(1.000, 0.4124)	8.4862	3.2427
3	(1.000, 0.4142)	8.4861	3.2426
5	(1.000, 0.4142)	8.4861	3.2426
10	(1.000, 0.4142)	8.4861	3.2426
20	(1.000, 0.4142)	8.4861	3.2426

TABLE 7.3

n	$\mathbf{x}_n$	λ	$\lambda + 3.2426$
0	(1, 1)		
1	(0.4142, −1.000)	−4.2426	−1
2	(−0.4142, 1.000)	−8.4852	−5.2426
3	(0.4142, −1.000)	−8.4852	−5.2426
5	(0.4142, −1.000)	−8.4852	−5.2426
10	(−0.4142, 1.000)	−8.4852	−5.2426
20	(−0.4142, 1.000)	−8.4852	−5.2426

(0.001), which is considerably smaller than the corresponding ratio for A. Thus, the vectors $\mathbf{x}_n$ will converge more rapidly in the latter case to the corresponding eigenvector. To illustrate this one more time we compute the eigenvalue with largest magnitude of the matrix $A-(3.2426)I$. This should give us a much better approximate value to the eigenvector $-1-3\sqrt{2}\approx -5.2426$, c.f., Table 7.3.

We've seen in the previous calculations that if the ratio λ_1/λ_2 is small, the convergence of $\mathbf{x}_n$ to an eigenvector is fairly rapid. We next discuss an effective way to force this ratio to be small. Suppose that A is an $n\times n$ matrix with real eigenvalues $\lambda_1, \ldots, \lambda_n$. Suppose also that previous calculations lead us to believe that $\lambda_n \approx c$, i.e., $|\lambda_n - c| < |\lambda_j - c|$ for $j = 1, 2, \ldots, n-1$. Let $B=(A-cI)^{-1}$. The eigenvalues of B are $(\lambda_1 - c)^{-1}, \ldots, (\lambda_n - c)^{-1}$. Moreover if c is close to λ_n, then $|1/(\lambda_n - c)|$ will be considerably larger than $|1/(\lambda_j - c)|$ for $j=1, \ldots, n-1$. Hence the ratios $[1/(\lambda_j - c)]/[1/(\lambda_n - c)]$, for $j=1, \ldots, n-1$, will be close to zero. If we now iterate with the matrix $B=(A-cI)^{-1}$, we would expect to have rapid convergence to the eigenvalue $1/(\lambda_n - c)$. When implementing this procedure we do not compute $(A-cI)^{-1}$ but solve the following equations for $\mathbf{y}_n$ and then rescale the $\mathbf{y}_n$ to get $\mathbf{x}_n$:

$$(A-cI)\mathbf{y}_1 = \mathbf{x}_0$$
$$(A-cI)\mathbf{y}_n = \mathbf{y}_{n-1}$$
$$\mathbf{x}_n = \frac{\mathbf{y}_n}{\|\mathbf{y}_n\|}$$

The preceding techniques are referred to as the power method and the inverse iteration method.

PROBLEM SET 7.3

1. Let $A=\begin{bmatrix} -10 & 1 \\ 0 & 1 \end{bmatrix}$. Let $\mathbf{x}_0 = (1,1)$. Using the iteration techniques described in this section, calculate approximations to the eigenvalues of A and their corresponding eigenvectors.

2. Let $A = \begin{bmatrix} 10 & 0 & 1 \\ 0 & 0 & 0 \\ 1 & 0 & 1 \end{bmatrix}$. Repeat problem 1 with $\mathbf{x}_0 = (1,1,1)$.

3. Let A and $\mathbf{x}_0$ be as in problem 1. How many iterations does it take for the approximation to the eigenvalue -10 to have an error of 10^{-4}? Same question for the eigenvalue 1.

4. Use (7.2) to estimate, in terms of the ratio (λ_1/λ_2), how many iterations it will take for $\mathbf{x}_n$ to approximate an eigenvector with a relative error of 10^{-5}. Assume that both c_1 and c_2 are nonzero.

SUPPLEMENTARY PROBLEMS

1. Write a computer program to solve systems of equations. Incorporate partial pivoting and updating in your scheme, and then solve

$$H\mathbf{x} = \mathbf{e}_j \qquad j = 1, 2, 3$$

where H is the 3×3 Hilbert matrix and $\mathbf{e}_j$ are the standard basis vectors of $\mathbb{R}^3$.

2. Write a computer program that uses the power method to calculate the largest eigenvalue of a square matrix. Then determine the largest eigenvalue of the 3×3 Hilbert matrix.

3. Quite often iteration schemes are used to solve systems of equations. For example, if the magnitudes of the diagonal elements of A are much larger than those of the off-diagonal elements, the following scheme, known as Jacobi's method, is helpful. Thus, let $A = [a_{jk}] = D + K$, where D is the $n \times n$ diagonal matrix $[a_{jk}\delta_{jk}]$, and K equals $A - D$. For example,

$$A = \begin{bmatrix} 6 & 1 \\ 1 & 9 \end{bmatrix} = \begin{bmatrix} 6 & 0 \\ 0 & 9 \end{bmatrix} + \begin{bmatrix} 0 & 1 \\ 1 & 0 \end{bmatrix}$$

To solve $A\mathbf{x} = \mathbf{b}$, we write this as

$$D\mathbf{x} = \mathbf{b} - K\mathbf{x}$$

An initial guess $\mathbf{x}_0$ is picked and then iterated according to the scheme

$$D\mathbf{x}_{k+1} = \mathbf{b} - K\mathbf{x}_k$$

This procedure assumes of course that D is an invertible matrix. The iterations are stopped when the magnitude of $\mathbf{x}_{k+1} - \mathbf{x}_k$ is smaller than some prescribed value. Use this idea to approximate the solution to

$$\begin{bmatrix} 6 & 1 \\ 1 & 9 \end{bmatrix}\begin{bmatrix} x_1 \\ x_2 \end{bmatrix} = \begin{bmatrix} -2 \\ 3 \end{bmatrix}$$

with $\mathbf{x}_0 = (0,0)$. Compare the difference between $\mathbf{x}_3$ and the actual solution.

4. Write a computer program to solve systems of equations that uses Jacobi's iteration scheme. When testing your scheme try various coefficient matrices. In particular, compare rates of convergence between matrices with relatively large diagonal entries versus those whose diagonal entries are of the same magnitude as the off-diagonal entries.

8
APPLICATIONS

This chapter will be somewhat disjointed in that the sections are independent of one another. Each is a discussion of a particular problem and how the material we've covered so far, or its extensions, can be used to analyze these problems. The point of these sections is not the particular problem but rather the demonstration of the applicability of the ideas and techniques discussed in the preceding chapters.

8.1 A SAMPLING THEOREM

Suppose that we are receiving some sort of signal that we can measure at various times; that is, there is a function $f(t)$ and we are able to determine $f(t_k)$, $1 \leq k \leq N$. Is it possible to calculate $f(t)$ from the finite number of samples $f(t_k)$? The answer, in general, is clearly no. Given the N data points $(t_k, f(t_k))$ there are an infinite number of different functions whose graphs pass through these points. We thus need to make some additional assumptions about the signal we are measuring, if we hope to be able to recover it from a finite number of samples.

We make two assumptions, first that we are receiving a periodic signal and second that the frequencies are bounded. These two assumptions are expressed mathematically by assuming that our signal $f(t)$ can be written as a finite linear combination of sines and cosines.

$$f(t) = \frac{a_0}{2} + \sum_{k=1}^{N} b_k \sin kt + \sum_{k=1}^{N} a_k \cos kt \tag{8.1}$$

Each of the functions $\sin kt$, $1 \le k \le N$, and $\cos kt$, $0 \le k \le N$, is periodic with period $2\pi/k$ and frequency $k/2\pi$. Thus, the largest frequency is $N/2\pi$. What we wish to show is that if we know $f(t_k)$ for $2N+1$ points t_k, then $f(t)$ is completely determined, i.e., the constants a_k, $k = 0, 1, \ldots, N$, and b_k, $k = 1, 2, \ldots, N$, can be calculated. Let V be the vector space of all functions of the form (8.1). Clearly a spanning set for V is the set $\{1, \cos t, \ldots, \cos Nt, \sin t, \ldots, \sin Nt\}$. It is not immediately clear that this set is linearly independent; so we may only assert that $\dim(V) \le 2N+1$. We next exhibit a function $\mathbf{g}(t)$ that has the property that $\mathbf{g}(0) = 1$ and $\mathbf{g}[2k\pi/(2N+1)] = 0$, $k = \pm 1, \pm 2, \ldots, \pm N$, and such that $\mathbf{g}(t)$ is also in V. Define $\mathbf{g}(t)$ by

$$\mathbf{g}(t) = \frac{1}{2N+1}\left\{1 + 2\sum_{k=1}^{N} \cos kt\right\} \tag{8.2}$$

Note that $\mathbf{g}(t)$ is in V and that $\mathbf{g}(0) = 1$. We next rewrite $\mathbf{g}(t)$ in the following manner:

$$\begin{aligned}
\mathbf{g}(t) = \frac{\mathbf{g}(t)\sin(t/2)}{\sin(t/2)} &= \frac{\sin(t/2) + \sum_{k=1}^{N} 2\cos kt \sin(t/2)}{(2N+1)\sin(t/2)} \\
&= \frac{\sin(t/2) + \sum_{k=1}^{N} [\sin(k+\frac{1}{2})t - \sin(k-\frac{1}{2})t]}{(2N+1)\sin(t/2)} \\
&= \frac{\sin(2N+1)t/2}{(2N+1)\sin(t/2)}
\end{aligned}$$

Thus,

$$\mathbf{g}(t) = \frac{1}{2N+1}\left\{1 + 2\sum_{k=1}^{N} \cos kt\right\} = \frac{\sin(2N+1)t/2}{(2N+1)\sin(t/2)} \tag{8.3}$$

Formula (8.3) immediately implies that $\mathbf{g}[2j\pi/(2N+1)] = 0$ for $j = \pm 1, \pm 2, \ldots, \pm N$. Let $t_k = 2k\pi/(2N+1)$, and define the functions $\mathbf{g}_k$ by

$$\mathbf{g}_k(t) = \mathbf{g}(t - t_k) \qquad k = -N, \ldots, N \tag{8.4}$$

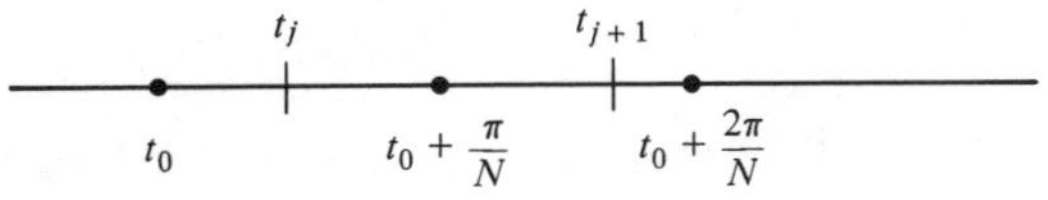

Figure 8.1

These $2N + 1$ functions are in the vector space V (the reader should verify this), and satisfy $\mathbf{g}_j(t_k) = \delta_{jk}$. The preceding discussion implies that these $2N + 1$ vectors are linearly independent. Since $\dim(V) \le 2N + 1$, we may now infer that $\dim(V) = 2N + 1$, and that the functions $\{\mathbf{g}_j\}$ form a basis of V. Thus there are constants c_j such that our signal function $\mathbf{f}(t)$ is equal to

$$\mathbf{f}(t) = \sum_{j=-N}^{N} c_j \mathbf{g}_j(t) \tag{8.5}$$

Moreover we have $\mathbf{f}(t_k) = \sum_{j=-N}^{N} c_j \mathbf{g}_j(t_k) = c_k$. Thus,

$$\mathbf{f}(t) = \sum_{j=-N}^{N} \mathbf{f}(t_j)\mathbf{g}_j(t)$$

Examining (8.1) we see that the largest frequency in f equals $N/2\pi$. Thus the shortest wavelength is $2\pi/N$. We now observe that the distance between any two consecutive t_j is equal to $2\pi/(2N + 1)$ (less than half the shortest wavelength). Hence, between the endpoints of one wave there are always two sampling points, at least one of which is not located at one of the three nodes. See Figure 8.1. Intuitively this should have been expected. For suppose the spacing of the t_j was exactly one-half of the shortest wavelength. Then $\mathbf{f}(t_j)$ could equal zero for each t_j, and we would then infer that $\mathbf{f}(t) = 0$ for all t.

There is also an infinite-dimensional version of this result, which is called the sampling theorem. Its exact statement may be found in many books dealing with communication theory, digital filters, or Fourier analysis.

PROBLEM SET 8.1

1. The two sets of functions $F = \{1, \sin t, \cos t, \ldots, \sin Nt, \cos Nt\}$ and $G = \{\mathbf{g}_j(t): -N \le j \le N\}$, where the $\mathbf{g}_j(t)$ are defined by (8.2) and (8.4), are both bases of V.
   a. Let $\mathbf{g}(t)$ be defined by (8.2). Show that $\mathbf{g}(t - c)$ is in V for any constant c. Hence, deduce that the set G is a subset of V.
   b. Write the function identically equal to **1** as a linear combination of the functions in G.
   c. Write the function $\mathbf{g}_1(t)$ as a linear combination of the functions in F.
2. Let $\mathbf{f}(t) = a_1 + a_2 \sin 3t + a_3 \cos t$.
   a. How many sample points are needed, and where should they be located in order to determine $\mathbf{f}(t)$?

b. Calculate a_1, a_2, and a_3 assuming that $\mathbf{f}(t_j)$ is known, where $t_j = 2j\pi/7$, for $j = -3, -2, -1, 0, 1, 2, 3$.

3. Show that the set F in problem 1 consists of orthogonal vectors, where $\langle \mathbf{f},\mathbf{g} \rangle = \int_0^{2\pi} \mathbf{f}(t)\mathbf{g}(t)dt$.

4. Use the result of problem 3 to compute the coordinates of $\mathbf{f}(t)$ with respect to the basis F.

5. Define the following inner product for V.

$$[\mathbf{f},\mathbf{h}] = \sum_{j=-N}^{N} \mathbf{f}(t_j)\mathbf{h}(t_j)$$

a. Show that [,] is an inner product in the sense that it satisfies b and c of Theorem 6.3, and that it also satisfies $[\mathbf{f},\mathbf{f}] \geq 0$ and $[\mathbf{f},\mathbf{f}] = 0$ only if $\mathbf{f} = 0$.

b. Show that G is an orthonormal basis of V with respect to the inner product [,].

c. Find the coordinates of **f** with respect to the basis G.

8.2 AGE-DEPENDENT POPULATIONS

Suppose we wish to study the change in the population size of some species for which reproduction is possible only during certain stages and not throughout an individual's life span.

As a simple example of this, let's assume that we can partition the life span of the females into four stages, each lasting the same amount of time. We further assume that reproduction is possible only during the second and third stages. For humans, the first stage might be from 0 to 14 years, the second stage from 15 to 29 years, the third stage from 30 to 44 years, and the last from 45 to 59 years.

Let f_{jk}, $j = 1, 2, 3, 4$; $k = 1, 2, \ldots$, represent the number of females in stage j at time k. We next suppose that the female population at time $k + 1$ is related to the female population at time k via the following equations:

$$\begin{aligned} f_{1(k+1)} &= b_2 f_{2k} + b_3 f_{3k} \\ f_{2(k+1)} &= a_1 f_{1k} \\ f_{3(k+1)} &= a_2 f_{2k} \\ f_{4(k+1)} &= a_3 f_{3k} \end{aligned} \tag{8.6}$$

The numbers b_2 and b_3 are the respective birthrates during the second and third stages. The terms a_1, a_2, and a_3 represent the nondeath rate of the respective stages. That is, if 100 females are in stage 1 at time k, then $100a_1$ of them will enter stage 2 at time $k + 1$.

Let $\mathbf{F}_k = [f_{1k} \;\; f_{2k} \;\; f_{3k} \;\; f_{4k}]^T$. Writing (8.6) as a matrix equation we have

$$\begin{aligned} \mathbf{F}_{k+1} &= \begin{bmatrix} 0 & b_2 & b_3 & 0 \\ a_1 & 0 & 0 & 0 \\ 0 & a_2 & 0 & 0 \\ 0 & 0 & a_3 & 0 \end{bmatrix} \mathbf{F}_k \\ &= B\mathbf{F}_k \end{aligned} \tag{8.7}$$

Thus, we have

$$\mathbf{F}_{k+1} = B^k\mathbf{F}_1 \tag{8.8}$$

In trying to understand the long-term behavior of the population we need to analyze the powers of B. Since the fourth stage does not contribute to the birthrate, it suffices to analyze A^k, where A is the first 3×3 principal minor of B. Moreover one can easily show that

$$B^k = \left[\begin{array}{ccc|c} & & & 0 \\ & A^k & & 0 \\ & & & 0 \\ \hline - & - & - & 0 \end{array}\right]$$

The form of B^k as well as (8.6) tells us that the first three stages are essentially a closed system. Individuals may enter the fourth stage, but the size of this stage does not influence the population levels of the earlier stages. For purposes of illustration, let's take a specific A.

$$A = \begin{bmatrix} 0 & \frac{1}{3} & \frac{1}{6} \\ \frac{1}{2} & 0 & 0 \\ 0 & \frac{1}{2} & 0 \end{bmatrix} \tag{8.9}$$

There is a theorem that guarantees that a matrix with the above form (at least two consecutive nonzero entries in the first row) has the property that A^n has only positive entries for some positive integer N. For this particular A, $N = 5$ works. There is also another set of theorems due to Perron, which state that if A is a square matrix such that A^n has only positive entries for some n, then A has a largest positive eigenvalue λ_0, and there is an eigenvector $\mathbf{z}_0 = (z_1, z_2, \ldots, z_n)$ associated with λ_0 such that $z_j > 0$ for each component of $\mathbf{z}_0$. Moreover λ_0 is a simple eigenvalue, i.e., λ_0 has multiplicity 1, and if λ is any other eigenvalue of A (complex eigenvalues allowed), then $|\lambda| < \lambda_0$. For our A, $\lambda_0 = \frac{1}{2}$, and $\mathbf{z}_0 = (1,1,1)$.

Let $\mathbf{x}_0$ be any initial distribution of the female population. We are interested in what happens to $A^k\mathbf{x}_0$ for large k. Since every eigenvalue of A has absolute value less than λ_0 and λ_0 has multiplicity 1, it can be shown that

$$\lim_{k\to\infty} \frac{A^k[\mathbf{x}_0]}{\lambda_0^k} = c\mathbf{z}_0 \tag{8.10}$$

for some constant c. Thus, asymptotically $A^k\mathbf{x}_0$ looks like $\lambda_0^k(c\mathbf{z}_0)$. The exact value of c is also fairly easy to determine. To see how this may be done, let $\mathbf{z}^*$ be an eigenvector of A^T corresponding to λ_0. By Perron's theorem, $\mathbf{z}^*$ can be picked so that all its components are also positive. We remind the reader of formula (6.23).

$$\langle A\mathbf{x},\mathbf{y}\rangle = \langle \mathbf{x}, A^T\mathbf{y}\rangle$$

Since all the components of $\mathbf{z}_0$ and $\mathbf{z}^*$ are positive, we have $\langle \mathbf{z}_0, \mathbf{z}^* \rangle > 0$. Moreover,

$$\begin{aligned} c\langle \mathbf{z}_0, \mathbf{z}^* \rangle = \langle c\mathbf{z}_0, \mathbf{z}^* \rangle &= \lim_{k\to\infty} \left\langle \frac{A^k \mathbf{x}_0}{\lambda_0^k}, \mathbf{z}^* \right\rangle \\ &= \lim_{k\to\infty} \left\langle \frac{\mathbf{x}_0}{\lambda_0^k}, (A^T)^k \mathbf{z}^* \right\rangle \\ &= \lim_{k\to\infty} \left\langle \frac{\mathbf{x}_0}{\lambda_0^k}, \lambda_0^k \mathbf{z}^* \right\rangle \\ &= \langle \mathbf{x}_0, \mathbf{z}^* \rangle \end{aligned}$$

Thus,

$$c = \frac{\langle \mathbf{x}_0, \mathbf{z}^* \rangle}{\langle \mathbf{z}_0, \mathbf{z}^* \rangle} \tag{8.11}$$

Note that if $\mathbf{x}_0$ is an actual population distribution, then $\langle \mathbf{x}_0, \mathbf{z}^* \rangle > 0$. Thus $c > 0$.

For the specific A in (8.9) we have $\lambda_0 = \frac{1}{2}$, $\mathbf{z}_0 = (1,1,1)$, and $\mathbf{z}^* = (3,3,1)$. Thus, if we start with $\mathbf{x}_0 = (a_1, a_2, a_3)$, where a_j represents the number of females in the jth stage, we have

$$\begin{aligned} \lim_{k\to\infty} (\tfrac{1}{2})^{-k} A^k \mathbf{x}_0 &= \frac{\langle \mathbf{x}_0, \mathbf{z}^* \rangle}{\langle \mathbf{z}_0, \mathbf{z}^* \rangle} (1,1,1) \\ &= \frac{3a_1 + 3a_2 + a_3}{7} (1,1,1) \end{aligned} \tag{8.12}$$

Thus, $A^k \mathbf{x}_0 \approx (\frac{1}{2})^k [(3a_1 + 3a_2 + a_3)/7] (1,1,1)$, and we see that for large k, $A^k \mathbf{x}_0$ tends to the zero vector. That is, the female, and hence the entire population, dies out.

Since $A^k \mathbf{x}_0$ looks like λ_0^k, the limiting behavior of $A^k \mathbf{x}_0$ is known without having to calculate c, if $\lambda_0 \neq 1$. Thus, only when $\lambda_0 = 1$ is it necessary to calculate c, as we did in (8.11), in order to determine the limiting behavior of $A^k x_0$.

PROBLEM SET 8.2

1. Let $A = \begin{bmatrix} 0 & \frac{1}{3} & \frac{1}{6} \\ \frac{1}{2} & 0 & 0 \\ 0 & \frac{1}{2} & 0 \end{bmatrix}$.

a. Show that every entry in A^5 is positive.

b. Compute the eigenvalues of A, and verify that $\frac{1}{2}$ is the largest positive eigenvalue, that it is simple, and $|d| < \frac{1}{2}$ for the other eigenvalues.

2. Let A be the matrix in problem 1. Find $\mathbf{z}_0$ and $\mathbf{z}^*$, where $A\mathbf{z}_0 = \frac{1}{2}\mathbf{z}_0$ and $A^T\mathbf{z}^* = \frac{1}{2}\mathbf{z}^*$. For $\mathbf{x} = (-1,1,-2)$ find c such that $\lim_{k\to\infty} A^k\mathbf{x}/(\frac{1}{2})^k = c\mathbf{z}_0$.

3. Let $A = \begin{bmatrix} 1 & 2 \\ 3 & 6 \end{bmatrix}$. The eigenvalues of A are 0 and 7. The eigenspace of 7 is $S[(1,3)]$. Let

$\mathbf{x} = (a, b)$. Show that $\lim_{k\to\infty} A^k\mathbf{x}/7^k = c(1,3)$ for some constant c. Show how to calculate c by using the fact that $(1,2)$ is an eigenvector of A^T corresponding to 7.

4. Let $A = \begin{bmatrix} d & 1 \\ 0 & d \end{bmatrix}$. Notice that d is an eigenvalue of A with multiplicity 2 and that $\dim[\ker(A - dI)]$ equals 1.

a. Show that $A^n = \begin{bmatrix} d^n & nd^{n-1} \\ 0 & d^n \end{bmatrix}$.

b. Find a vector $\mathbf{x}$ such that $\lim_{n\to\infty} A^n\mathbf{x}/d^n$ does not exist.

5. Let $A = \begin{bmatrix} 0 & a & b \\ c & 0 & 0 \\ 0 & d & 0 \end{bmatrix}$, where a, b, c, and d are positive.

a. Show that every entry of A^5 is positive. What does this imply about the eigenvalues of A? What can you say about $\lim_{n\to\infty} A^n\mathbf{x}$?

b. Show that 1 is the largest positive eigenvalue of A if and only if $c(a + bd) = 1$.

6. Suppose that some species has three distinct stages in its life cycle. Let f_{jk}, $j = 1, 2, 3$, $k = 0, 1, 2, \ldots$, denote the number of females at time k in the jth stage. Suppose the following equations are valid:

$$\begin{aligned} f_{1(k+1)} &= 2f_{2k} + 4f_{3k} \\ f_{2(k+1)} &= \tfrac{3}{8}f_{1k} \\ f_{3(k+1)} &= \tfrac{1}{6}f_{2k} \end{aligned}$$

Determine the asymptotic behavior of the population sizes as a function of the initial population distribution.

8.3 STOCHASTIC MATRICES

Suppose that we have some system that must be in exactly one of a finite number of states $S_1, S_2, \ldots, S_n$ at any given time t_k, $k = 0, 1, 2, \ldots$, where $t_k < t_{k+1}$ for every k. We also assume that the system is not deterministic. That is, if at time t_k the system is in state S_j, then we cannot guarantee which state the system will be in at time t_{k+1}. We do, however, assume that we know the various conditional probabilities for the system to move from one state to another. By $P(S_j^{t_m}|S_k^{t_{m-1}})$ we denote the conditional probability that the system will be in state S_j at time t_m given that the system was in state S_k at time t_{m-1}. We assume that these conditional probabilities satisfy

$$P(S_j^{t_m}|S_k^{t_{m-1}}) = P(S_j^{t_2}|S_k^{t_1}) = P(S_j|S_k) = p_{kj} \tag{8.13}$$

That is, the system's probability of passing from state S_k to state S_j in one time period does not depend on when the system was in state S_k.

A second assumption is that these probabilities do not depend upon where the system was at time t_{m-2}. In other words, where the system will be in the next instant depends only upon its present state and not on how it came to be there.

A system that satisfies these assumptions is called a Markov chain. We remind the reader that the symbols p_{jk} denote the system's conditional probability of being in state S_k, if it was in state S_j at the preceding instant.

We wish to discuss two problems. The first is, given an initial probabilistic description of the system, i.e., if we know the probabilities that the system is in a particular state, can we determine the probability of the system's being in a given state at any future instant? The second is to determine what happens to these probabilities as t_m gets arbitrarily large.

Let $P = [p_{jk}]$ be the $n \times n$ matrix whose entries are the given conditional probabilities. We note two features of the matrix P. First, $0 \leq p_{jk} \leq 1$ for each entry, and second

$$p_{j1} + p_{j2} + \cdots + p_{jn} = \sum_{k=1}^{n} p_{jk} = 1 \qquad j = 1, 2, \ldots, n \tag{8.14}$$

This last equality follows from the fact that if the system is in state S_j at one instant, then it must go to one of the n possible states at the next instant.

Definition 8.1. Let $P = [p_{jk}]$ be an $n \times n$ matrix such that the entries p_{jk} are nonnegative. If each row of P also satisfies (8.14), P is called a stochastic matrix.

The reader should be aware that some authors define a stochastic matrix somewhat differently than here. Their matrices are stochastic if the matrix columns sum to 1. There is no essential difference since if a matrix P is stochastic in our sense, its transpose will be stochastic using the other definition, and conversely.

EXAMPLE 1

a. $P = \begin{bmatrix} \frac{1}{2} & \frac{1}{2} \\ \frac{1}{4} & \frac{3}{4} \end{bmatrix}$ is a stochastic matrix.

b. $P = \begin{bmatrix} \frac{3}{2} & -\frac{1}{2} \\ \frac{1}{4} & \frac{3}{4} \end{bmatrix}$ is not a stochastic matrix, since one of the entries is negative.

c. $P = \begin{bmatrix} \frac{1}{2} & \frac{1}{2} & \frac{1}{2} \\ 0 & 1 & 0 \\ \frac{1}{3} & \frac{1}{3} & \frac{1}{3} \end{bmatrix}$ is not a stochastic matrix, since the first row sums to $\frac{3}{2}$, not 1. ■

Notice that the columns of a stochastic matrix P need not sum to 1. The most that we can say about the column sums of an $n \times n$ stochastic matrix is that they are nonnegative and no greater than n.

It is easy to show, and we shall, that if λ is an eigenvalue of a stochastic matrix, then $|\lambda| \leq 1$. It is also easy to see that $\lambda = 1$ is an eigenvalue of any stochastic matrix P.

EXAMPLE 2 Let P equal

$$\begin{bmatrix} \frac{1}{2} & 0 & \frac{1}{2} \\ \frac{1}{4} & \frac{1}{8} & \frac{5}{8} \\ \frac{1}{3} & \frac{1}{3} & \frac{1}{3} \end{bmatrix}$$

Show that P is a stochastic matrix and that $(1,1,1)^T$ is an eigenvector of P corresponding to the eigenvalue 1.

Solution. Since each entry in P is nonnegative and each row sums to 1, P is a stochastic matrix. Moreover

$$P\begin{bmatrix} 1 \\ 1 \\ 1 \end{bmatrix} = \begin{bmatrix} \frac{1}{2} & 0 & \frac{1}{2} \\ \frac{1}{4} & \frac{1}{8} & \frac{5}{8} \\ \frac{1}{3} & \frac{1}{3} & \frac{1}{3} \end{bmatrix}\begin{bmatrix} 1 \\ 1 \\ 1 \end{bmatrix} = \begin{bmatrix} 1 \\ 1 \\ 1 \end{bmatrix}$$

Thus, 1 is an eigenvalue of P with eigenvector $(1,1,1)^T$. ■

Theorem 8.1. Let $P = [p_{jk}]$ be an $n \times n$ stochastic matrix. Then 1 is an eigenvalue of P with $\mathbf{z}_0 = (1, \ldots, 1)^T$ a corresponding eigenvector. Moreover if λ is any eigenvalue of P, then $|\lambda| \leq 1$.

Proof. To see that $\mathbf{z}_0$ is an eigenvector of P we merely note that the jth entry of $P\mathbf{z}_0$ is just the sum of the jth row of P. Since P is a stochastic matrix, this sum equals 1, the jth component of $\mathbf{z}_0$. Hence $P\mathbf{z}_0 = \mathbf{z}_0$. Suppose now that λ is an eigenvalue of P with $\mathbf{x} = (x_1, x_2, \ldots, x_n)$ a corresponding eigenvector. We then have $\lambda\mathbf{x} = P\mathbf{x}$. In terms of the coordinates of $\mathbf{x}$,

$$\lambda \mathbf{x}_j = \sum_{k=1}^{n} p_{jk}\mathbf{x}_k \qquad j = 1, 2, \ldots, n \tag{8.15}$$

Let q be an index such that $|x_j| \leq |x_q|$ for every j. Since $\mathbf{x}$ is an eigenvector, we know that $\mathbf{x}$ is not the zero vector. Hence, $|x_q| > 0$. From (8.15) we have

$$\begin{aligned} |\lambda||x_q| &\leq \sum_{k=1}^{n} |p_{qk}||x_k| \leq \sum_{k=1}^{n} |p_{qk}||x_q| \\ &= \left(\sum_{k=1}^{n} p_{qk}\right)|x_q| = |x_q| \end{aligned}$$

Since $|x_q|$ is positive, we must have that $|\lambda| \leq 1$.

We remark that this proof also demonstrates that if λ is an eigenvalue of an $n \times n$ matrix $A = [a_{jk}]$, then

$$|\lambda| \le \max_{1\le j\le n} \sum_{k=1}^{n} |a_{jk}|$$

That is, the largest value of the sums of the absolute values of the entries in each row of A is larger than any eigenvalue of A.

We now develop the formula that gives the conditional probability that the system will be in state S_j at time t_{m+2} if it was in state S_k at time t_m; i.e., two time intervals elapse. Let $p_{kj}^{(2)}$ denote this conditional probability. Then

$$p_{kj}^{(2)} = p_{k1}p_{1j} + p_{k2}p_{2j} + \cdots + p_{kn}p_{nj} \tag{8.16}$$

But this is just the k,j entry of the matrix P^2. In fact, if $p_{kj}^{(q)}$ denotes the conditional probability that the system will be in state S_j q time intervals after it is in state S_k, then $p_{kj}^{(q)}$ is the k,j entry of the matrix P^q. Now suppose that at time t_0 the system is in state S_j with probability p_j. Then at time t_1 the probability that the system will be in state S_k is

$$p_1p_{1k} + p_2p_{2k} + \cdots + p_np_{nk} \qquad k = 1, 2, \ldots, n \tag{8.17}$$

Thus, if $(p_1, p_2, \ldots, p_n)$ is that vector whose components are the probabilities that the system is in state j, then the kth column of

$$(p_1, \ldots, p_n)P$$

gives the probability that the system is in state S_k at time t_1. In general if $\mathbf{p} = (p_1, \ldots, p_n)$ is the probability distribution at time t_m, then the probability distribution at time t_{m+q} is $\mathbf{p}P^q$.

EXAMPLE 3 Let $P = \begin{bmatrix} \frac{1}{2} & \frac{1}{2} \\ \frac{1}{4} & \frac{3}{4} \end{bmatrix}$ be the stochastic matrix for a system that must be in one of two possible states. If (p_1, p_2) is the initial probability distribution, determine the probability distribution of the system at time t_m.

If $\mathbf{p}^{(m)} = (p_1^{(m)}, p_2^{(m)})$ denotes these probabilities, then

$$\mathbf{p}^{(m)} = (p_1, p_2)\begin{bmatrix} \frac{1}{2} & \frac{1}{2} \\ \frac{1}{4} & \frac{3}{4} \end{bmatrix}^m$$

$$= (p_1, p_2)\begin{bmatrix} -2 & 1 \\ 1 & 1 \end{bmatrix}\begin{bmatrix} \frac{1}{4} & 0 \\ 0 & 1 \end{bmatrix}^m\begin{bmatrix} -\frac{1}{3} & \frac{1}{3} \\ \frac{1}{3} & \frac{2}{3} \end{bmatrix}$$

$$= (p_1, p_2)\begin{bmatrix} \dfrac{1 + 2(\frac{1}{4})^m}{3} & \dfrac{2 - 2(\frac{1}{4})^m}{3} \\[2ex] \dfrac{1 - (\frac{1}{4})^m}{3} & \dfrac{2 + (\frac{1}{4})^m}{3} \end{bmatrix} \tag{8.18}$$

As the number of time intervals increase, i.e., as m tends to infinity, we see that

$$\lim_{m\to\infty} \mathbf{p}^m = (p_1, p_2)\begin{bmatrix} \frac{1}{3} & \frac{2}{3} \\ \frac{1}{3} & \frac{2}{3} \end{bmatrix}$$
$$= ([p_1 + p_2]/3, 2[p_1 + p_2]/3)$$
$$= (\tfrac{1}{3}, \tfrac{2}{3})$$

The last equality follows from the fact that $p_1 + p_2$ must equal 1. Thus, for this particular system the initial probability distribution does not affect the limiting probabilities. In this example we see that the system will wind up in state S_1 with probability $\frac{1}{3}$ and in state S_2 with probability $\frac{2}{3}$. ■

One would like to know when it is true that $\lim_{m\to\infty} P^m$ will exist for a given stochastic matrix P. The following definition and theorem give a complete answer to this question.

Definition 8.2. A stochastic matrix P is said to be regular if whenever λ is an eigenvalue of P such that $|\lambda| = 1$, then $\lambda = 1$.

Theorem 8.2. Let P be a stochastic matrix. Then $\lim_{m\to\infty} P^m$ exists if and only if P is a regular stochastic matrix.

Suppose now that $\lim_{m\to\infty} P^m = P^\infty$. Then we have the following equation:

$$O_n = P^\infty - P^\infty = \lim_{m\to\infty} P^m - \lim_{m\to\infty} P^{m+1}$$
$$= (I_n - P)\lim_{m\to\infty} P^m = (I_n - P)P^\infty$$

Thus each column of P^∞ is an eigenvector of the matrix P corresponding to the eigenvalue 1.

Notice in Example 3, the columns of P^∞ are $[\frac{1}{3},\frac{1}{3}]^T$ and $[\frac{2}{3},\frac{2}{3}]^T$. Both of these columns are eigenvectors corresponding to the eigenvalue 1. Thus if P is a regular stochastic matrix and the eigenvalue 1 has multiplicity 1, then every column of P^∞ must be a scalar multiple of the eigenvector $(1,1, \ldots ,1)$. In this case we see, as in the preceding example, that the limiting probabilities will not depend on the initial probability distributions. If, however, the eigenspace corresponding to the eigenvalue 1 has dimension greater than 1, then the limiting probabilities will depend on the initial probability distributions.

For example, if $P = I_2$, i.e., the probability that the system changes from one state to a different one is zero, then $P^\infty = P = I_2$. Hence, if (p_1, p_2) is the initial probability distribution, then it is also the limiting probability distribution.

EXAMPLE 4 Suppose we have a system of four states. Suppose further that the probability of leaving any state is 1 and each of the other states is then equally likely to occur. That is

$$p_{jk} = \begin{cases} \frac{1}{3} & \text{if } j \neq k \\ 0 & \text{if } j = k \end{cases}$$

Thus, if P is the probability matrix for this system

$$P = \begin{bmatrix} 0 & \frac{1}{3} & \frac{1}{3} & \frac{1}{3} \\ \frac{1}{3} & 0 & \frac{1}{3} & \frac{1}{3} \\ \frac{1}{3} & \frac{1}{3} & 0 & \frac{1}{3} \\ \frac{1}{3} & \frac{1}{3} & \frac{1}{3} & 0 \end{bmatrix}$$

Note that P is symmetric and every entry of P^2 is positive. Since P is stochastic, 1 is the largest eigenvalue, and since P^2 has positive entries, the multiplicity of this eigenvalue is 1. Moreover if λ is any other eigenvalue we have $|\lambda| < 1$ (Perron's theorems). Using these facts, we show that regardless of the initial probability distribution the limiting probability equals $\frac{1}{4}(1,1,1,1)$. Let $\mathbf{p} = (p_1, p_2, p_3, p_4)$ be the initial probability distribution. Then

$$\mathbf{p}_\infty = \lim_{n\to\infty} \mathbf{p}P^n = c\mathbf{z}_0$$

where $\mathbf{z}_0 = (1,1,1,1)$. As in the preceding section we have

$$\begin{aligned} 4c = c\langle \mathbf{z}_0, \mathbf{z}_0 \rangle &= \langle c\mathbf{z}_0, \mathbf{z}_0 \rangle = \lim_{n\to\infty} \langle \mathbf{p}P^n, \mathbf{z}_0 \rangle \\ &= \lim_{n\to\infty} \langle \mathbf{p}, \mathbf{z}_0 P^n \rangle = \langle \mathbf{p}, \mathbf{z}_0 \rangle \\ &= \sum_{j=1}^{4} p_j = 1 \end{aligned}$$

Thus, c equals $\frac{1}{4}$, and we have $\mathbf{p}_\infty = \frac{1}{4}(1,1,1,1)$. ∎

PROBLEM SET 8.3

1. Which of the following matrices is stochastic?

a. $\begin{bmatrix} 1 & 0 \\ 0 & 0 \end{bmatrix}$ b. $\begin{bmatrix} 0 & 1 \\ 1 & 0 \end{bmatrix}$ c. $\begin{bmatrix} \frac{1}{2} & \frac{1}{4} \\ \frac{1}{2} & \frac{3}{4} \end{bmatrix}$

d. $\begin{bmatrix} 1 & 0 & 0 \\ 0 & \frac{1}{8} & \frac{7}{8} \\ 1 & 0 & 0 \end{bmatrix}$ e. $\begin{bmatrix} 1 & 0 & 0 \\ 0 & \frac{1}{2} & 0 \\ 0 & \frac{1}{2} & 1 \end{bmatrix}$

2. Let $A = \begin{bmatrix} a & b \\ c & d \end{bmatrix}$ be an arbitrary 2×2 matrix. If λ is an eigenvalue of A, show that $|\lambda| \leq k$, where k is the larger of the two numbers $|a| + |b|$ and $|c| + |d|$. Use this fact to deduce that if $a = 6$, $b = 2$, $c = 7$, and $d = 0$, then $|\lambda| \leq 8$ for any eigenvalue λ.

3. Show that every 2×2 stochastic matrix P, with one exception, is regular. Hence $\lim_{n\to\infty} P^n$ must exist, again with one exception. Hint: If P is a 2×2 stochastic matrix, then P is regular if and only if -1 is not an eigenvalue of P. The exceptional matrix is $\begin{bmatrix} 0 & 1 \\ 1 & 0 \end{bmatrix}$.

4. Let $P = \begin{bmatrix} \frac{1}{2} & \frac{1}{2} \\ \frac{3}{4} & \frac{1}{4} \end{bmatrix}$. Compute $\lim_{n\to\infty} P^n$.

5. Suppose we have a system consisting of two states S_1 and S_2. Let the conditional probabilities p_{11} and p_{22} equal $\frac{1}{5}$ and $\frac{2}{3}$, respectively.
   a. $p_{12} = ?$, $p_{21} = ?$
   b. If the system at time t_0 has probability distribution $\mathbf{p} = (\frac{1}{2},\frac{1}{2})$, what is the probability the system will be in state S_1 at time t_1? At time t_3? Same question for state S_2.
   c. Is the stochastic matrix P regular?

6. If P and Q are $n \times n$ stochastic matrices, show that PQ is also a stochastic matrix.

7. Let

$$P = \begin{bmatrix} 0 & \frac{1}{3} & \frac{1}{3} & \frac{1}{3} \\ \frac{1}{3} & 0 & \frac{1}{3} & \frac{1}{3} \\ \frac{1}{3} & \frac{1}{3} & 0 & \frac{1}{3} \\ \frac{1}{3} & \frac{1}{3} & \frac{1}{3} & 0 \end{bmatrix}$$

   a. Show that P is a symmetric stochastic matrix.
   b. Show that every entry of P^2 is positive.

8. Let $P = [p_{jk}]$ be an $n \times n$ matrix $(n > 2)$ such that p_{jj} equals zero, and $p_{jk} = 1/(n-1)$ if $j \neq k$.
   a. Show that P is a symmetric stochastic matrix and that every entry of P^2 is positive.
   b. Show that if $\mathbf{p}$ is any initial probability distribution, then $\mathbf{p}_\infty = 1/n\,(1,1, \ldots ,1)$. Thus, no matter what our original distribution, we are equally likely to wind up in any one of the n given states.

9. Let $P = [p_{jk}]$ be an $n \times n$ matrix $(n \geq 2)$ such that

$$p_{jk} = \begin{cases} a, & j = k \\ (1-a)/(n-1), & j \neq k \end{cases}$$

   where $0 < a < 1$.
   a. Show that P is symmetric and stochastic. Note also that each p_{jk} is positive.
   b. If $\mathbf{p} = (p_1, \ldots ,p_n)$ is any probability distribution, i.e., $p_j \geq 0$ and $\sum_{j=1}^{n} p_j$ equals 1, then $\mathbf{p}_\infty = \lim_{k\to\infty} \mathbf{p}P^k = 1/n(1,1, \ldots ,1)$. Thus, as in Example 4, we again are equally likely to wind up in any of the given states regardless of the initial probability distribution.
   c. What happens if $a = 1$?

8.4 SYSTEMS OF DIFFERENTIAL EQUATIONS

Suppose we have a linear system of two first-order, constant coefficient, differential equations involving two functions $x_1(t)$ and $x_2(t)$. That is, we have equations of the form

$$\begin{aligned} x_1' &= a_{11}x_1 + a_{12}x_2 \\ x_2' &= a_{21}x_1 + a_{22}x_2 \end{aligned} \tag{8.19}$$

where the a_{ij} are constant. Let A be the 2×2 matrix

$$\begin{bmatrix} a_{11} & a_{12} \\ a_{21} & a_{22} \end{bmatrix}$$

Let $\mathbf{x}(t)$ equal $[x_1(t), x_2(t)]^T$, and write (8.19) as

$$\mathbf{x}' = A\mathbf{x} \tag{8.20}$$

Clearly this formulation of (8.19) is also possible for a linear system of n first-order differential equations involving n unknown functions.

In this section we discuss one way of obtaining an approximate value of $\mathbf{x}(t)$ if $\mathbf{x}(0)$ is known. The first step is to remember that $x_j'(t) = \lim_{h\to 0}(1/h)[x_j(t+h) - x_j(t)]$. Thus, it follows that $\mathbf{x}'(t) = \lim_{h\to 0}(1/h)[\mathbf{x}(t+h) - \mathbf{x}(t)]$. For small h, we have $\mathbf{x}'(t) \approx (1/h)[\mathbf{x}(t+h) - \mathbf{x}(t)]$. Hence, from (8.20) we have

$$\mathbf{x}(t+h) \approx \mathbf{x}(t) + hA\mathbf{x}(t) = (I + hA)\mathbf{x}(t) \tag{8.21}$$

For any k we then have

$$\mathbf{x}(t+kh) \approx (I+hA)\mathbf{x}(t+(k-1)h) \approx \cdots \approx (I+hA)^k\mathbf{x}(t) \tag{8.22}$$

Thus, suppose we want to calculate $\mathbf{x}(t_0)$, $t_0 > 0$. Let N be any positive integer and set $h = t_0/N$. We then have the following approximation to the solution of (8.20) at the points $h, 2h, \ldots, Nh = t_0$.

$$\mathbf{x}(kh) \approx (I+hA)^k\mathbf{x}(0) \qquad k = 0, 1, \ldots, N = \frac{t_0}{h}$$

EXAMPLE 1 Find an approximate value at the point t_0 for the solution to the differential equation

$$\begin{aligned} x_1' &= 3x_1 + x_2 \\ x_2' &= 2x_1 + 4x_2 \end{aligned}$$

assuming that $x_1(0) = 3$ and $x_2(0) = 4$.

Solution. Let N be any positive integer. From (8.22) we have

$$\begin{bmatrix} x_1(t_0) \\ x_2(t_0) \end{bmatrix} \approx [I + hA]^N \begin{bmatrix} 3 \\ 4 \end{bmatrix}$$

where $A = \begin{bmatrix} 3 & 1 \\ 2 & 4 \end{bmatrix}$ and $h = t_0/N$. Since A equals

$$\frac{1}{3}\begin{bmatrix} 1 & 1 \\ -1 & 2 \end{bmatrix}\begin{bmatrix} 2 & 0 \\ 0 & 5 \end{bmatrix}\begin{bmatrix} 2 & -1 \\ 1 & 1 \end{bmatrix}$$

we have $(I + hA)^N$ equal to

$$\frac{1}{3}\begin{bmatrix} 1 & 1 \\ -1 & 2 \end{bmatrix}\left(\begin{bmatrix} 1 & 0 \\ 0 & 1 \end{bmatrix} + h\begin{bmatrix} 2 & 0 \\ 0 & 5 \end{bmatrix}\right)^N \begin{bmatrix} 2 & -1 \\ 1 & 1 \end{bmatrix}$$

$$= \frac{1}{3}\begin{bmatrix} 1 & 1 \\ -1 & 2 \end{bmatrix}\begin{bmatrix} (1+2h)^N & 0 \\ 0 & (1+5h)^N \end{bmatrix}\begin{bmatrix} 2 & -1 \\ 1 & 1 \end{bmatrix}$$

Thus,

$$\begin{bmatrix} x_1(t_0) \\ x_2(t_0) \end{bmatrix} \approx \frac{1}{3}\begin{bmatrix} 1 & 1 \\ -1 & 2 \end{bmatrix}\begin{bmatrix} (1+2h)^N & 0 \\ 0 & (1+5h)^N \end{bmatrix}\begin{bmatrix} 2 & -1 \\ 1 & 1 \end{bmatrix}\begin{bmatrix} 3 \\ 4 \end{bmatrix}$$

where $h = t_0/N$. If one assumes that the approximations will converge to the solution as N becomes infinite, then

$$\begin{bmatrix} x_1(t_0) \\ x_2(t_0) \end{bmatrix} = \lim_{N\to\infty} \frac{1}{3}\begin{bmatrix} 1 & 1 \\ -1 & 2 \end{bmatrix}\begin{bmatrix} (1+2t_0/N)^N & 0 \\ 0 & (1+5t_0/N)^N \end{bmatrix}\begin{bmatrix} 2 & -1 \\ 1 & 1 \end{bmatrix}\begin{bmatrix} 3 \\ 4 \end{bmatrix}$$

If we remember that $\lim_{N\to\infty}(1 + a/N)^N = \exp(a)$, then we see that

$$\begin{bmatrix} x_1(t_0) \\ x_2(t_0) \end{bmatrix} = \begin{bmatrix} \frac{1}{3} & \frac{1}{3} \\ -\frac{1}{3} & \frac{2}{3} \end{bmatrix}\begin{bmatrix} \exp(2t_0) & 0 \\ 0 & \exp(5t_0) \end{bmatrix}\begin{bmatrix} 2 & -1 \\ 1 & 1 \end{bmatrix}\begin{bmatrix} 3 \\ 4 \end{bmatrix}$$ ■

Let's redo this example but in a more general fashion. We start out wanting an approximate solution to $\mathbf{x}' = A\mathbf{x}$ at t_0. Suppose that A is similar to a diagonal matrix D, i.e., $A = P^{-1}DP$. Then $\mathbf{x}(t_0) = \mathbf{x}(Nh) \approx (I + hA)^N\mathbf{x}(0)$. But $(I + hA) = P^{-1}P + hP^{-1}DP = P^{-1}(I + hD)P$, and $I + hD$ is a diagonal matrix with diagonal entries $1 + hd_j$, where d_j is the jth diagonal entry of D. Thus, $\lim_{N\to\infty}(I + hA)^N = \lim_{N\to\infty} P^{-1}(I + hD)^N P$. Hence, if D equals the diagonal matrix

$$D = \begin{bmatrix} d_1 & 0 & \cdots & & \\ 0 & d_2 & 0 & \cdots & 0 \\ \cdots & \cdots & \cdots & \cdots & \cdots \\ 0 & & & & d_n \end{bmatrix}$$

we then have

$$\lim_{N\to\infty}(I + hD)^N = \begin{bmatrix} \exp(d_1 t_0) & & 0 \\ 0 & \exp(d_2 t_0) & 0 \\ \cdots & \cdots & \cdots \\ 0 & & \exp(d_n t_0) \end{bmatrix}$$

Thus,

$$\mathbf{x}(t_0) = P^{-1}\begin{bmatrix} \exp(d_1t_0) & 0 & 0 \\ 0 & \exp(d_2t_0) & 0 \\ \cdots & \cdots & \cdots \\ 0 & 0 & \exp(d_nt_0) \end{bmatrix} P\mathbf{x}_0$$

The matrix $P^{-1}[\exp(d_jt_0)\delta_{jk}]P$ can be interpreted in a very interesting fashion. Let A be an $n \times n$ matrix. If $p(x)$ is a polynomial, then we've defined $p(A)$. Suppose now that $p(x)$ is an infinite series in x; that is, $p(x)$ equals $\sum_{n=0}^{\infty} p_nx^n$, and the series converges for all x. We then define $p(A) = \sum_{n=0}^{\infty} p_nA^n$. One particular infinite series is $p(x) = \exp(x) = \sum_{n=0}^{\infty} x^n/n!$

Definition 8.3. Let A be any square matrix. We define $\exp[A]$ as the matrix

$$\exp[A] = \sum_{n=0}^{\infty} \frac{A^n}{n!} \tag{8.23}$$

We will show that if D is a diagonal matrix, then $\exp[D]$ is the diagonal matrix whose diagonal entries are the exponentials of the diagonal entries of D. Thus, the matrix product above equals $\exp[t_0A]$.

EXAMPLE 2 Let $A = \begin{bmatrix} d_1 & 0 \\ 0 & d_2 \end{bmatrix}$. Compute $\exp[A]$.

$$\exp[A] = \sum_{n=0}^{\infty} \frac{A^n}{n!} = \sum_{n=0}^{\infty} \begin{bmatrix} d_1 & 0 \\ 0 & d_2 \end{bmatrix}^n \frac{1}{n!}$$

$$= \sum_{n=0}^{\infty} \begin{bmatrix} \frac{d_1^n}{n!} & 0 \\ 0 & \frac{d_2^n}{n!} \end{bmatrix} = \begin{bmatrix} \exp(d_1) & 0 \\ 0 & \exp(d_2) \end{bmatrix}$$

■

Theorem 8.3. Let A be an $n \times n$ matrix. Then the solution to the system (8.20), which also satisfies $\mathbf{x}(0) = \mathbf{x}_0$, is given by

$$\mathbf{x}(t) = \exp[tA]\mathbf{x}_0 \tag{8.24}$$

EXAMPLE 3 Solve the following system of differential equations:

$$\begin{aligned} x_1' &= 2x_1 \qquad & x_1(0) &= -1 \\ x_2' &= -x_2 \qquad & x_2(0) &= 1 \end{aligned} \tag{8.25}$$

Solution. Writing (8.25) as a vector equation, we have

$$\mathbf{x}' = \begin{bmatrix} 2 & 0 \\ 0 & -1 \end{bmatrix} \mathbf{x}$$

The solution, as given by (8.25), is

$$\mathbf{x}(t) = \begin{bmatrix} \exp(2t) & 0 \\ 0 & \exp(-t) \end{bmatrix} \begin{bmatrix} -1 \\ 1 \end{bmatrix}$$
$$= \begin{bmatrix} -\exp(2t) \\ \exp(-t) \end{bmatrix} \mathbf{x}$$

The reader should check that the functions $x_1(t) = -\exp(2t)$ and $x_2(t) = \exp(-t)$ actually solve the given problem. ■

Theorem 8.3 enables us to compute the solution to a linear system of differential equations if we know how to compute $\exp[B]$ for any square matrix B. If B is similar to a diagonal matrix D, i.e., $B = P^{-1}DP$, then it can be shown that $\exp[B] = P^{-1}\exp[D]P$. Thus the problem of computing $\exp[B]$ is reduced to the case when B is not similar to a diagonal matrix. In more advanced texts, it is shown how to compute $\exp[B]$, by finding the Jordan normal form of the matrix B. The interested reader should consult the bibliography for appropriate references.

PROBLEM SET 8.4

1. For each of the following matrices, compute $\exp[A]$.

a. $\begin{bmatrix} 2 & 0 \\ 0 & -3 \end{bmatrix}$ **b.** $\begin{bmatrix} 4 & 6 \\ 0 & 3 \end{bmatrix}$ **c.** $\begin{bmatrix} 2 & 1 \\ 0 & 2 \end{bmatrix}$

2. Solve the following systems of differential equations with the given initial conditions:

a. $x_1' = 2x_1,\ x_1(0) = 10,\ x_2' = -3x_1,\ x_2(0) = -3$

b. $x_1' = 4x_1 + 6x_2,\ x_1(0) = -2,\ x_2' = 3x_2,\ x_2(0) = 1$

3. Let A and B be two similar square matrices.

a. Show that tA is similar to tB.

b. Show that $\exp[tA]$ is similar to $\exp[tB]$.

4. Let λ be an eigenvalue of A with eigenvector $\mathbf{x}$. Use (8.23) to show that $\exp[\lambda]$ is an eigenvalue of $\exp[A]$ with eigenvector $\mathbf{x}$.

5. Solve the following system of equations:

$$\begin{aligned} x_1' &= 8x_1 - x_2 + x_3, & x_1(0) &= 1 \\ x_2' &= -x_1 + 8x_2 + x_3, & x_2(0) &= 1 \\ x_3' &= x_1 + x_2 + 8x_3, & x_3(0) &= 0 \end{aligned}$$

6. Let $A = \begin{bmatrix} 1 & 0 & 0 \\ 0 & 2 & 1 \\ 0 & 0 & 2 \end{bmatrix}$.

a. Compute exp[A].
b. Solve the system $\mathbf{x}' = A\mathbf{x}$, $\mathbf{x}(0) = [1,1,1]^T$.

8.5 GEOMETRY OF LINEAR TRANSFORMATIONS

Let A be a nonsingular 2×2 matrix. In this section we discuss what the mapping $\mathbf{x} \to A\mathbf{x}$ does to various geometrical objects. For example, if K is an ellipse in $\mathbb{R}^2$, what is $A(K)$, where $A(K)$ is the set of images of elements in K?

Suppose first that K is a straight line in $\mathbb{R}^2$. Then there are constants a_1, a_2, and c such that

$$K = K_c = \{(x_1, x_2) \ : \ a_1x_1 + a_2x_2 = c\}$$

Writing this in vector notation we have

$$K_c = \{\mathbf{x} \ : \ \langle \mathbf{x},\mathbf{a} \rangle = c\}$$

where $\mathbf{a} = (a_1, a_2)$ and $\mathbf{x} = (x_1, x_2)$. See Figure 8.2. The distance between the lines K_1 and K_0 need not equal 1; cf. problem 1. To determine $A(K)$, we use the fact that A is an invertible matrix. Thus,

$$\begin{aligned} A(K) &= \{\mathbf{y} \ : \ \mathbf{y} = A\mathbf{x} \text{ for some } \mathbf{x} \text{ in } K\} \\ &= \{\mathbf{y} \ : \ A^{-1}\mathbf{y} \text{ is in } K\} \\ &= \{\mathbf{y} \ : \ \langle A^{-1}\mathbf{y},\mathbf{a} \rangle = c\} \\ &= \{\mathbf{y} \ : \ \langle \mathbf{y},(A^{-1})^T\mathbf{a} \rangle = c\} \end{aligned}$$

Thus, if K is a line perpendicular to the direction $\mathbf{a}$, then $A(K)$ is a line perpendicular to the direction $(A^{-1})^T\mathbf{a}$.

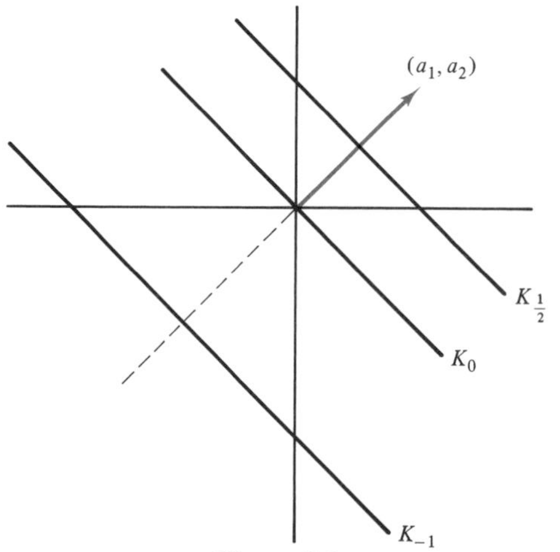

Figure 8.2

EXAMPLE 1 Let $K = \{(x_1, x_2) : 2x_1 + 3x_2 = -1\}$, and let A be the matrix

$$A = \begin{bmatrix} 6 & -2 \\ 3 & 1 \end{bmatrix}$$

$$A^{-1} = \left(\frac{1}{12}\right)\begin{bmatrix} 1 & 2 \\ -3 & 6 \end{bmatrix} \quad (A^{-1})^T\mathbf{a} = \left(\frac{1}{12}\right)\begin{bmatrix} 1 & -3 \\ 2 & 6 \end{bmatrix}\begin{bmatrix} 2 \\ 3 \end{bmatrix} = \left(\frac{1}{12}\right)\begin{bmatrix} -7 \\ 22 \end{bmatrix}$$

$$A(K) = \{\mathbf{y} : \langle \mathbf{y}, (A^{-1})^T\mathbf{a} \rangle = -1\}$$

$$A(K) = \{\mathbf{y} : -(\tfrac{7}{12})y_1 + (\tfrac{22}{12})y_2 = -1\}$$

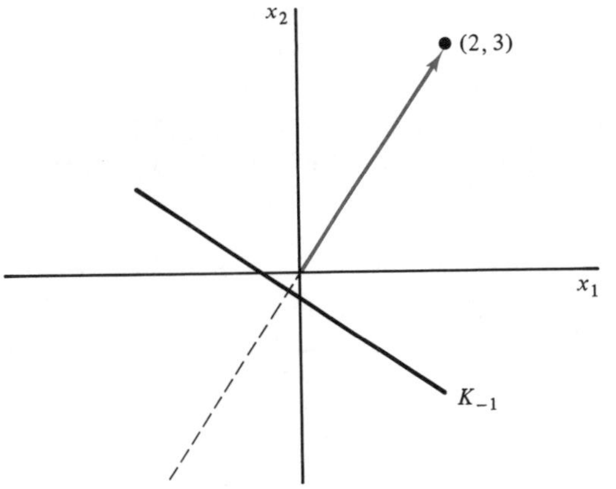

(a)

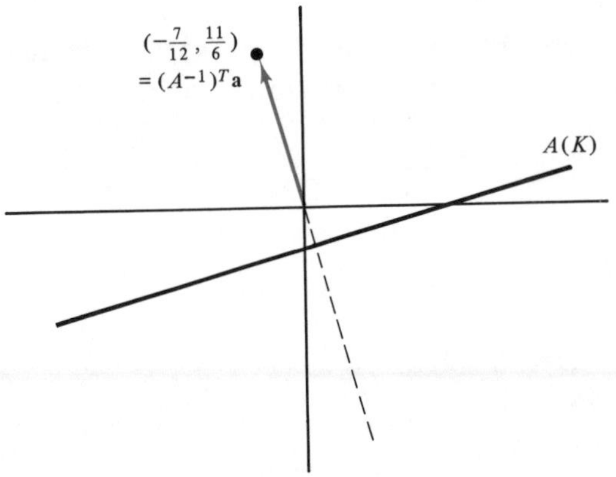

(b)

Figure 8.3 ■

In $\mathbb{R}^3$ one can characterize planes in the same manner that lines are described in $\mathbb{R}^2$. That is, if K is some plane in $\mathbb{R}^3$, then there are constants a_1, a_2, a_3, and c such that

$$K = K_c = \left\{(x_1, x_2, x_3) \; : \; \sum_{j=1}^{3} x_j a_j = c\right\}$$
$$= \{\mathbf{x} \; : \; \langle \mathbf{x}, \mathbf{a} \rangle = c\}$$

If A is a 3×3 nonsingular matrix, then

$$\begin{aligned} A(K) &= \{\mathbf{y} \; : \; A^{-1}\mathbf{y} \text{ is in } K\} \\ &= \{\mathbf{y} \; : \; \langle A^{-1}\mathbf{y}, \mathbf{a} \rangle = c\} \\ &= \{\mathbf{y} \; : \; \langle \mathbf{y}, (A^{-1})^T \mathbf{a} \rangle = c\} \end{aligned}$$

Thus, A transforms planes perpendicular to the direction $\mathbf{a}$ into planes perpendicular to the direction $(A^{-1})^T\mathbf{a}$.

We next discuss the conic sections and what a linear transformation does to them. There are three types of conics: ellipses, hyperbolas, and parabolas. Each of these is the locus of a particular type of quadratic equation, that is, an equation of the form

$$ax_1^2 + 2bx_1x_2 + cx_2^2 + 1_1x_1 + 1_2x_2 = k \tag{8.26}$$

We assume that not all the coefficients a, b, and c are zero. It is well known that the type of conic (8.26) describes is completely determined by the sign of $ac - b^2$. That is,

$$ac - b^2 \begin{cases} >0, \text{ ellipse} \\ =0, \text{ parabola} \\ <0, \text{ hyperbola} \end{cases}$$

We want to write (8.26) as a vector equation, but before doing so we note that the second-degree terms in (8.26) can be written in the following form. Let Q be the 2×2 matrix

$$Q = \begin{bmatrix} a & b \\ b & c \end{bmatrix}$$

Then,

$$\begin{aligned} ax_1^2 + 2bx_1x_2 + cx_2^2 &= (ax_1 + bx_2)x_1 + (bx_1 + cx_2)x_2 \\ &= \langle (ax_1 + bx_2, bx_1 + cx_2), (x_1, x_2) \rangle \\ &= \langle Q\mathbf{x}, \mathbf{x} \rangle \end{aligned}$$

Note that Q is a symmetric matrix and that $\det(Q) = ac - b^2$. In fact it is because we want Q to be symmetric that the factor 2 appears in the mixed term x_1x_2 of (8.26).

If we let $\mathbf{1} = (1_1, 1_2)$, then (8.26) can be written

$$\langle Q\mathbf{x}, \mathbf{x} \rangle + \langle \mathbf{x}, \mathbf{1} \rangle = k \tag{8.27}$$

EXAMPLE 2 Determine Q, $\mathbf{l}$, and k for each of the following quadratic equations, and also the type of conic the equation describes.

a. $2x_1^2 - 4x_1x_2 + 7x_2^2 + 6x_1 = 2.$

$$Q = \begin{bmatrix} 2 & -2 \\ -2 & 7 \end{bmatrix} \qquad \mathbf{l} = (6,0) \qquad k = 2$$

$\det(Q) = 10 > 0$ implies the conic is an ellipse.

b. $3x_1^2 + x_1x_2 + 7x_1 - 8x_2 = 0.$

$$Q = \begin{bmatrix} 3 & \frac{1}{2} \\ \frac{1}{2} & 0 \end{bmatrix} \qquad \mathbf{l} = (7,-8) \qquad k = 0$$

$\det(Q) = -\frac{1}{4} < 0$ implies the conic is a hyperbola.

c. $6x_1^2 - 6x_1x_2 + \frac{3}{2}x_2^2 = 1 - x_1.$

$$Q = \begin{bmatrix} 6 & -3 \\ -3 & \frac{3}{2} \end{bmatrix} \qquad \mathbf{l} = (1,0) \qquad k = 1$$

$\det(Q) = 0$ implies the conic is a parabola. ∎

Let A be any 2×2 nonsingular matrix. Let

$$K = \{\mathbf{x} : \langle Q\mathbf{x},\mathbf{x}\rangle + \langle \mathbf{x},\mathbf{l}\rangle = k\}$$

Then

$$\begin{aligned} A(K) &= \{\mathbf{y} : \mathbf{y} = A\mathbf{x} \text{ for some } \mathbf{x} \text{ in } K\} \\ &= \{\mathbf{y} : A^{-1}\mathbf{y} \text{ is in } K\} \\ &= \{\mathbf{y} : \langle QA^{-1}\mathbf{y}, A^{-1}\mathbf{y}\rangle + \langle A^{-1}\mathbf{y},\mathbf{l}\rangle = k\} \end{aligned}$$

Thus,

$$A(K) = \{\mathbf{y} : \langle (A^{-1})^T QA^{-1}\mathbf{y}, \mathbf{y}\rangle + \langle \mathbf{y}, (A^{-1})^T\mathbf{l}\rangle = k\} \tag{8.28}$$

Let R be the matrix $(A^{-1})^T QA^{-1}$. Using the facts that $(AB)^T = B^TA^T$ and $(A^{-1})^T = (A^T)^{-1}$, it is easy to show that R is a symmetric matrix. Moreover, since we have assumed that A is nonsingular and that Q is not the zero matrix, R is not equal to the zero matrix. Thus, $A(K)$ is also a conic section, but which type?

$$\begin{aligned} \det(R) &= \det((A^{-1})^T QA^{-1}) = \det((A^{-1})^T)\det(Q)\det(A^{-1}) \\ &= \det(Q)[\det(A^{-1})]^2 \end{aligned}$$

Thus, the sign of $\det(R)$ is the same as the sign of $\det(Q)$. Hence, $A(K)$ is the same type of conic section as K.

We now describe in more detail the conic sections $A(K)$. To do this we use the fact that R is symmetric and thus there is an orthonormal basis of $\mathbb{R}^2$ consisting of eigenvectors of $\mathbb{R}$. Let $F = \{\mathbf{f}_1, \mathbf{f}_2\}$ be this basis and let λ_1 and λ_2 be the associated eigenvalues. Let $\mathbf{y} = y_1\mathbf{f}_1 + y_2\mathbf{f}_2$. Let $(A^{-1})^T\mathbf{l} = a_1\mathbf{f}_1 + a_2\mathbf{f}_2$. Then

$$\begin{aligned} k &= \langle (A^{-1})^T Q A^{-1}\mathbf{y}, \mathbf{y}\rangle + \langle \mathbf{y}, (A^{-1})^T\mathbf{l}\rangle \\ &= \langle R(y_1\mathbf{f}_1 + y_2\mathbf{f}_2), y_1\mathbf{f}_1 + y_2\mathbf{f}_2\rangle + \langle y_1\mathbf{f}_1 + y_2\mathbf{f}_2, a_1\mathbf{f}_1 + a_2\mathbf{f}_2\rangle \\ &= \lambda_1 y_1^2 + \lambda_2 y_2^2 + a_1 y_1 + a_2 y_2 \end{aligned} \tag{8.29}$$

Notice that in terms of the coordinates of $\mathbf{y}$ with respect to the basis F, there is no mixed product $y_1 y_2$. Thus, a simple translation of axes will now put this quadratic expression into standard form.

EXAMPLE 3 Let $K = \{(x_1, x_2) : x_1^2 = x_2\}$. For each of the matrices below determine $A(K)$.

a. Let $A = \begin{bmatrix} 1 & -1 \\ 1 & 1 \end{bmatrix}$. The quadratic equation that describes the set K gives rise to the following matrix Q and vector $\mathbf{l}$.

$$Q = \begin{bmatrix} 1 & 0 \\ 0 & 0 \end{bmatrix} \qquad \mathbf{l} = (0, -1)$$

Thus $R = (A^{-1})^T Q A^{-1} = \left(\frac{1}{4}\right)\begin{bmatrix} 1 & -1 \\ 1 & 1 \end{bmatrix}\begin{bmatrix} 1 & 0 \\ 0 & 0 \end{bmatrix}\begin{bmatrix} 1 & 1 \\ -1 & 1 \end{bmatrix}$

$$= \left(\frac{1}{4}\right)\begin{bmatrix} 1 & 1 \\ 1 & 1 \end{bmatrix}$$

The eigenvalues of R and their corresponding eigenvectors are:

$$\lambda_1 = 0 \qquad \mathbf{f}_1 = (1, -1)/\sqrt{2} \qquad \lambda_2 = \tfrac{1}{2} \qquad \mathbf{f}_2 = (1, 1)/\sqrt{2}$$

We also have $(A^{-1})^T\mathbf{l} = (\frac{1}{2})(1, -1) = (1/\sqrt{2})f_1$. From (8.29) we have

$$\begin{aligned} A(K) &= \{y_1\mathbf{f}_1 + y_2\mathbf{f}_2 : \lambda_1 y_1^2 + \lambda_2 y_2^2 + (1/\sqrt{2})y_1 = 0\} \\ &= \{y_1\mathbf{f}_1 + y_2\mathbf{f}_2 : (\tfrac{1}{2})y_2^2 + (1/\sqrt{2})y_1 = 0\} \end{aligned}$$

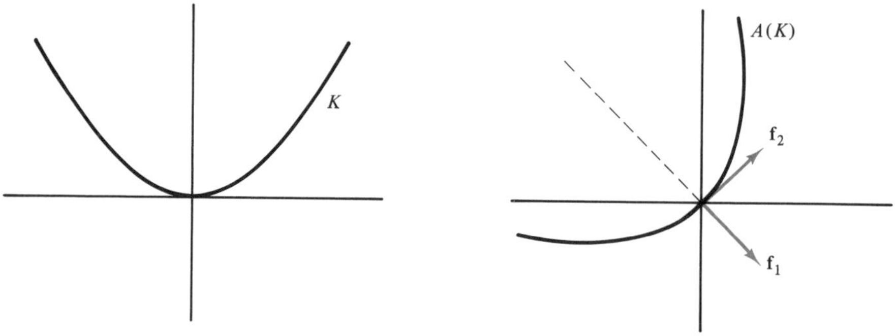

Figure 8.4

See Figure 8.4. Thus, $A(K)$ is a parabola with vertex at the origin which opens up about the line through the vertex in the $-\mathbf{f}_1$ direction. It is clear from the picture that A rotates the parabola 45 degrees in a counterclockwise direction. This linear transformation also magnifies by a factor of $\sqrt{2}$. One way to see this is to write

$$A = \begin{bmatrix} \sqrt{2} & 0 \\ 0 & \sqrt{2} \end{bmatrix} \begin{bmatrix} \frac{1}{\sqrt{2}} & \frac{-1}{\sqrt{2}} \\ \frac{1}{\sqrt{2}} & \frac{1}{\sqrt{2}} \end{bmatrix} = \begin{bmatrix} \sqrt{2} & 0 \\ 0 & \sqrt{2} \end{bmatrix} \begin{bmatrix} \cos 45 & -\sin 45 \\ \sin 45 & \cos 45 \end{bmatrix}$$

the second factor being a rotation of 45 degrees and the first an elongation in all directions by a factor of $\sqrt{2}$.

b. Let $A = \begin{bmatrix} 1 & -1 \\ 0 & 1 \end{bmatrix}$. Then $A^{-1} = \begin{bmatrix} 1 & 1 \\ 0 & 1 \end{bmatrix}$ and $(A^{-1})^T = \begin{bmatrix} 1 & 0 \\ 1 & 1 \end{bmatrix}$. Thus, $R = \begin{bmatrix} 1 & 1 \\ 1 & 1 \end{bmatrix}$. The eigenvalues of R are $\lambda_1 = 0$ and $\lambda_2 = 2$. The corresponding eigenvectors are $\mathbf{f}_1 = (1,-1)/\sqrt{2}$ and $\mathbf{f}_2 = (1,1)/\sqrt{2}$. We compute $(A^{-1})^T\mathbf{l} = (0,-1) = (1/\sqrt{2})\mathbf{f}_1 - (1/\sqrt{2})\mathbf{f}_2$. Thus,

$$\begin{aligned} A(K) &= \{y_1\mathbf{f}_1 + y_2\mathbf{f}_2 : 2y_2^2 + y_1/\sqrt{2} - y_2/\sqrt{2} = 0\} \\ &= \left\{y_1\mathbf{f}_1 + y_2\mathbf{f}_2 : 2[y_2 - (\tfrac{1}{4}\sqrt{2})]^2 = 1/\sqrt{2}\left[\left(\frac{1}{8\sqrt{2}}\right) - y_1\right]\right\} \end{aligned}$$

Hence, $A(K)$ is a parabola with vertex at $\mathbf{y}_0 = [1(8\sqrt{2})](\mathbf{f}_1 + 2\mathbf{f}_2)$ which opens up about the half line starting at $\mathbf{y}_0$ in the direction of $-\mathbf{f}_1$.

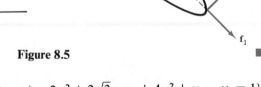

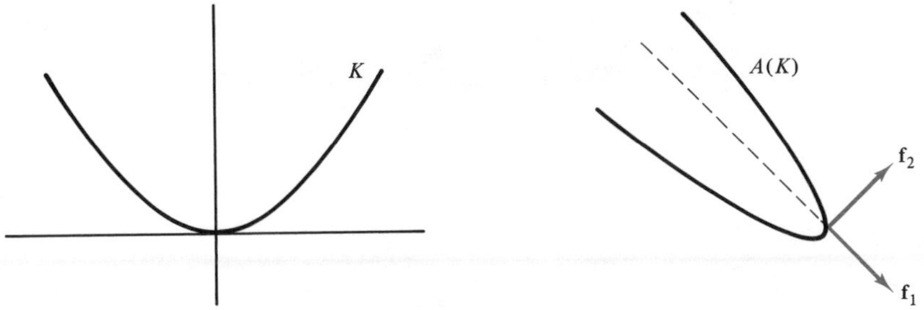

Figure 8.5 ■

EXAMPLE 4 Let $K = \{(x_1, x_2) : 2x_1^2 + 2\sqrt{3}x_1x_2 + 4x_2^2 + x_1 - x_2 = 1\}$. Sketch K by first finding the eigenvalues and eigenvectors of the matrix Q associated with the quadratic equation that describes K.

$$Q = \begin{bmatrix} 2 & 3 \\ 3 & 4 \end{bmatrix} \qquad \mathbf{l} = (1,-1)$$

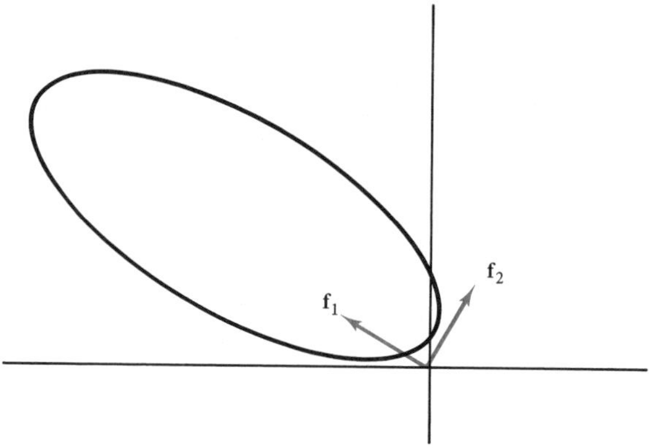

Figure 8.6

The eigenvalues and eigenvectors of Q are

$$\lambda_1 = 1 \qquad \mathbf{f}_1 = (-\sqrt{3},1)/2 \qquad \lambda_2 = 5 \qquad \mathbf{f}_2 = (1,\sqrt{3})/2$$

Thus,

$$\begin{aligned} 1 &= \langle Q(y_1\mathbf{f}_1 + y_2\mathbf{f}_2), y_1\mathbf{f}_1 + y_2\mathbf{f}_2\rangle + \langle y_1\mathbf{f}_1 + y_2\mathbf{f}_2, \mathbf{l}\rangle \\ &= y_1^2 + 5y_2^2 + \frac{-1-\sqrt{3}}{2} y_1 + \frac{1-\sqrt{3}}{2} y_2 \end{aligned}$$

Completing the square we have

$$\begin{aligned} &\left(y_1 - \frac{1+\sqrt{3}}{2}\right)^2 + 5\left(y_2 + \frac{1-\sqrt{3}}{4}\right)^2 \\ &\qquad = 1 + \left(\frac{1+\sqrt{3}}{4}\right)^2 + 5\left(\frac{1-\sqrt{3}}{4}\right)^2 \end{aligned}$$

See Figure 8.6. ■

PROBLEM SET 8.5

1. Let $K_c = \{\mathbf{x} : \langle \mathbf{x},\mathbf{a}\rangle = c\}$, where $\mathbf{a}$ is some fixed vector in $\mathbb{R}^2$.
   a. Show that $K_c = K_0 + c\mathbf{a}/\|\mathbf{a}\|^2$. That is, K_c is equal to K_0 translated by the vector $c\mathbf{a}/\|\mathbf{a}\|^2$.
   b. Determine the distance between K_{c_1} and K_{c_2} in terms of c_1, c_2, and $\mathbf{a}$.
2. Let $V = \mathbb{R}^n$. For a fixed vector $\mathbf{a}$ in V, define $K_c = \{\mathbf{x} : \langle \mathbf{x},\mathbf{a}\rangle = c\}$. K_c is called a hyperplane. Let A be an $n \times n$ invertible matrix.
   a. Show $K_c = K_0 + c\mathbf{a}/\|\mathbf{a}\|^2$.
   b. Show that $A(K_c) = \{\mathbf{y} : \langle \mathbf{y}, (A^{-1})^T\mathbf{a}\rangle = c\}$.

3. Let $A = \begin{bmatrix} 6 & -2 \\ 4 & 1 \end{bmatrix}$. Let $K_c = \{\mathbf{x} : \langle \mathbf{x}, (1,1) \rangle = c\}$.
   a. Sketch K_c for various values of c.
   b. Sketch $A(K_1)$ and $A(K_{-1})$.
   c. How is the distance between $A(K_1)$ and $A(K_{-1})$ related to the distance between K_1 and K_{-1}?

4. Let $A = \begin{bmatrix} 1 & -1 \\ 1 & 1 \end{bmatrix}$. For each of the following conic sections K, sketch K and $A(K)$.
   a. $x_1^2 + 2x_2^2 = 1$ b. $x_1^2 - x_2^2 = 1$ c. $x_1 - x_2^2 = 1$.

5. Let $A = \begin{bmatrix} 1 & -1 \\ 0 & 1 \end{bmatrix}$. For each of the conic sections in problem 4, sketch $A(K)$.

6. For each of the conic sections below determine the quadratic matrix Q and the associated orthonormal basis of $\mathbb{R}^2$, and then express the conic in terms of the orthonormal basis coordinates.
   a. $2x_1^2 + 6x_1x_2 - x_2^2 = 1$
   b. $3x_1^2 - 4x_1x_2 + 6x_2^2 + 4x_1 = 3$
   c. $4x_1^2 - 8x_1x_2 + 4x_2^2 = 1$

7. Let $A = \begin{bmatrix} 0 & 1 \\ -1 & 0 \end{bmatrix}$.
   a. Show that $\langle A\mathbf{x},\mathbf{x} \rangle = 0$ for every $\mathbf{x}$ in $\mathbb{R}^2$.
   b. Let $A = [a_{jk}]$ be any 2×2 matrix for which $\langle A\mathbf{x},\mathbf{x} \rangle = 0$ for all $\mathbf{x}$. Show that the diagonal entries of A must equal zero, and that $a_{12} = -a_{21}$, that is, A is a skew symmetric matrix ($A = -A^T$).
   c. Let A be a 2×2 symmetric matrix for which $\langle A\mathbf{x},\mathbf{x} \rangle = 0$ for all $\mathbf{x}$. Show that A must equal the 2×2 zero matrix.

8. Let A be an $n \times n$ matrix. Show that $\langle A\mathbf{x},\mathbf{x} \rangle = 0$ for all $\mathbf{x}$ in $\mathbb{R}^n$, if and only if A is a skew symmetric matrix.

9. Let A be any $n \times n$ matrix. The equality

$$A = \frac{A + A^T}{2} + \frac{A - A^T}{2}$$

   shows that we can write A as the sum of a symmetric and a skew symmetric matrix.
   a. Verify that the two matrices above are symmetric and skew symmetric, respectively.
   b. Show that if $A = B + C$, where B is symmetric and C is skew symmetric, then $B = (A + A^T)/2$ and $C = (A - A^T)/2$.
   c. Given any $n \times n$ matrix A, show that there is a unique symmetric matrix Q such that $\langle Q\mathbf{x},\mathbf{x} \rangle = \langle A\mathbf{x},\mathbf{x} \rangle$ for all $\mathbf{x}$ in $\mathbb{R}^n$.

Appendix A

INDUCTION

Proof by induction is a method that enables one to prove, at one time, an infinite number of consecutively indexed statements. Suppose that these statements are called $P_1, P_2, \ldots$. We verify that each one of them is true via the following scheme:

1. Show that P_1 is true.
2. Show that if P_k is true, then P_{k+1} must also be true for $k = 1, 2, \ldots$.

Clearly if 1 and 2 are valid, then each P_n must be true. For assume that one of the P_n is not true. Let P_N be the first false statement. N must be greater than **1**, since we have shown that P_1 is true. Thus P_{N-1} is true, but then **2** says that $P_{(N-1)+1}$ must be true. Since we can have no first false statement, all the statements must be true. The following two examples demonstrate this technique.

EXAMPLE 1 Show that $1 + n$ equals $n + 1$ for $n = 1, 2, \ldots$. Thus P_n is the statement $1 + n = n + 1$.

1. Since $1 + 1$ clearly equals $1 + 1$, P_1 is true.
2. Suppose that P_k is true. That is, $1 + k = k + 1$. P_{k+1} is the statement $1 + (k+1) = (k+1) + 1$. But $1 + (k+1)$ equals $(1 + k) + 1$, and since we are assuming that P_k is true, this equals $(k+1) + 1$. Hence, P_{k+1} follows from P_k. ■

EXAMPLE 2 Let A be an $n \times n$ matrix. Let $\mathbf{x}_k$ be a sequence of vectors in $\mathbb{R}^n$ that satisfy

$$\mathbf{x}_{k+1} = A\mathbf{x}_k \qquad \text{for } k = 0, 1, \ldots$$

Show that $\mathbf{x}_k = A^k\mathbf{x}_0$.

Solution. P_k is the statement $A^k\mathbf{x}_0 = \mathbf{x}_k$. Hence P_1 is the statement $\mathbf{x}_1 = A\mathbf{x}_0$, and this is clearly true. We thus need to prove P_{k+1} follows from P_k. But

$$\begin{aligned} \mathbf{x}_{k+1} = A\mathbf{x}_k &= A[A^k\mathbf{x}_0] \\ &= A^{k+1}\mathbf{x}_0 \end{aligned}$$

and this is P_{k+1}. ∎

Appendix B
SUMMATION NOTATION

Throughout the text we have used the notation $\sum_{j=1}^{n} a_j$ and $\sum_{j=1}^{n} \sum_{k=1}^{m} a_{jk}$ to indicate finite sums of vectors or numbers. We now explain this notation, and discuss some of its properties.

First, let a_j, $j = 1, 2, \ldots$, be a sequence of numbers or vectors. We want a compact expression for the sum

$$a_1 + a_2 + \cdots + a_n$$

We thus define $\sum_{j=1}^{n} a_j$ inductively by

$$\sum_{j=1}^{1} a_j = a_1 \tag{B.1a}$$

$$\sum_{j=1}^{n+1} a_j = \sum_{j=1}^{n} a_j + a_{n+1} \tag{B.1b}$$

Thus, if $a_1 = 5$, $a_2 = 4$, $a_3 = 3$, $a_4 = 2$, $a_5 = 1$, then

$$\sum_{j=1}^{1} a_j = a_1 = 5$$

$$\sum_{j=1}^{2} a_j = \sum_{j=1}^{1} a_j + a_2 = 5 + 4 = 9$$

$$\sum_{j=1}^{3} a_j = \sum_{j=1}^{2} a_j + a_3 = 9 + 3 = 12$$

If we have a doubly indexed sequence a_{jk}, sums over both indices are defined as follows:

$$\sum_{j=1}^{n}\sum_{k=1}^{m} a_{jk} = \sum_{j=1}^{n} b_j$$

where b_j equals $\Sigma_{k=1}^{m} a_{jk}$.

Theorem B.1

a. $\sum_{j=1}^{n}(a_j + b_j) = \sum_{j=1}^{n} a_j + \sum_{j=1}^{n} b_j$

b. $\sum_{j=1}^{n} ca_j = c\sum_{j=1}^{n} a_j$

c. $\sum_{j=1}^{n}\sum_{k=1}^{m} a_{jk} = \sum_{k=1}^{m}\sum_{j=1}^{n} a_{jk}$

Proof. We prove a and c by induction and leave b to the reader.

a. $\sum_{j=1}^{1}(a_j + b_j) = a_1 + b_1 = \sum_{j=1}^{1} a_j + \sum_{j=1}^{1} b_j$. Thus statement P_1 is verified. We next prove P_{n+1} assuming P_n.

$$\begin{aligned}\sum_{j=1}^{n+1}(a_j + b_j) &= \sum_{j=1}^{n}(a_j + b_j) + (a_{n+1} + b_{n+1})\\ &= \sum_{j=1}^{n} a_j + \sum_{j=1}^{n} b_j + a_{n+1} + b_{n+1}\\ &= \sum_{j=1}^{n} a_j + a_{n+1} + \sum_{j=1}^{n} b_j + b_{n+1}\\ &= \sum_{j=1}^{n+1} a_j + \sum_{j=1}^{n+1} b_j\end{aligned}$$

c. We prove this by induction on n. For $n = 1$ we have

$$\sum_{j=1}^{1}\sum_{k=1}^{m} a_{jk} = \sum_{k=1}^{m} a_{1k} = \sum_{k=1}^{m}\sum_{j=1}^{1} a_{jk} \qquad \text{by (B.1a)}$$

Now assume that the theorem is true for n and show it is true for $n + 1$.

$$\sum_{j=1}^{n+1}\sum_{k=1}^{m} a_{jk} = \sum_{j=1}^{n}\sum_{k=1}^{m} a_{jk} + \sum_{k=1}^{m} a_{(n+1)k} \qquad \text{by (B.1}b\text{)}$$

$$= \sum_{k=1}^{m} \sum_{j=1}^{n} a_{jk} + \sum_{k=1}^{m} a_{(n+1)k} \qquad \text{by induction hypothesis}$$

$$= \sum_{k=1}^{m} \left(\sum_{j=1}^{n} a_{jk} + a_{(n+1)k} \right) \qquad \text{by } \mathbf{a}$$

$$= \sum_{k=1}^{m} \sum_{j=1}^{n+1} a_{jk} \qquad \text{by (B.1b)}$$

Thus, for all n we have $\sum_{j=1}^{n} \sum_{k=1}^{m} a_{jk} = \sum_{k=1}^{m} \sum_{j=1}^{n} a_{jk}$.

Appendix C

Permutations

Definition C.1. A permutation σ (of N things) is a one-to-one function that maps the first N integers into themselves. Another way to say this is, a permutation is a rearrangement of the first N integers. The collection of all such permutations is denoted by S_N. A convenient way to display a permutation σ is by listing $\sigma(i)$ below the integer i. Thus, if σ is the permutation in S_3 for which $\sigma(1) = 2$, $\sigma(2) = 1$, and $\sigma(3) = 3$, then

$$\sigma = \begin{pmatrix} 1 & 2 & 3 \\ 2 & 1 & 3 \end{pmatrix}$$

For any integer N, there are $N!$ permutations. If σ is in S_N, and

$$\sigma = \begin{pmatrix} 1 & 2 & \cdots & N \\ j_1 & j_2 & \cdots & j_N \end{pmatrix}$$

then the collection $\{j_1, \ldots, j_N\}$ must list exactly once, each of the first N integers.

If σ_1 and σ_2 are two permutations in S_N, we define their product $\sigma_1\sigma_2$ to be that permutation σ in S_N which satisfies $\sigma(i) = \sigma_1[\sigma_2(i)]$. In other words the product of two permutations is just functional composition, and is therefore an associative operation. That is, $(\sigma_1\sigma_2)\sigma_3 = \sigma_1(\sigma_2\sigma_3)$.

EXAMPLE 1 Let $\sigma_1 = \begin{pmatrix} 1 & 2 & 3 & 4 \\ 4 & 1 & 3 & 2 \end{pmatrix}$, $\sigma_2 = \begin{pmatrix} 1 & 2 & 3 & 4 \\ 2 & 3 & 4 & 1 \end{pmatrix}$. Then $\sigma_1\sigma_2 = \begin{pmatrix} 1 & 2 & 3 & 4 \\ 1 & 3 & 2 & 4 \end{pmatrix}$ while $\sigma_2\sigma_1 = \begin{pmatrix} 1 & 2 & 3 & 4 \\ 1 & 2 & 4 & 3 \end{pmatrix}$. Notice that $\sigma_1\sigma_2$ does not equal

$\sigma_2\sigma_1$. We let σ_{I} denote that permutation which maps each integer to itself, i.e., $\sigma_{\mathrm{I}}(j) = j$ for each j. Clearly $\sigma\sigma_I = \sigma_I\sigma = \sigma$ for each σ in S_N. To each σ in S_N we may associate a permutation σ^{-1}, also in S_N, which satisfies $\sigma\sigma^{-1} = \sigma^{-1}\sigma = \sigma_I$. σ^{-1} is called σ inverse. ■

EXAMPLE 2

a. $\sigma = \begin{pmatrix} 1 & 2 & 3 & 4 \\ 3 & 1 & 2 & 4 \end{pmatrix} \qquad \sigma^{-1} = \begin{pmatrix} 1 & 2 & 3 & 4 \\ 2 & 3 & 1 & 4 \end{pmatrix}$

b. $\sigma = \begin{pmatrix} 1 & 2 & 3 & 4 & 5 \\ 4 & 1 & 3 & 5 & 2 \end{pmatrix} \qquad \sigma^{-1} = \begin{pmatrix} 1 & 2 & 3 & 4 & 5 \\ 2 & 5 & 3 & 1 & 4 \end{pmatrix}$

To construct σ^{-1} one just reads the table describing σ from bottom to top. Thus, if $\sigma(1) = j$, then $\sigma^{-1}(j) = 1$.

Definition C.2. A permutation σ is called a transposition if σ leaves every integer but two fixed, and interchanges those two. Thus,

$\sigma = \begin{pmatrix} 1 & 2 & 3 & 4 & 5 \\ 1 & 2 & 4 & 3 & 5 \end{pmatrix}$ is a transposition, since $\sigma(i) = i$ unless $i = 3$ or 4,

and $\sigma(3) = 4$, $\sigma(4) = 3$. A more convenient way to describe a transposition is to list only the interchanged pair of integers. Thus, the above transposition would be written as $\sigma = (34)$.

Theorem C.1. Let $S_N = \{\sigma_1, \ldots, \sigma_{N!}\}$. Let σ be any permutation in S_N. Then the set $\{\sigma\sigma_1, \sigma\sigma_2, \ldots, \sigma\sigma_{N!}\}$ equals S_N.

Proof. Since the above set consists of $N!$ different symbols, each of which represents a permutation in S_N, it will suffice to show that no two of them are equal. Thus suppose that $\sigma\sigma_j = \sigma\sigma_k$. Then

$$\sigma_j = \sigma^{-1}\sigma\sigma_j = \sigma^{-1}\sigma\sigma_k = \sigma_k$$

Hence $i = j$. Thus, as τ varies over S_N so too will $\sigma\tau$ vary over S_N.

Given a permutation σ we say that σ has an inversion if for $i < j$, $\sigma(i) > \sigma(j)$.

Definition C.3. A permutation is said to be odd or even according to whether it has an odd or even number of inversions.

EXAMPLE 3

a. $\sigma = \begin{pmatrix} 1 & 2 & 3 & 4 \\ 3 & 1 & 4 & 2 \end{pmatrix}$. σ has $2 + 0 + 1 = 3$ inversions, because

$\sigma(1) = 3$ is greater that $\sigma(2)$ and $\sigma(4)$.

$\sigma(2) = 1$ produces no inversions.

$\sigma(3) = 4$ is greater than $\sigma(4)$.

Thus, σ is a odd permutation

b. $\sigma = \begin{pmatrix} 1 & 2 & 3 & 4 & 5 & 6 \\ 5 & 4 & 6 & 3 & 2 & 1 \end{pmatrix}$. σ has $4 + 3 + 3 + 2 + 1 = 13$ inversions. thus, σ is an odd permutation.

To count the number of inversions we start at the first entry of the second row of σ and count the number of integers smaller than it. We then do the same for each of the other integers.

c. $\sigma = \begin{pmatrix} 1 & 2 & 3 & 4 & 5 & 6 & 7 \\ 2 & 1 & 4 & 7 & 6 & 5 & 3 \end{pmatrix}$. σ has $1 + 0 + 1 + 3 + 2 + 1 = 8$ inversions. Thus σ is an even permutation. ■

Theorem C.2. A transposition is an odd permutation.

Proof. Suppose σ is in S_N and $\sigma = (i_1, i_2)$. Thus,

$$\sigma = \begin{pmatrix} 1 & \cdots & i_1 & \cdots & i_2 & \cdots & N \\ 1 & \cdots & i_2 & \cdots & i_1 & \cdots & N \end{pmatrix}$$

Clearly the only pairs of integers j and k for which σ has an inversion must lie between i_1 and i_2. For each k such that $i_1 < k \le i_2$, $\sigma(i_1) = i_2 > \sigma(k)$ and there are $i_2 - i_1$ such inversions. For each pair j and k for which $i_1 < j < k < i_2$ there are no inversions. For each j such that $i_1 \le j < i_2$, $\sigma(j) > i_1 = \sigma(i_2)$ and there are $(i_2 - 1) - i_1$ such inversions. Thus, the total number of inversions is $(i_2 - i_1) + [(i_2 - 1) - i_1]$ and this number is odd. Hence, every transposition is an odd permutation.

Definition C.4. We define the sign of σ, denoted by $|\sigma|$, as follows:

$$|\sigma| = \begin{cases} 1 & \text{if } \sigma \text{ is even} \\ -1 & \text{if } \sigma \text{ is odd} \end{cases}$$

Theorem C.3. Let σ and τ be two permutations in S_N. Then $|\tau\sigma| = |\tau||\sigma|$.

Proof. The idea is to show that the number of inversions in σ plus the number of inversions in τ has the same parity as the number of inversions in $\tau\sigma$. This, along with even plus even equals even, odd plus odd equals even, and even plus odd equals odd then proves the theorem. Thus, suppose

$$\sigma = \begin{pmatrix} 1 & 2 & & N \\ \sigma(1) & \sigma(2) & \cdots & \sigma(N) \end{pmatrix} \quad \text{and} \quad \tau\sigma = \begin{pmatrix} \sigma(1) & \cdots & \sigma(N) \\ \tau\sigma(1) & \cdots & \tau\sigma(N) \end{pmatrix}$$

There are four cases to consider:

$i < j$, $\sigma(i) < \sigma(j)$, and $\tau\sigma(i) < \tau\sigma(j)$
σ and τ have no inversions and neither does $\tau\sigma$

$i < j$, $\sigma(i) < \sigma(j)$, and $\tau\sigma(i) > \tau\sigma(j)$
σ has no inversion, τ has an inversion, $\tau\sigma$ has an inversion

$i < j$, $\sigma(i) > \sigma(j)$, and $\tau\sigma(i) > \tau\sigma(j)$
σ has an inversion, τ has no inversion, $\tau\sigma$ has an inversion

$i < j$, $\sigma(i) > \sigma(j)$, and $\tau\sigma(i) < \tau\sigma(j)$
σ has an inversion, τ has an inversion, $\tau\sigma$ has no inversion

In all four cases the number of inversions in σ plus the number of inversions in τ differs from the number of inversions of $\tau\sigma$ by an even number.

EXAMPLE 4 Let $\sigma = \begin{pmatrix} 1 & 2 & 3 & 4 & 5 \\ 3 & 2 & 4 & 1 & 5 \end{pmatrix}$, $\tau = \begin{pmatrix} 1 & 2 & 3 & 4 & 5 \\ 4 & 3 & 1 & 5 & 2 \end{pmatrix}$. Then $\tau\sigma = \begin{pmatrix} 1 & 2 & 3 & 4 & 5 \\ 1 & 3 & 5 & 4 & 2 \end{pmatrix}$.

σ has $2 + 1 + 1 + 0 = 4$ inversions, $|\sigma| = 1$
τ has $3 + 2 + 0 + 1 = 6$ inversions, $|\tau| = 1$
$\tau\sigma$ has $0 + 1 + 2 + 1 = 4$ inversions, $|\tau\sigma| = 1 = |\tau||\sigma|$ ■

Appendix D

DETERMINANTS

Definition D.1. Let $A = [a_{jk}]$ be an $n \times n$ matrix. The determinant of A, $\det(A)$, is defined as

$$\det(A) = \sum_{\sigma} |\sigma| a_{1\sigma(1)} a_{2\sigma(2)} \cdots a_{n\sigma(n)}$$

where the sum is taken over all permutations in S_N. Each summand consists of a product of n terms, and each factor in the product comes from a unique row and column of A. This product is then multiplied by plus or minus 1 depending upon whether σ is even or odd respectively.

EXAMPLE 1 Let $A = \begin{bmatrix} a_{11} & a_{12} \\ a_{21} & a_{22} \end{bmatrix}$. S_2 has two permutations σ_I and $\tau = (12)$. Thus,

$$\begin{aligned} \det(A) &= |\sigma_I| a_{1\sigma_I(1)} a_{2\sigma_I(2)} + |\tau| a_{1\tau(1)} a_{2\tau(2)} \\ &= a_{11}a_{22} - a_{12}a_{21} \end{aligned}$$

■

EXAMPLE 2 Let $A = \begin{bmatrix} 2 & 3 & 1 & -2 \\ 4 & -3 & 6 & 0 \\ 1 & -1 & 2 & 6 \\ 5 & 8 & -2 & 7 \end{bmatrix}$. Compute the product associated with each of the following permutations in S_4:

a. $\sigma = \begin{pmatrix} 1 & 2 & 3 & 4 \\ 3 & 1 & 2 & 4 \end{pmatrix}$. σ has two inversions. Thus $|\sigma| = 1$.

$$\begin{aligned} a_{1\sigma(1)} a_{2\sigma(2)} a_{3\sigma(3)} a_{4\sigma(4)} &= a_{13} a_{21} a_{32} a_{44} \\ &= (1)(4)(-1)(7) = -28 \end{aligned}$$

b. $\sigma = \begin{pmatrix} 1 & 2 & 3 & 4 \\ 4 & 1 & 2 & 3 \end{pmatrix}$. σ has three inversions. Thus $|\sigma| = -1$.

$$\begin{aligned} -a_{1\sigma(1)}a_{2\sigma(2)}a_{3\sigma(3)}a_{4\sigma(4)} &= -a_{14}a_{21}a_{32}a_{43} \\ &= -(-2)(4)(-1)(-2) = 16 \end{aligned}$$

■

Theorem D.1. $\det(A) = \det(A^T)$.

Proof. Let $A = [a_{jk}]$. Then $A^T = [b_{jk}]$, where $b_{jk} = a_{kj}$. Thus,

$$\det(A^T) = \sum_{\sigma} |\sigma| b_{1\sigma(1)} \cdots b_{n\sigma(n)} = \sum_{\sigma} |\sigma| a_{\sigma(1)1} \cdots a_{\sigma(n)n}$$

Since the two terms $a_{\sigma(1)1} \cdots a_{\sigma(n)n}$ and $a_{1\sigma^{-1}(1)} \cdots a_{n\sigma^{-1}(n)}$ have exactly the same factors but in perhaps a different order, and since $|\sigma| = |\sigma^{-1}|$, we have

$$\det(A^T) = \sum_{\sigma} |\sigma^{-1}| a_{1\sigma^{-1}(1)} \cdots a_{n\sigma^{-1}(n)}$$

Moreover as σ varies throughout S_N, so too does σ^{-1}. Thus, $\det(A^T) = \det(A)$.

Theorem D.2. Let A_1 be a matrix obtained by interchanging two rows (columns) of A. Then $\det(A_1)$ equals $-\det(A)$.

Proof. Let $A = [a_{jk}]$. Suppose rows p and q of A are interchanged. Then $A_1 = [b_{jk}]$, where

$$b_{jk} = \begin{cases} a_{jk} & j \neq p \text{ or } q \\ a_{qk} & j = p \\ a_{pk} & j = q \end{cases}$$

Let τ be the transposition (pq). Then

$$\begin{aligned} \det(A_1) &= \sum_{\sigma} |\sigma| b_{1\sigma(1)} \cdots b_{p\sigma(p)} \cdots b_{q\sigma(q)} \cdots b_{n\sigma(n)} \\ &= \sum_{\sigma} |\sigma| a_{1\sigma(1)} \cdots a_{q\sigma(p)} \cdots a_{p\sigma(q)} \cdots a_{n\sigma(n)} \\ &= \sum_{\sigma} |\sigma| a_{1\sigma\tau(1)} \cdots a_{q\sigma\tau(q)} \cdots a_{p\sigma\tau(p)} \cdots a_{n\sigma\tau(n)} \end{aligned}$$

Since $|\tau| = -1$ and as σ varies over S_N so too will $|\sigma\tau|$, we have

$$\det(A_1) = -\sum_{\sigma} |\sigma\tau| a_{1\sigma\tau(1)} \cdots a_{n\sigma\tau(n)} = -\det(A)$$

If A_1 is obtained by interchanging two columns of A, then $\det(A_1) = -\det(A)$ follows from $\det(A) = \det(A^T)$.

Corollary D.1. If two rows (columns) of A are identical, $\det(A)$ equals zero.

Proof. Suppose rows p and q of A are the same. Then if we interchange these two rows to get the matrix A_1, we have $A_1 = A$. Thus, $\det(A_1) = \det(A) = -\det(A)$. Hence $\det(A) = 0$.

Theorem D.3. If the matrix A_1 is obtained by multiplying a row (column) of A by the constant c, then $\det(A_1) = c \det(A)$.

Proof. Let $A = [a_{jk}]$. Suppose row p of A is multiplied by c. Then $A_1 = [b_{jk}]$, where

$$b_{jk} = \begin{cases} a_{jk} & j \neq p \\ ca_{pk} & j = p \end{cases}$$

Thus,

$$\begin{aligned} \det(A_1) &= \sum_\sigma |\sigma| b_{1\sigma(1)} \cdots b_{p\sigma(p)} \cdots b_{n\sigma(n)} \\ &= \sum_\sigma |\sigma| a_{1\sigma(1)} \cdots (ca_{p\,\sigma(p)}) \cdots a_{n\sigma(n)} \\ &= c \sum_\sigma |\sigma| a_{1\sigma(1)} \cdots a_{p\sigma(p)} \cdots a_{n\sigma(n)} \\ &= c \det(A) \end{aligned}$$

That this result is true if a column of A is multiplied by a constant follows from Theorem D.1.

Corollary D.2. If A has a row or column of zeros, $\det(A)$ equals zero.

Proof. Suppose row p of A consists entirely of zeros. Multiplying this row by the constant 0 does not change A. Thus, $\det(A) = 0 \det(A) = 0$.

Theorem D.4. If the matrix A_1 is obtained by multiplying row (column) p of A by a constant c and then adding this to row (column) q ($p \neq q$), then $\det(A_1)$ equals $\det(A)$.

Proof. Let $A = [a_{jk}]$. Then $A_1 = [b_{jk}]$, where

$$b_{jk} = \begin{cases} a_{jk} & j \neq q \\ ca_{pk} + a_{qk} & j = q \end{cases}$$

$$\begin{aligned} \det(A_1) &= \sum_\sigma |\sigma| b_{1\sigma(1)} \cdots b_{p\sigma(p)} \cdots b_{q\sigma(q)} \cdots b_{n\sigma(n)} \\ &= \sum_\sigma |\sigma| a_{1\sigma(1)} \cdots a_{p\sigma(p)} \cdots (ca_{p\sigma(q)} + a_{q\sigma(q)}) \cdots a_{n\sigma(n)} \\ &= c \sum_\sigma |\sigma| a_{1\sigma(1)} \cdots a_{p\sigma(p)} \cdots a_{p\sigma(q)} \cdots a_{n\sigma(n)} \\ &\quad + \sum_\sigma |\sigma| a_{1\sigma(1)} \cdots a_{p\sigma(p)} \cdots a_{q\sigma(q)} \cdots a_{n\sigma(n)} \end{aligned}$$

The first sum is c times the determinant of a matrix with rows p and q identical. Hence, this sum is zero. The second sum is just the determinant of A. Thus, $\det(A_1)$ equals $\det(A)$. Once again $\det(A) = \det(A^T)$ shows that this theorem is true for the corresponding column operation.

Theorem D.5. The following are equivalent:

a. $\det(A) \neq 0$.
b. The rows (columns) of A are linearly independent.
c. A is invertible.

Proof. In Chapter 1 we showed that A is invertible if and only if A is row equivalent to I_n. It is easy to see that A is not row equivalent to I_n if and only if some linear combination of the rows equals the zero vector. Thus **b** and **c** are equivalent. To see that **a** and **c** are equivalent one merely notes that $\det(I_n) = 1$, and then uses Theorems D.2, D.3, and D.4.

Theorem D.6. $\det(AB) = \det(A)\det(B)$.

Proof. We first prove this for A an elementary row matrix. Suppose A is an elementary row matrix obtained by interchanging two rows of I_n. Then $\det(A) = -\det(I_n) = -1$. Since AB is the matrix obtained by interchanging two rows of B, we have $\det(AB) = -\det(B) = \det(A)\det(B)$. If A is one of the other two types of elementary row matrices, Theorems D.3 and D.4 show, in a similar manner, that $\det(AB) = \det(A)\det(B)$. Next, suppose that A is a nonsingular matrix. Then $A = E_1E_2 \cdots E_p$, where each E_k is an elementary row matrix. Thus,

$$\begin{aligned}\det(AB) &= \det(E_1)\det(E_2 \cdots E_pB) = \cdots \\ &= \det(E_1) \cdots \det(E_p)\det(B) \\ &= \det(E_1 \cdots E_p)\det(B) \\ &= \det(A)\det(B)\end{aligned}$$

Suppose now that A is a singular matrix. Then AB is also a singular matrix, and by Theorem D.5 we have $\det(A) = 0$ and $\det(AB) = 0$. Thus, $\det(AB) = \det(A)\det(B)$.

Theorem D.7. If A_{jk} denotes the cofactor of a_{jk}, i.e., $A_{jk} = (-1)^{j+k}\det(M_{jk})$, then

$$\sum_{k=1}^{n} a_{ik}A_{jk} = \delta_{ij}\det(A) \qquad 1 \leq i \leq n,\ 1 \leq j \leq n \tag{D.1}$$

$$\sum_{i=1}^{n} a_{ij}A_{ik} = \delta_{jk}\det(A) \qquad 1 \leq j \leq n,\ 1 \leq k \leq n \tag{D.2}$$

Proof. Formula (D.2) follows from (D.1) by using $\det(A) = \det(A^T)$ and then applying (D.1) to the matrix A^T. It will suffice to prove (D.1) for $i = 1$, since the more general case can be reduced to this by using Theorem D.2 $(i-1)$ times. For $i = 1$, we first show that

$$\sum_{k=1}^{n} a_{1k}A_{1k} = \det(A)$$

$$\det(A) = \sum_{\sigma\in S_n} |\sigma| a_{1\sigma(1)}a_{2\sigma(2)} \cdots a_{n\sigma(n)}$$
$$= \sum_{k=1}^{n} a_{1k} \sum_{\sigma(1)=k} |\sigma| a_{2\sigma(2)} \cdots a_{n\sigma(n)}$$

If $\sigma(1) = k$, then each of the products $a_{2\sigma(2)} \cdots a_{n\sigma(n)}$ corresponds to a unique term in the formula for $\det(M_{1k})$. In fact we want to show that

$$\sum_{\sigma(1)=k} |\sigma| a_{2\sigma(2)} \cdots a_{n\sigma(n)} = (-1)^{1+k}\det(M_{1k})$$

Thus, let $b_{ij} = \begin{cases} a_{i+1,j} & 1 \le j < k \\ a_{i+1,j+1} & k \le j \le n-1 \end{cases}$

Then,

$$\det(M_{1k}) = \sum_{\tau\in S_{n-1}} |\tau| b_{1\tau(1)} \cdots b_{n-1\tau(n-1)}$$
$$= \sum_{\tau\in S_{n-1}} |\tau| b_{\tau^{-1}(1)1} \cdots b_{\tau^{-1}(n-1)(n-1)}$$
$$= \sum_{\tau\in S_{n-1}} |\tau| a_{(\tau^{-1}(1)+1)1} \cdots a_{\tau^{-1}(k-1)+1(k-1)}a_{\tau^{-1}(k)+1(k+1)} \cdots a_{(\tau^{-1}(n-1)+1)n}$$

If σ^{-1} is defined as follows:

$$\begin{pmatrix} 1 & 2 \cdots & k-1 & k & k+1 & \cdots & n \\ \tau^{-1}(1)+1 & & \tau^{-1}(k-1)+1 & 1 & \tau^{-1}(k)+1 & & \tau^{-1}(n-1)+1 \end{pmatrix}$$

then σ^{-1} is in S_n and the number of inversions of σ^{-1} equals the number of inversions of $\tau^{-1} + (k-1)$. Thus

$$|\sigma| = |\sigma^{-1}| = |\tau^{-1}|(-1)^{k-1} = |\tau^{-1}|(-1)^{k+1} = |\tau|(-1)^{k+1}$$

Moreover as τ varies through S_{n-1}, the corresponding σ will vary over those permutations of S_n for which $\sigma(1) = k$. Thus,

$$\det(M_{1k}) = \sum_{\substack{\sigma\in S_n \\ \sigma(1)=k}} (-1)^{1+k}|\sigma| a_{2\sigma(2)} \cdots a_{n\sigma(n)}$$

and we have

$$\det(A) = \sum_{k=1}^{n} a_{1k}(-1)^{1+k}\det(M_{1k}) = \sum_{k=1}^{n} a_{1k}A_{1k}$$

To see that

$$\sum_{k=1}^{n} a_{1k}A_{jk} = 0$$

if j is not equal to 1, we note that this is the expansion across the first row of a matrix for which the first and jth rows are the same. By Corollary D.1 and the above, this sum must equal zero.

Theorem D.8

$$A\text{adj}(A) = \text{adj}(A)A = \det(A)I_n$$

Proof. $\text{adj}(A) = [A_{jk}]^T$. Thus, the i,j entry of the matrix $A\text{adj}(A)$ equals

$$\sum_{k=1}^{n} a_{ik}A_{jk} = \delta_{jk} \det(A)$$

Hence, every term not on the main diagonal of A adj(A) equals zero, and every term on the main diagonal of this product equals det(A). To see that [adj(A)]A equals $\det(A)I_n$, use formula (D.2).

BIBLIOGRAPHY

Differential Equations, Dynamical Systems, and Linear Algebra, M. W. Hirsch and S. Smale, Academic Press, 1974.

Finite Dimensional Vector Spaces, second edition, Paul R. Halmos, D. Van Nostrand Co., Inc., 1958.

Fourier Series and Integrals, H. Dym and H. P. McKean, Academic Press, 1972.

Introduction to Matrix Computations, Gilbert W. Stewart, Academic Press, 1973.

Linear Algebra and Its Applications, second edition, Gilbert Strang, Academic Press, 1980.

Linear Algebra and Matrix Theory, Evar D. Nering, John Wiley & Sons, Inc., 1964.

Mathematical Models for the Growth of Human Populations, J. H. Pollard, Cambridge University Press, 1973.

Matrix Computations, G. H. Golub and C. F. Van Loan, Johns Hopkins University Press, 1983.

Modules in Applied Mathematics, Vol. 3, *Discrete and System Models,* Springer-Verlag, 1982.

ANSWERS TO ODD-NUMBERED PROBLEMS

PROBLEM SET 1.1

1. **a.** $x_1 = -2x_2$. **b.** No solution.
3. **a.** $x_1 = -1 + 2x_3$, $x_2 = 3 - 3x_3$. **b.** $x_1 = 1.4$, $x_2 = -0.6$, $x_3 = 1.2$.
5. $x_1 = 1$, $x_2 = -1$, $x_3 = 0$, $x_4 = 3$.
7. Abe is 6. (Dave is 13, Matt is 16.)
9. $\frac{7}{22}$ or 0.3182 hour (about 19.091 minutes). The slower travels about 14.32 miles, the faster, about 20.68 miles ($\frac{315}{22}$ and $\frac{455}{22}$, respectively).
11. $x_1 = \frac{2}{7} - x_3$, $x_2 = -\frac{3}{7} + x_3$, $x_4 = \frac{2}{7}$.
13. 10 tons of mortar and 12 tons of cement.
15. **a.** Plane $x_1 + x_2 + x_3 = 0$. **b.** $x_1 + x_3 = 0$, $x_2 = 0$. Line $x_3 = -x_1$ in x_1, x_3 plane. **c.** Origin.

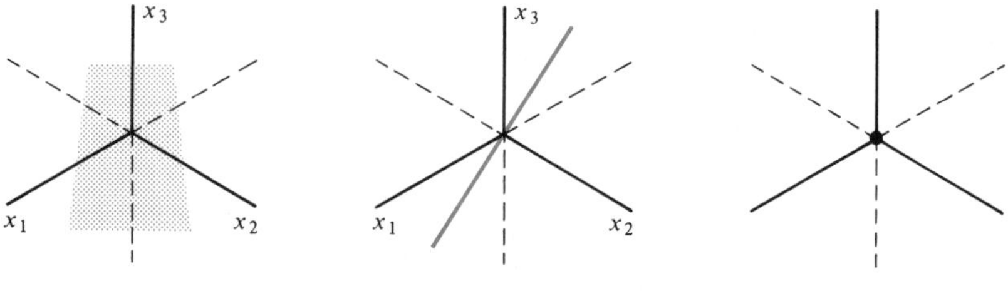

17. a. No. **b.** No.

19. a. $x_1 = -x_2$. **b.** $x_1 = 1 - x_2$. **c.** Parallel lines.

21. $x_{2n+1} = 2^n x_1$, $y_{2n+1} = 2^n y_1$, $x_{2n} = 2^{n-1}(x_1 + y_1)$, $y_{2n} = 2^{n-1}(x_1 - y_1)$

PROBLEM SET 1.2

1. a. $3A = \begin{bmatrix} 3 & 6 & 9 \\ -3 & 0 & -6 \end{bmatrix}$. **b.** $A - B = \begin{bmatrix} 2 & -4 & 7 \\ 1 & 3 & -2 \end{bmatrix}$.

c. $2A + 3B = \begin{bmatrix} -1 & 22 & -6 \\ -8 & -9 & -4 \end{bmatrix}$.

3. a. $\begin{bmatrix} -1 & 6 \\ 4 & -9 \end{bmatrix}$. **b.** $\begin{bmatrix} -1 & -2 \\ 12 & -9 \end{bmatrix}$. **c.** $\begin{bmatrix} 3 & 4 \\ 8 & -9 \end{bmatrix}$.

5. a. $\begin{bmatrix} 3 & 6 \\ 10 & 21 \end{bmatrix}$. **b.** $\begin{bmatrix} 15 & 6 \\ 22 & 9 \end{bmatrix}$.

7. a. $A^2 = \begin{bmatrix} 1 & 0 \\ 0 & 4 \end{bmatrix} A^3 = \begin{bmatrix} 1 & 0 \\ 0 & 8 \end{bmatrix}$. **b.** $A^2 = \begin{bmatrix} 1 & 3 \\ 0 & 4 \end{bmatrix} A^3 = \begin{bmatrix} 1 & 7 \\ 0 & 8 \end{bmatrix}$.

c. $A^2 = \begin{bmatrix} a^2 + bc & ab + bd \\ ac + cd & bc + d^2 \end{bmatrix}$

$$A^3 = \begin{bmatrix} a^3 + 2abc + bcd & a^2b + b^2c + abd + bd^2 \\ a^2c + acd + ac^2 + c^2d & abc + 2bcd + d^3 \end{bmatrix}$$

9. If $A^2 = \begin{bmatrix} a^2 + bc & ab + bd \\ ac + cd & bc + d^2 \end{bmatrix} = \begin{bmatrix} 0 & 1 \\ 0 & 0 \end{bmatrix}$, then $a^2 + bc = c(a + d) = bc + d^2 = 0$, and $b(a + d) = 1$. Since $b(a + d) = 1$ and $c(a + d) = 0$, it follows that $c = 0$. But then, $a^2 = d^2 = 0$, so $a = d = 0$, and $a + d = 0$, a contradiction. Thus, there can be no such a, b, c, d.

11. If $AX = B$, then X must be 2×3, so let $X = \begin{bmatrix} a & b & c \\ d & e & f \end{bmatrix}$.

$AX = \begin{bmatrix} 2a - d & 2b - e & 2c - f \\ 4a + d & 4b + e & 4c + f \end{bmatrix} = \begin{bmatrix} 2 & 3 & 6 \\ 1 & -1 & 0 \end{bmatrix}$. Thus, we solve the sets $\begin{bmatrix} 2a - d = 2 \\ 4a + d = 1 \end{bmatrix}$, $\begin{bmatrix} 2b - e = 3 \\ 4b + e = -1 \end{bmatrix}$, $\begin{bmatrix} 2c - f = 6 \\ 4c + f = 0 \end{bmatrix}$, or $\begin{bmatrix} 2 & -1 & : & 2 & 3 & 6 \\ 4 & 1 & : & 1 & -1 & 0 \end{bmatrix}$.

We get $\begin{bmatrix} 1 & 0 & : & \frac{1}{2} & \frac{1}{3} & 6 \\ 0 & 1 & : & -1 & -\frac{7}{3} & -4 \end{bmatrix}$ so that $X = \begin{bmatrix} \frac{1}{2} & \frac{1}{3} & 1 \\ -1 & -\frac{7}{3} & -4 \end{bmatrix}$. There is no Y with $YA = B$, since A has 2 columns and B has 3.

13. Proof by induction on p: case $p = 0$. If $p = 0$, then $A^pA^q = A^0A^q = I_nA^q = A^q = A^{0+q} = A^{p+q}$. Suppose that $A^pA^q = A^{p+q}$. Then $A^{p+1}A^q = (AA^p)A^q = A(A^pA^q) = A(A^{p+q})$ [by assumption] $= A^{p+q+1}$ [by definition of A^{n+1}]. The theorem follows.

15. $(AB)^T = [9 \quad 18] = B^TA^T$.

17. (2) Let $A = [a_{ij}]$ be $m \times n$, $B = [b_{jk}]$ be $n \times p$, and $c = [c_{kl}]$ be $p \times q$. Then $(AB)C = ([a_{ij}][b_{jk}])C = ([\Sigma_{j=1}^n a_{ij}b_{jk}])[c_{kl}] = (\Sigma_{j=1}^n \Sigma_{k=1}^p a_{ij}b_{jk}c_{kl}] = [\Sigma_{j=1}^n a_{ij}(\Sigma_{k=1}^p b_{jk}c_{kl})] = [a_{ij}][\Sigma_{k=1}^p b_{jk}c_{kl}] = A([b_{jk}][c_{kl}]) = A(BC)$.

(3) $a(AB) = a([a_{ij}][b_{jk}]) = a\ [\Sigma_{j=1}^{n} a_{ij}b_{jk}] = [\Sigma_{j=1}^{n} aa_{ij}b_{jk}] = [aa_{ij}][b_{jk}] = (aA)B = [\Sigma_{j=1}^{n} a_{ij}(ab_{jk})] = [a_{ij}][ab_{jk}] = A(aB)$, where a is a real number, A is $m \times n$ and B is $n \times k$.

19. $[-1 \quad 6]B = [-19 \quad 1]$, $[4 \quad 3]B = [22 \quad 23]$, and $AB = \begin{bmatrix} -19 & 1 \\ 22 & 23 \end{bmatrix}$.

21. A is 3×3 with the last row all 0's, so that $a_{31} = a_{32} = a_{33} = 0$. If B is any other 3×3 matrix, then the last row of AB consists of the numbers $\Sigma_{j=1}^{3} a_{3j}b_{j1}$, $\Sigma_{j=1}^{3} a_{3j}b_{j2}$, $\Sigma_{j=1}^{3} a_{3j}b_{j3}$, all of which must be 0, since all the a_{3j} are.

23. b and c are symmetric.

25. If A is $m \times n$, then A^T is $n \times m$, so that $(A^T)^T$ is $m \times n$ again. $(A^T)^T = [a_{ij}^{TT}] = [a_{ji}^T] = [a_{ij}]$.

27. Let A be 2×2 and B be 2×3. If $A = [a_{ij}]$ and $B = [b_{jk}]$, then $A^T = [a_{ji}^T]$, and $B^T = [b_{kj}^T]$, where $a_{ji}^T = a_{ij}$ and $b_{kj}^T = b_{jk}$. $(AB)^T = ([a_{ij}][b_{jk}])^T = ([a_{i1}b_{1k} + a_{i2}b_{2k}])^T = [(a_{i1}b_{1k} + a_{i2}b_{2k})^T]$. So the kith element of $(AB)^{\mathrm{T}}$ is $(a_{i1}b_{1k} + a_{i2}b_{2k})^T$, which equals $a_{1i}^T b_{k1}^T + a_{2i}^T b_{k2}^T = b_{k1}^T a_{1i}^T + b_{k2}^T a_{2i}^T =$ the kith element of $B^T A^T$.

29. If A and B are symmetric, then $A = A^T$ and $B = B^T$, so $a_{ij} = a_{ji}$ and $b_{ij} = b_{ji}$, where the indices are the same because A, B are the same size. $(A + B)^T = [(a_{ij} + b_{ij})^T] = [(a_{ji} + b_{ji})^T] = [a_{ij} + b_{ij}]$.

31. If A and B have the same size ($n \times n$) and are upper triangular, then $a_{ij} = b_{ij} = 0$ when $i > j$. $A + B = [a_{ij} + b_{ij}]$, and $a_{ij} + b_{ij} = 0$ when $i > j$. $AB = [\Sigma_{k=1}^{n} a_{ik}b_{kj}]$. Suppose $i > j$; k goes from 1 to n, and each k is either $> j$ or $k \leq j < i$. In the first case, $b_{kj} = 0$ since B is upper triangular; in the second, $a_{ik} = 0$ since A is upper triangular. In any case, each term of the sum (when $i > j$) is 0, so AB is also upper triangular.

33. By Theorem 2.5, $(AA^T)^T = A^{TT}A^T$ and $A^{TT}A^T = AA^T$, by exercise 25. The same results give $(A^TA)^T = A^TA^{TT} = A^TA$, so both AA^T and A^TA are symmetric.

35. $AB = \begin{bmatrix} 5 & -1 \\ 7.5 & -1.5 \end{bmatrix} = AC.$

PROBLEM SET 1.3

1. a. The second and and the fourth. **b.** The first and the third are; the middle one is not.

3. a. $\begin{bmatrix} 1 & 0 \\ 0 & 1 \end{bmatrix}, \begin{bmatrix} 1 & -2 \\ 0 & 0 \end{bmatrix}$. **b.** Both are equivalent to I_3.

c. $\begin{bmatrix} 1 & 0 & 0 & 0.6 \\ 0 & 1 & 1 & 2.4 \\ 0 & 0 & 0 & -0.8 \end{bmatrix}$.

5. a. $\begin{bmatrix} 1 & 0 \\ 0 & 1 \\ 0 & 0 \\ 0 & 0 \end{bmatrix}$. **b.** $[1 \quad -\frac{3}{4} \quad -\frac{2}{3} \quad 2]$. **c.** $\begin{bmatrix} 1 \\ 0 \\ 0 \\ 0 \end{bmatrix}$

d. $\begin{bmatrix} 1 & 0 & -\frac{8}{3} & \frac{2}{3} & -3 \\ 0 & 1 & 16 & -1 & 16 \end{bmatrix}$.

7. **a.** $\begin{bmatrix}1 & 2\\0 & 1\end{bmatrix}\begin{bmatrix}0 & 1\\1 & 0\end{bmatrix}=\begin{bmatrix}2 & 1\\1 & 0\end{bmatrix}$. **b.** $\begin{bmatrix}1 & 3\\0 & 1\end{bmatrix}\begin{bmatrix}1 & 0\\2 & 1\end{bmatrix}\begin{bmatrix}1 & 0\\-1 & 1\end{bmatrix}=\begin{bmatrix}4 & 3\\1 & 1\end{bmatrix}$.

9. $-2C1$ to $C2$.

11. If $E=[e_{mk}]$ is the elementary matrix that corresponds to interchanging row i with row j, then all entries of E are 0 except $e_{ij}=e_{ji}=1$ and $e_{mm}=1$ when m is not equal to i or j. If $A=[a_{kl}]$, then the $m1$th element of EA is $e_{m1}a_{11}+e_{m2}a_{21}+e_{m3}a_{31}+\cdots+e_{mn}a_{n1}$. If m is not equal to i or j, then the only nonzero term of this sum is $e_{mm}a_{ml}=a_{ml}$, so that the mth row of EA is the same as the mth row of A. If, however, $m=i$, then $e_{ij}a_{jl}=a_{jl}$ is the only nonzero term in the sum, so that the ith row of EA is the same as the jth row of A. If $m=j$, then the sum $=e_{ji}a_{il}=a_{il}$, and the jth row of EA is the same as the ith row of A. AE is the matrix obtained when columns i and j of A are interchanged.

13. $\begin{bmatrix}d_1e_1 & 0 & 0\\ & d_2e_2 & 0\\0 & 0 & d_3e_3\end{bmatrix}$

15. The first column is a_{11} followed by 0,0.

17. No. If $B=\begin{bmatrix}w & x\\y & z\end{bmatrix}$ is a candidate, then $BA=\begin{bmatrix}aw+cx & bw+dx\\ay+cz & by+dz\end{bmatrix}=\begin{bmatrix}a+c & b+d\\c & d\end{bmatrix}$ so that w and x satisfy $a(w-1)+c(x-1)=0$ and $b(w-1)+d(x-1)=0$ simultaneously. Since we cannot assume any relationship among a, b, c, and d, we have $w=x=1$. Also, y and z satisfy $ay+c(z-1)=0$ and $by+d(z-1)=0$. Again, we are forced to conclude that $y=0$ and $z=1$.

19. $\begin{bmatrix}a & b\\0 & d\end{bmatrix}$.

PROBLEM SET 1.4

1. **a.** $x_1=2891/3127=0.9245$, $x_2=5133/6254=0.8208$, $x_3=61/106=0.5755$. **b.** $x_1=5.5$, $x_2=1$, $x_3=-0.5$. **c.** No solution.

3. **a.** $7x_1=5$, $3x_1=2$, no solution. **b.** $4x_1+6x_2=-2$, $-3x_1-2x_2=1$, solution: $x_1=-0.2$, $x_2=-0.2$. **c.** $-2x_1=7$, $4x_1=-1$, $3x_1=2$, no solution.

5. **a.** $4x_1+6x_2=0$, $-3x_1+x_3=0$, solution: $x_1=\frac{1}{3}x_3$, $x_2=-\frac{2}{9}x_3$.
   b.
$$\begin{aligned}x_1-x_2\qquad\quad +2x_4&=0\\ x_1+x_2+x_3+\ x_4&=0\\ x_2+x_3-2x_4&=0\end{aligned}$$
   solution: $x_1=-3x_4$, $x_2=-x_4$, $x_3=3x_4$.

7. $a_{1k+1}=a_{1k}+0.05a_{1k}=(1.05)a_{1k}$, so $a_{1k}=(1.05)^{k-1}a_{11}$. $a_{2k+1}=a_{2k}+0.05a_{2k}+0.05a_{1k}=(1.05)a_{2k}+0.05a_{1k}$, so $a_{2k}=(1.05)^{k-1}a_{21}+(k-1)(0.05)(1.05)^{k-2}a_{11}$.

9. **a.** $T=\begin{bmatrix}1 & 0\\0 & 1\end{bmatrix}$ and $T^n=T=I_2$.

   b. $T=\begin{bmatrix}0 & 1\\1 & 0\end{bmatrix}$ and $T^n=\begin{cases}T, \text{ when } n \text{ is odd}\\ I_2, \text{ when } n \text{ is even.}\end{cases}$

   c. $T=\begin{bmatrix}1 & 1\\0 & 1\end{bmatrix}$ and $T^n=\begin{bmatrix}1 & n\\0 & 1\end{bmatrix}$.

11. $T = \begin{bmatrix} 1 & 1 & 0 \\ 0 & 1 & 0 \\ 1 & 1 & 0 \end{bmatrix}$. The proof that $T^n = \begin{bmatrix} 1 & n & 0 \\ 0 & 1 & 0 \\ 1 & n & 0 \end{bmatrix}$ is by induction. Case $n = 1$: obvious. Inductive step: Assume that $T^n = \begin{bmatrix} 1 & n & 0 \\ 0 & 1 & 0 \\ 1 & n & 0 \end{bmatrix}$. Show $T^{n+1} = \begin{bmatrix} 1 & n+1 & 0 \\ 0 & 1 & 0 \\ 1 & n+1 & 0 \end{bmatrix}$. But $T^{n+1} = TT^n = \begin{bmatrix} 1 & 1 & 0 \\ 0 & 1 & 0 \\ 1 & 1 & 0 \end{bmatrix}\begin{bmatrix} 1 & n & 0 \\ 0 & 1 & 0 \\ 1 & n & 0 \end{bmatrix} = \begin{bmatrix} 1 & n+1 & 0 \\ 0 & 1 & 0 \\ 1 & n+1 & 0 \end{bmatrix}$.

13. Suppose that $Ax_1 = 0$, $Ax_2 = 0$, and that c_1 and c_2 are constants. Then $A(c_1x_1 + c_2x_2) = A(c_1x_1) + A(c_2x_2) = c_1Ax_1 + c_2Ax_2 = c_10 + c_20 = 0$.

15. Let $A = [a_{jk}]$ be an $m \times n$ matrix, and let $C_k = [a_{1k}a_{2k} \; . \; . \; . \; a_{mk}]^T$ be the kth column of A. Suppose $X = [x_1x_2 \; . \; . \; . \; x_n]$. Then $AX = [\Sigma_{k=1}^n a_{jk}x_k] =$

$$[\Sigma_{k=1}^n a_{jk}x_k] = \begin{bmatrix} a_{11}x_1 + a_{12}x_2 + \cdots + a_{1n}x_n \\ a_{21}x_1 + a_{22}x_2 + \cdots + a_{2n}x_n \\ \cdot \quad \cdot \quad \cdot \\ \cdot \quad \cdot \quad \cdot \\ \cdot \quad \cdot \quad \cdot \\ a_{m1}x_1 + a_{m2}x_2 + \cdots + a_{mn}x_n \end{bmatrix} = x_1\begin{bmatrix} a_{11} \\ a_{21} \\ \cdot \\ \cdot \\ \cdot \\ a_{m1} \end{bmatrix} + x_2\begin{bmatrix} a_{12} \\ a_{22} \\ \cdot \\ \cdot \\ \cdot \\ a_{m2} \end{bmatrix} + \cdots + x_n\begin{bmatrix} a_{1n} \\ a_{2n} \\ \cdot \\ \cdot \\ \cdot \\ a_{mn} \end{bmatrix} = x_1C_1 + x_2C_2 + \cdots + x_nC_n.$$

PROBLEM SET 1.5

1. Only **a** has an inverse, and $A^{-1} = \frac{1}{5}\begin{bmatrix} 4 & -3 \\ -1 & 2 \end{bmatrix}$.

3. By definition, $A^{-k} = (A^{-1})^k$. By exercise 13 of section 1.2, $A^pA^q = A^{p+q}$ for p and q nonegative integers.

Case 1: p and q are both negative, so $-p, -q > 0$. Then $A^pA^q = (A^{-1})^{-p}(A^{-1})^{-q} = (A^{-1})^{-p+-q}$ (by exercise 13 quoted above) $= (A^{-1})^{-(p+q)} = A^{p+q}$ by definition.

Case 2: $p >= 0$ and $q < 0$.

Subcase a: $p + q = 0$, so $q = -p$, and $A^pA^q = A^pA^{-p} = A^p(A^{-1})^p = A^{p-1}AA^{-1}(A^{-1})^{p-1} = A^{p-1}I(A^{-1})^{p-1} = A^{p-1}(A^{-1})^{p-1} = \cdots = AA = I$.

Subcase b: $p + q > 0$, so $p = (p + q) + -q =$ the sum of two positive numbers. $A^pA^q = A^{(p+q)+-q}A^q = (A^{p+q}A^{-q})A^q =$ (by exercise 13) $A^{p+q}(A^{-q}A^q) = A^{p+q}I =$ (by subcase a) A^{p+q}.

Subcase c: $p + q < 0$, so $q = (p + q) + -p =$ the sum of two negative numbers. $A^pA^q = A^pA^{-p+(p+q)} = A^p(A^{-p}A^{p+q} =$ (by case 1) $(A^pA^{-p})A^{p+q} = IA^{p+q}$ (by subcase a) $= A^{p+q}$.

5. **a.** $\begin{bmatrix} x_{k+1} \\ y_{k+1} \end{bmatrix} = \begin{bmatrix} 2x_k + y_k \\ x_k + 3y_k \end{bmatrix} = \begin{bmatrix} 2 & 1 \\ 1 & 3 \end{bmatrix} \begin{bmatrix} x_k \\ y_k \end{bmatrix}.$

b. $\begin{bmatrix} x_{k+1} \\ y_{k+1} \end{bmatrix} = \begin{bmatrix} 2 & 1 \\ 1 & 3 \end{bmatrix} \begin{bmatrix} x_k \\ y_k \end{bmatrix} = \begin{bmatrix} 2 & 1 \\ 1 & 3 \end{bmatrix} \begin{bmatrix} 2 & 1 \\ 1 & 3 \end{bmatrix} \begin{bmatrix} x_{k-1} \\ y_{k-1} \end{bmatrix} = \begin{bmatrix} 2 & 1 \\ 1 & 3 \end{bmatrix}^2 \begin{bmatrix} x_{k-1} \\ y_{k-1} \end{bmatrix}.$

c. Proof by induction on k:

Case $k = 0$: $\begin{bmatrix} x_0 \\ y_0 \end{bmatrix} = A^0 \begin{bmatrix} x_0 \\ y_0 \end{bmatrix} = I_2 \begin{bmatrix} x_0 \\ y_0 \end{bmatrix}.$

Inductive step: Assume that $\begin{bmatrix} x_k \\ y_k \end{bmatrix} = A^k \begin{bmatrix} x_0 \\ y_0 \end{bmatrix}$. Then, $\begin{bmatrix} x_{k+1} \\ y_{k+1} \end{bmatrix} = A \begin{bmatrix} x_k \\ y_k \end{bmatrix} =$ $AA^k \begin{bmatrix} x_0 \\ y_0 \end{bmatrix} = A^{k+1} \begin{bmatrix} x_0 \\ y_0 \end{bmatrix}$. The theorem follows by induction.

d. Since $\begin{bmatrix} x_3 \\ y_3 \end{bmatrix} = A^3 \begin{bmatrix} x_0 \\ y_0 \end{bmatrix}$, it follows that $\begin{bmatrix} x_0 \\ y_0 \end{bmatrix} = A^{-3} \begin{bmatrix} x_3 \\ y_3 \end{bmatrix}.$

e. $A^3 = \begin{bmatrix} 15 & 20 \\ 20 & 35 \end{bmatrix}$, and $A^{-3} = \frac{1}{25} \begin{bmatrix} 7 & -4 \\ -4 & 3 \end{bmatrix}.$

7. For 5 units of sector 1, $(0.1)5 = 0.5$ units from sector 1.
For 8 units of sector 2, $(0.2)8 = 1.6$ units from sector 1.
For 9 units of sector 3, $(0.1)9 = 0.9$ units from sector 1.
Or, to get all three simultaneously, $0.5 + 1.6 + 0.9 = 3$ units from sector 1.

9. $X = (I - A)^{-1}D = \left(\begin{bmatrix} 1 & 0 & 0 \\ 0 & 1 & 0 \\ 0 & 0 & 1 \end{bmatrix} - \begin{bmatrix} \frac{1}{2} & \frac{1}{2} & 0 \\ \frac{1}{2} & 0 & \frac{1}{2} \\ 0 & \frac{1}{2} & \frac{1}{3} \end{bmatrix} \right)^{-1} \begin{bmatrix} 2 \\ 4 \\ 1 \end{bmatrix} =$

$\begin{bmatrix} \frac{1}{2} & -\frac{1}{2} & 0 \\ -\frac{1}{2} & 1 & -\frac{1}{2} \\ 0 & -\frac{1}{2} & \frac{2}{3} \end{bmatrix}^{-1} \begin{bmatrix} 2 \\ 4 \\ 1 \end{bmatrix} = \begin{bmatrix} 10 & 8 & 6 \\ 8 & 8 & 6 \\ 6 & 6 & 6 \end{bmatrix} \begin{bmatrix} 2 \\ 4 \\ 1 \end{bmatrix} = \begin{bmatrix} 58 \\ 54 \\ 42 \end{bmatrix}.$

Thus, 54 units are needed from sector 2.

11. If $E = [e_{ij}]$ with all $e_{ij} = 0$ except e_{jj}, and with $e_{jj} = 1$ except for $e_{ii} = c =$ a nonzero number, then E^{-1} looks exactly like E, except for its i,i entry, which is $1/c$.

13. **a.** The six are:

$$\begin{bmatrix} 1 & 0 & 0 \\ 0 & 1 & 0 \\ 0 & 0 & 1 \end{bmatrix} \begin{bmatrix} 1 & 0 & 0 \\ 0 & 0 & 1 \\ 0 & 1 & 0 \end{bmatrix} \begin{bmatrix} 0 & 1 & 0 \\ 1 & 0 & 0 \\ 0 & 0 & 1 \end{bmatrix} \begin{bmatrix} 0 & 1 & 0 \\ 0 & 0 & 1 \\ 1 & 0 & 0 \end{bmatrix} \begin{bmatrix} 0 & 0 & 1 \\ 1 & 0 & 0 \\ 0 & 1 & 0 \end{bmatrix} \begin{bmatrix} 0 & 0 & 1 \\ 0 & 1 & 0 \\ 1 & 0 & 0 \end{bmatrix}$$

b. There are n rows in the $n \times n$ identity matrix. To build an $n \times n$ permutation matrix, there are n choices for the first row and for each of these, there are $(n - 1)$ choices for the second row, and for each pair, $(n - 2)$ choices for the third row, etc., down to 1 choice for the last row. This gives $n(n - 1) \ldots (2)(1) = n!$ permutation matrices.

c. Proof by induction on $k =$ the number of rows "in the right place."
Case $k = 1$: If the first row of the permutation matrix does not have its first entry equal to 1, search down the first column until you find the row that does, and interchange that row with the first.
Inductive step: Assume that the first k rows of the matrix have 1's in columns 1 through k, respectively. We show that we can put the $k + 1$st row into place. Search down the $k + 1$st column from row $k + 1$ to n until a 1 is encountered. There is one because a permutation matrix has exactly one 1 in each row and

column, and none of the first k rows has a 1 in the $k+1$st column. Switch the row with 1 in the $k+1$st column with the row that is currently the $k+1$st. Thus, every $n \times n$ permutation matrix can be turned into the appropriate identity matrix by interchanging rows.

15. If $D = \begin{bmatrix} d_{11} & 0 \\ 0 & d_{22} \end{bmatrix}$, then $D^{-1} = \begin{bmatrix} 1/d_{11} & 0 \\ 0 & 1/d_{22} \end{bmatrix}$, so D^{-1} exists precisely when both entries on the main diagonal are nonzero.

17. Let A be 3×3 and upper triangular, so that $a_{ij} = 0$ if $i > j$. If a_{11}, a_{22}, and a_{33} are nonzero, then the ith row can be divided by a_{ii}, so that the diagonal entries are all 1. Then, starting with the third column, we subtract the new a_{i3} times row 3 from row i ($i = 1,2$); then we subtract the resulting a_{12} times row 2 from row 1 to get the identity. So A is invertible. If, on the other hand, one of the a_{ii} is zero, then using the technique above, A can be transformed into a matrix with at least one row of zeros. (If $a_{33} = 0$, then third row has all zeros; if $a_{22} = 0$ and a_{33} does not, then subtract a_{23}/a_{33} times the third row from the second to get a new second row with all zeros; if $a_{11} = 0$ and a_{22} and a_{33} do not, then use the second and third rows to get a new first row with all zeros.) If B is any 3×3 matrix, then AB must have a row of zeros, since A does. Thus, for no B does $AB = I$, and A is singular.

19. Suppose A is symmetric and invertible. Then $A = A^T$. But by the previous problem, $(A^{-1})^T = (A^T)^{-1} = A^{-1}$, so A^{-1} is symmetric.

SUPPLEMENTARY EXERCISES

3. $x_1 = -\frac{1}{2} - \frac{2}{3}x_4$, $x_2 = 1 + \frac{4}{9}x_4$, $x_3 = -1 - x_4$

5. **a.** $x_1 = \frac{10}{3} + x_3$, $x_2 = \frac{2}{3}$

b. No solution.

c. $x_1 = \frac{3}{2}(1 - x_3)$, $x_2 = \frac{1}{2}(1 + x_3)$

7. From $B = EA$ and $E = E^{-1}$ we have $B^{-1} = A^{-1}E$. Multiplying on the right by E interchanges the appropriate columns of A^{-1}. If two columns of A are interchanged, then the corresponding two rows of A^{-1} are interchanged.

9. $ax_1 + bx_3 = 1$, $ax_2 + bx_4 = 0$, $cx_1 + dx_3 = 0$, $cx_2 + dx_4 = 1$. The solution to this system is $x_1 = d/(ad - bc)$, $x_2 = -b/(ad - bc)$, $x_3 = -c/(ad - bc)$, $x_4 = a/(ad - bc)$. To see that $BA = I$, compute their product.

11. Let X be the $n \times 1$ matrix whose column consists of the numbers $x_1, x_2, \ldots, x_n$. If $Y = [y_1, y_2, \ldots, y_n]$, then $Y = AX$. Thus, $X = A^{-1}Y$, and $x_j = \Sigma_{k=1}^{n} b_{jk}y_k$.

13. Clearly, if A is invertible, then $AB = O_{nn}$ implies that $B = A^{-1}O_n = O_n$. Suppose now that A is not invertible; then there is a sequence of elementary row matrices $E_1, \ldots, E_p$ such that the matrix $C = E_1 \ldots E_pA$ has only zeros in its last row. We note that CB, for any B, is a matrix whose last row has only zeros in it. Thus, in solving the equation $CB = O_n$, we have a system of $n(n-1)$ equations with n^2 unknowns, and we must have a nontrivial solution. But then we have $AB = E_p^{-1} \ldots E_1^{-1}O_n = O_n$ with B not equal to the $n \times n$ zero matrix.

15. Let $A(t) = [a_{jk}]$ and $B(t) = [b_{jk}]$. Then $(A + B)' = [a_{jk} + b_{jk}]' = [a'_{jk} + b'_{jk}] = [a'_{jk}] + [b'_{jk}] = A' + B'$. $(AB)' = [(\Sigma_{k=1}^{n} a_{jk}b_{kl})'] = [\Sigma_{k=1}^{n} a'_{jk}b_{kl} + \Sigma_{k=1}^{n} a_{jk}b'_{kl}] = A'B + AB'$.

PROBLEM SET 2.1

1. a.

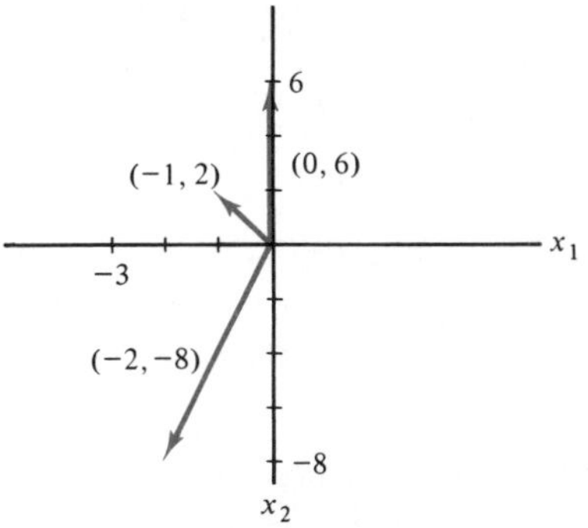

b.

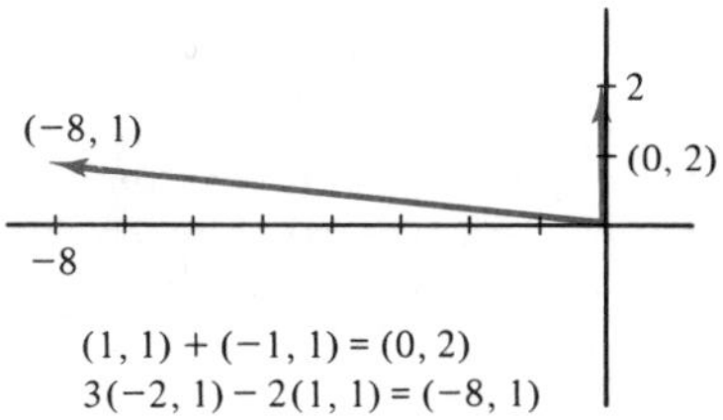

3. These are all vectors of the form $(t, -t)$, where t is real.

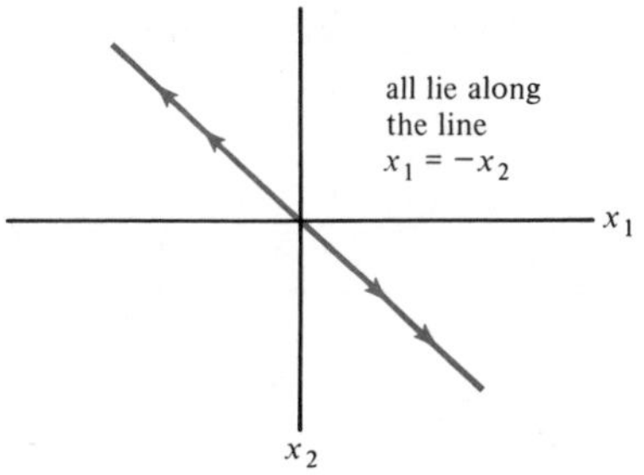

5. All vectors in $\mathbb{R}^2$ can be written as $c_1\mathbf{A} + c_2\mathbf{B}$.

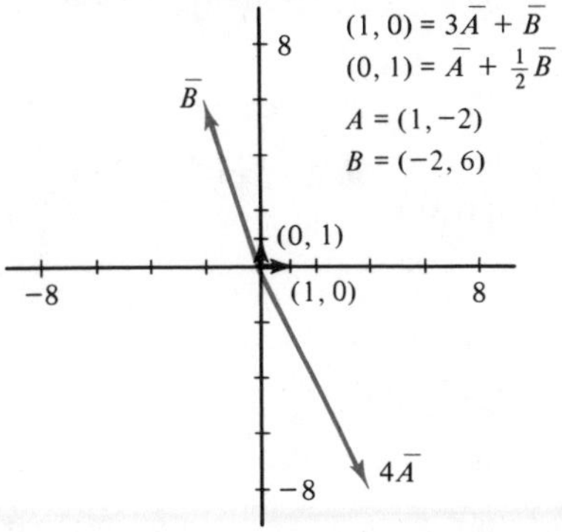

7. $(t, -t, 2t)$, and they all lie along the same line in 3-space.

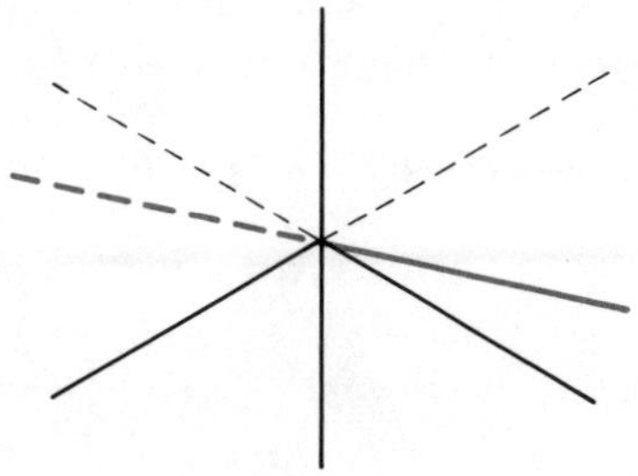

9. **a.** The line segment from (0,1) to (1,0).
b. The square with vertices (0,0), (1,0), (1,1), (0,1), and its interior.
c. The line though (1,0) and (0,1).
d. Portion of the first quadrant above the line $y = 1$, including all points on $y = 1$ from (0,1) to the right, and all points on the y-axis from (0,1) on up.
e. The entire plane.

11. **a.** The line segment from (1,1,1) to (1,1,0).
b. Quadrilateral with vertices (0,0,0), (1,1,0), (1,1,1), (2,2,1).
c. The line through (1,1,0) and (1,1,1).

PROBLEM SET 2.2

1. For numbers 3 through 10, repeat the proof of Theorem 2.1.
(1) If $(a_1,a_2,0)$ and $(b_1,b_2,0)$ are in V, then so is their sum, $(a_1 + b_1,a_2 + b_2,0)$.
(2) If a is a real number and if $(a_1,a_2,0)$ is in V, then so is $a(a_1,a_2,0) = (aa_1,aa_2,0)$.

3. Not a vector space.
(4) Associativity fails.
$[(x_1,x_2) + (y_1,y_2)] + (z_1,z_2) = (x_2 + y_2,x_1 + y_1) + (z_1,z_2) = (x_1 + y_1 + z_2,x_2 + y_2 + z_1)$, whereas $(x_1,x_2) + [(y_1,y_2) + (z_1,z_2)] = (x_1,x_2) + (y_2 + z_2,y_1 + z_1) = (x_2 + y_1 + z_1,x_1 + y_2 + z_2)$.
(8) Distributive law fails.
$2(x_1,x_2) = (1 + 1)(x_1,x_2) = 1(x_1,x_2) + 1(x_1,x_2) = (x_1,x_2) + (x_1,x_2) = (x_2 + x_2,x_1 + x_1) = (2x_2,2x_1)$, whereas $2(x_1,x_2) = (2x_1,2x_2)$.

5. The verifications for (3) through (6) are like the proof of Theorem 2.1. (1) works analogously to the verification in number 1 of this section, only a_2 and b_2 are 0 rather than a_3 and b_3.
(2) Given any real number a and a vector $(x_1,0,x_3)$ in V, $a(x_1,0,x_3) = (ax_1,0,x_3)$ is back in V.
(7) $a[(x_1,0,x_3) + (y_1,0,y_3)] = a(x_1 + y_1,0,x_3 + y_3) = (a(x_1 + y_1),0,x_3 + y_3) = (ax_1 + ay_1,0,x_3 + y_3) = a(x_1,0,x_3) + a(y_1,0,y_3)$.
(8) fails.
$(a + b)(x_1,0,x_3) = [(a + b)x_1,0,x_3] = (ax_1 + bx_1,0,x_3)$, whereas $a(x_1,0,x_3) + b(x_1,0,x_3) = (ax_1,0,x_3) + (bx_1,0,x_3) = (ax_1 + bx_1,0,2x_3)$.
(9) $(ab)(x_1,0,x_3) = (abx_1,0,x_3) = a(bx_1,0,x_3) = a[b(x_1,0,x_3)]$.
(10) $1(x_1,0,x_3) = (1x_1,0,x_3) = (x_1,0,x_3)$.

7. (1) and (2) are verified in the text.
(3) $(\mathbf{f} + \mathbf{g})(\mathbf{t}) = \mathbf{f}(t) + \mathbf{g}(t) = \mathbf{g}(t) + \mathbf{f}(t) = (\mathbf{g} + \mathbf{f})(t)$, since $\mathbf{f}(t)$ and $\mathbf{g}(t)$ are real numbers. So $(\mathbf{f} + \mathbf{g}) = (\mathbf{g} + \mathbf{f})$.
(4) $[(\mathbf{f} + \mathbf{g}) + \mathbf{h}](t) = (\mathbf{f} + \mathbf{g})(t) + \mathbf{h}(t) = [\mathbf{f}(t) + \mathbf{g}(t)] + \mathbf{h}(t) = \mathbf{f}(t) + [\mathbf{g}(t) + \mathbf{h}(t)] = \mathbf{f}(t) + (\mathbf{g} + \mathbf{h})(t) = [\mathbf{f} + (\mathbf{g} + \mathbf{h})](t)$ since $\mathbf{f}(t)$, $\mathbf{g}(t)$, and $\mathbf{h}(t)$ are real numbers. So $(\mathbf{f} + \mathbf{g}) + \mathbf{h} = \mathbf{f} + (\mathbf{g} + \mathbf{h})$.
(5) Let $\mathbf{0}\,(t) = 0$ for all t with $0 <= t <= 1$. Then $(\mathbf{f} + \mathbf{0})(t) = \mathbf{f}(t) + \mathbf{0}(t) = \mathbf{f}(t) + \mathbf{0} = \mathbf{f}(t)$, so $\mathbf{f} + \mathbf{0} = \mathbf{f}$.
(6) Given $\mathbf{f}$, let $-\mathbf{f}(t) = -\mathbf{f}(t)$. Then $(\mathbf{f} + -\mathbf{f})(t) = \mathbf{f}(t) + -\mathbf{f}(t) = \mathbf{f}(t) + -\mathbf{f}(t) = 0 = \mathbf{0}(t)$, so that $\mathbf{f} + -\mathbf{f} = \mathbf{0}$.
(7) $a(\mathbf{f} + \mathbf{g})(t) = a[\mathbf{f}(t) + \mathbf{g}(t)] = a\mathbf{f}(t) + a\mathbf{g}(t) = (a\mathbf{f} + a\mathbf{g})(t)$, and $a(\mathbf{f} + \mathbf{g}) = a\mathbf{f} + a\mathbf{g}$.

(8) $(a+b)\mathbf{f}(t) = a\mathbf{f}(t) + b\mathbf{f}(t) = (a\mathbf{f} + b\mathbf{f})(t)$, so $(a+b)\mathbf{f} = a\mathbf{f} + b\mathbf{f}$.
(9) $(ab)\mathbf{f}(t) = a[b\mathbf{f}(t)] = a(b\mathbf{f})(t)$, or $(ab)\mathbf{f} = a(b\mathbf{f})$.
(10) $1\mathbf{f}(t) = \mathbf{f}(t)$, and $1\mathbf{f} = \mathbf{f}$, therefore.

9. $V_0 =$ the set of 2×2 matrices with 1,1 entry 0 is a vector space. $V_1 =$ the set of 2×2 matrices with 1,1 entry nonzero, is not. There are several problems with V_1. First, closure: multiplication by 0 does not work, and adding a matrix with 1,1 entry say 2 to one with 1,1 entry -2 will also take you out of V_1. V_1 does not have the zero matrix, so (5) and (6) fail, as do (7), (8), and (9) because of the problems with multiplication by a real number.

11. V is not a vector space, since the sum of two things in V is not back in V. Note that V is not closed under scalar multiplication either.

13. $c = 0$ only. Otherwise, for example, the zero matrix would not be included; also, if c is nonzero, two such matrices cannot be added to get another in the set; closure under multiplication by real numbers also fails.

15. V is a vector space. If you add two matrices, each of whose entries sum to zero, you get a matrix whose entries sum to zero. The same is true of multiplication by a real number. (3) and (4) are the same for all matrices. The zero matrix has entries that sum to zero, so it is in V; and if A has entries that sum to 0, then $-A$ has entries that sum to $-0 = 0$. The rest of the properties are the same for matrices in general.

17. V is again a vector space: If $\mathbf{p}$ and $\mathbf{q}$ are in V, then so is $\mathbf{p}+\mathbf{q}$, since $(\mathbf{p}+\mathbf{q})(1) + (\mathbf{p}+\mathbf{q})(6) - (\mathbf{p}+\mathbf{q})(2) = \mathbf{p}(1) + \mathbf{q}(1) + \mathbf{p}(6) + \mathbf{q}(6) - \mathbf{p}(2) - \mathbf{q}(2) = 0$.
If $\mathbf{p}$ is and V and c is a real number, the $c\mathbf{p}$ is in V, since $(c\mathbf{p})(1) + (c\mathbf{p})(6) - (c\mathbf{p})(2) = c\mathbf{p}(1) + c\mathbf{p}(6) - c\mathbf{p}(2) = c[\mathbf{p}(1) + \mathbf{p}(6) - \mathbf{p}(2)] = 0$. Thus $0\mathbf{p}$ and $-\mathbf{p} = (-1)\mathbf{p}$ are back in V, so (5) and (6) are satisfied. The rest of the axioms are satisfied because they are satisfied in P_{17}. The only significance of 17 is that a positive whole number is needed. Again, if $\mathbf{p}(1) + \mathbf{p}(6) - \mathbf{p}(2) = 1$, then V is not closed under addition or multiplication by a real number. Hence, V would not be a vector space.

PROBLEM SET 2.3

1. $(x,y) = x(1,0) + y(0,1)$.

3. Since both $(1,0,1)$ and $(1,2,-1)$ satisfy $x_1 - x_2 - x_3 = 0$ we have $S[(1,0,1),(1,2,-1)] \subseteq \{(x_1,x_2,x_3): x_1 - x_2 - x_3 = 0\}$. Conversely if $x_1 - x_2 - x_3 = 0$, then $(x_1,x_2,x_3) = (x_1,x_2,x_1 - x_2) = \frac{2x_1 - x_2}{2}(1,0,1) + \frac{x_2}{2}(1,2,-1)$. Thus, $\{(x_1,x_2,x_3)\colon x_1 - x_2 - x_3 = 0\} \subseteq S[(1,0,1),(1,2,-1)]$. Hence the two sets are equal.

5. Only $(0,-1,1)$ is in $S[A]$. $(0,-1,1) = (1,-1,0) + (-1,0,1)$.

7. $S[\mathbf{x}] = \{c\mathbf{x}: c \text{ real}\} = \{(cx_1,cx_2): c \text{ real}\}$. Clearly, $(0,0)$ and (x_1,x_2) are in the set. Furthermore, the slope of the line through (cx_1,cx_2) and $(0,0)$ (assuming that c is not 0) is $(x_2)/(x_1)$, which is the same as the slope of the line through (x_1,x_2) and $(0,0)$, so all points (cx_1,cx_2) lie on the same line. Conversely, if (a,b) is on the line through $(0,0)$ and (x_1,x_2), then the slope of the line through (a,b) and $(0,0)$ should be the same as the slope of the line through (x_1,x_2) and $(0,0)$, so $b/a = x_2/x_1$, or there is a real number c with $a = cx_1$ and $b = cx_2$.

9. **a.** (1) and (2) of Theorem 3.4 are satisfied in W.
b. V satisfies (1) and (2) also, so it is a subspace of itself.

11. If $\mathbf{x}$ is in $S[(1,1,0)]$, then $\mathbf{x} = c(1,1,0)$. But $c(1,1,0) = c(1,1,0) + 0(0,0,1)$, a linear combination of $(1,1,0)$ and $(0,0,1)$. Thus, if $\mathbf{x}$ is in $S[A_1]$, $\mathbf{x}$ is in $S[A_2]$, as well. $S[A_1]$ is a line in 3-space, while $S[A_2]$ is a plane. Both go through the origin.

13. Since the solution to the homogeneous system is $x_1 = -3x_3, x_2 = -\frac{3}{2}x_3$, a generator is $(-3,-\frac{3}{2},1)$ so is $(6,-3,-2)$.

15. **a.** Yes. The four exhibited together with any other 2×2 matrix.
b. No. At least four are necessary. No matter what we choose for $M_1 = \begin{bmatrix} a_1 & b_1 \\ c_1 & d_1 \end{bmatrix}$, $M_2 = \begin{bmatrix} a_2 & b_2 \\ c_2 & d_2 \end{bmatrix}$, and $M_3 = \begin{bmatrix} a_3 & b_3 \\ c_3 & d_3 \end{bmatrix}$, we can always find an $M = \begin{bmatrix} w & x \\ y & z \end{bmatrix}$ so that the following system has no solution.

$$\begin{aligned} a_1x_1 + a_2x_2 + a_3x_3 &= w \\ b_1x_1 + b_2x_2 + b_3x_3 &= x \\ c_1x_1 + c_2x_2 + c_3x_3 &= y \\ d_1x_1 + d_2x_2 + d_3x_3 &= z \end{aligned}$$

Thus, there are no real x_i's with $x_1M_1 + x_2M_2 + x_3M_3 = M$.

17. **a.** Given cI_n and dI_n in W and a, a real number, then $cI_n + dI_n = (c+d)I_n$ and $a(cI_n) = (ac)I_n$ are back in W.
b. Given $[d_j\delta_{jk}]$ and $[c_j\delta_{jk}]$ in W and a real number a, both $[d_j\delta_{jk}] + [c_j\delta_{jk}] = [(d_j + c_j)\delta_{jk}]$ and $a[d_j\delta_{jk}] = [ad_j\delta_{jk}]$ are back in W.
c. Given $A = [a_{ij}]$ and $B = [b_{ij}]$ with $a_{ij} = b_{ij} = 0$ for $i > j$, and a, a real number, both $A + B = [a_{ij} + b_{ij}]$ and $aA = [aa_{ij}]$ are back in W. This is because $a_{ij} + b_{ij} = 0 + 0 = 0$ for $i > j$ and $aa_{ij} = a0 = 0$ for $i > j$.
d. Substitute $i < j$ in c. Or, if A and B are in W and a is real, then A^T and B^T are upper triangular. Thus, $A^T + B^T = (A+B)^T$ and $aA^T = (aA)^T$ are upper triangular by c. But then $(A+B)^{TT} = A + B$ and $(aA)^{TT} = aA$ are back in W.

19. The argument of 17 d works. Let A^T and B^T be in W^T and let a be real. Then A and B are in W. Since W is a subspace, $A + B$ and aA are back in W; but then $(A+B) = A^T + B^T$ and $(aA)^T = aA^T$ are back in W^T.

21. Let $\mathbf{x}$ and $\mathbf{y}$ be in $W = W_1 \cap W_2$ and let a be real. Then $\mathbf{x}$ and $\mathbf{y}$ are in W_1 and W_2, both of which are subspaces. So, $\mathbf{x} + \mathbf{y}$ and $a\mathbf{x}$ are in W_1 and W_2; therefore, they are in W.

23. Let $\mathbf{x}_1$ and $\mathbf{x}_2$ be in $W_1 + W_2$; then there exist vectors $\mathbf{u}_1, \mathbf{u}_2$ in W_1 and $\mathbf{v}_1, \mathbf{v}_2$ in W_2 with $\mathbf{x}_1 = \mathbf{u}_1 + \mathbf{v}_1$ and $\mathbf{x}_2 = \mathbf{u}_1 + \mathbf{v}_2$. W_1 and W_2 are subspaces, so $\mathbf{u}_1 + \mathbf{u}_2$ is back in W_1 and $\mathbf{v}_1 + \mathbf{v}_2$ is back in W_2; therefore, $\mathbf{x}_1 + \mathbf{x}_2 = \mathbf{u}_1 + \mathbf{v}_1 + \mathbf{u}_2 + \mathbf{v}_2 = (\mathbf{u}_1 + \mathbf{u}_2) + (\mathbf{v}_1 + \mathbf{v}_2) =$ an element of $W_1 + W_2$. If a is real, then $a\mathbf{u}_1$ is in W_1 and $a\mathbf{v}_1$ is in W_2 since W_1 and W_2 are subspaces. Therefore, $a\mathbf{x}_1 = a(\mathbf{u}_1 + \mathbf{v}_1) = a\mathbf{u}_1 + a\mathbf{v}_1 =$ an element of $W_1 + W_2$.

Now, suppose that W is a subspace of V that contains both W_1 and W_2; then if $\mathbf{x}$ is in $W_1 + W_2$, there exist $\mathbf{u}$ and $\mathbf{v}$ in W_1 and W_2, respectively, with $\mathbf{x} = \mathbf{u} + \mathbf{v}$. But $\mathbf{u}$ is in W because W_1 is, and $\mathbf{v}$ is in W because W_2 is. Thus, $\mathbf{u} + \mathbf{v}$ is in W since W is a subspace. Therefore, $\mathbf{x} = \mathbf{u} + \mathbf{v}$ is in W, so $W_1 + W_2 \subseteq W$.

25. No. If it were a subspace, then $a\mathbf{v}$ would be back in it wherever a is real and $\mathbf{v}$ is in W. So, if $\mathbf{v} = 1$ in W and $a = \frac{1}{2}$ in R, then $a\mathbf{v} = \frac{1}{2}$ should be in W, but it is not.

27. $c = 0$ only, in which case $x_1 = x_2 = 0$.

29. a. Since the solution of $A\mathbf{x} = \mathbf{0}$ is $x_1 = -3x_2 + \frac{1}{2}x_3$, $(x_1,x_2,x_3) = (-3x_2 + \frac{1}{2}x_3, x_2, x_3) = x_2(-3,1,0) + x_3(\frac{1}{2},0,1)$ and generators are $(-3,1,0)$ and $(\frac{1}{2},0,1)$, or $(-3,1,0)$ and $(1,0,2)$. Alternatively, since $2x_1 + 6x_2 - x_3 = 0$, $x_3 = 2x_1 + 6x_2$; thus, $(x_1,x_2,x_3) = (x_1,x_2,2x_1 + 6x_2) = x_1(1,0,2) + x_2(0,1,6)$. So $(1,0,2)$ and $(0,1,6)$ form another spanning set.

b. The solution to $A\mathbf{x} = \mathbf{0}$ is $x_1 = \frac{25}{2}x_3$ and $x_2 = -4x_3$. Thus, $(x_1,x_2,x_3) = (\frac{25}{2}x_3, -4x_3, x_3) = x_3(\frac{25}{2},-4,1)$. We may take either $(\frac{25}{2},-4,1)$ or $(25,-8,2)$ to be the sole element of a spanning set.

c. The solution to $A\mathbf{x} = \mathbf{0}$ is $x_1 = x_2 = x_3 = 0$, so the only element is $(0,0,0)$.

PROBLEM SET 2.4

1. $(1,0,1) = \frac{1}{2}(2,0,0) + (-1)(0,0,-1)$.

3. Let $\mathbf{x}$ be in $S[A]$. Then there exist $c_1, \ldots, c_k$, real numbers with $\mathbf{x} = c_1\mathbf{x}_1 + \cdots + c_k\mathbf{x}_k$, where $\mathbf{x}_1, \ldots, \mathbf{x}_k$ are the elements of A. If $\mathbf{y}$ is any vector in $S[A,\mathbf{x}]$, then there are real numbers $d_1, \ldots, d_k, d$ such that $\mathbf{y} = d_1\mathbf{x}_1 + \cdots + d_k\mathbf{x}_k + d\mathbf{x}$. But then, $\mathbf{y} = d_1\mathbf{x}_1 + \cdots + d_k\mathbf{x}_k + d(c_1\mathbf{x}_1 + \cdots + c_k\mathbf{x}_k) = (d_1 + dc_1)\mathbf{x}_1 + \cdots + (d_k + dc_k)\mathbf{x}_k$, an element of $S[A]$. Therefore, $S[a,\mathbf{x}] = S[A]$.

5. Only b is linearly dependent.

a. If $a(1,2,3,0) + b(-6,1,0,0) + c(1,2,0,3) = \mathbf{0}$, then we solve

$$\begin{bmatrix} 1 & -6 & 1 & : & 0 \\ 2 & 1 & 2 & : & 0 \\ 3 & 0 & 0 & : & 0 \\ 0 & 0 & 3 & : & 0 \end{bmatrix} \text{ and get } a = b = c = 0.$$

b. $0(1,-1,6) + 2(1,1,1) + (-2)(-1,0,1) = (4,2,0)$.

c. As in a, the system $\begin{bmatrix} 2 & 1 & -1 & : & 0 \\ 1 & 1 & 1 & : & 0 \\ 4 & 1 & 1 & : & 0 \end{bmatrix}$ has only the trivial solution.

7. If A is linearly dependent, then there are vectors $\mathbf{x}_1, \ldots, \mathbf{x}_k$ in A and real numbers $c_1, \ldots, c_k$ with not all $c_j = 0$, and $\mathbf{0} = c_1\mathbf{x}_1 + c_2\mathbf{x}_2 + \cdots + c_k\mathbf{x}_k$. If B is any set that contains A, then this same relation holds in B, so that B is also dependent.

9. Suppose that $\{\mathbf{x}_i = (a_{1i},a_{2i},a_{3i})$: for $i = 1,2,3,4\}$ is a set of four vectors in $\mathbb{R}^3$ and suppose that $c_1\mathbf{x}_1 + c_2\mathbf{x}_2 + c_3\mathbf{x}_3 + c_4\mathbf{x}_4 = \mathbf{0}$. Then to find values for the c's, we solve the system:

$$\begin{bmatrix} a_{11} & a_{12} & a_{13} & a_{14} & : & 0 \\ a_{21} & a_{22} & a_{23} & a_{24} & : & 0 \\ a_{31} & a_{32} & a_{33} & a_{34} & : & 0 \end{bmatrix}$$

Since there can never be exactly one solution for such a system, there are solutions other than $c_1 = c_2 = c_3 = c_4 = 0$, so the set of vectors is dependent.

11. Let $c_1(1 + t) + c_2(1 - t^2) + c_3(t^2) = \mathbf{0}$. Then $c_1 + c_2 = 0$, $c_1 = 0$, and $-c_2 + c_3 = 0$, or $c_1 = c_2 = c_3 = 0$. The set is independent. $S[A] = P_2$.

13. A is a spanning set for $\mathbb{R}^2$ since A contains $\{(1,0),(0,1)\}$, which is a spanning set for $\mathbb{R}^2$. (See problem 3.) It is not independent, for let t be any angle other than 0, 90, degrees. Then $\mathbf{x} = (\cos t, \sin t)$ is in A, $\mathbf{x}$ is not equal to either $(1,0)$ or $(0,1)$, and

$\mathbf{x} = (\cos t)(1,0) + (\sin t)(0,1)$, a linear combination of other vectors in A. Therefore, A is dependent.

15. If $a \sin t + b \sin 2t + c \cos t = 0$, then this equation holds for all values of t. In particular, if $t = 0$, we have that $c = 0$; if $t = \pi/2$ (in radians = 90 degrees), then we have $a = 0$. If $t = \pi/4$ (in radians = 45 degrees), then $b = 0$, and the set is independent. Since all the polynomials are in $C[0,1]$, and there is no way we can write them as linear combinations of $\sin t$, $\sin 2t$, and $\cos t$, the set cannot possible span $C[0,1]$.

17. If $p(t)$ is in V, then $p(t) = p_0 + p_1 t + p_2 t^2 + p_3 t^3$ and $p'(t) = p_1 + 2p_2 t + 3p_3 t^2$; $p'(0) = p_1 = 0$, so $p(t) = p_0 + p_2 t^2 + p_3 t^3$. Since $p(1) - p(0) = 0$, $p_0 + p_2 + p_3 - p_0 = 0$, and $p_3 = -p_2$, so $p(t) = p_0 + p_2 t^2 + -p_2 t^3 = p_0(1) + p_2(t^2 - t^3)$. A spanning set is $\{1, t^2 - t^3\}$. This spanning set is linearly independent, since if $p_0 + p_2(t^2 - t^3) = \mathbf{0}$, then $p_0 = 0$, and so does p_2.

19. a. We suppose $A = \{\mathbf{x}_1, \mathbf{x}_2, \ldots, \mathbf{x}_k\}$ and that whenever $c_1, \ldots, c_k$ are real numbers with $c_1\mathbf{x}_1 + \cdots + c_k\mathbf{x}_k = \mathbf{0}$, then $c_1 = c_2 = \cdots = c_k = 0$. Now suppose that $\{\mathbf{y}_1, \ldots, \mathbf{y}_p\}$ is a finite set of elements of A with $d_1\mathbf{y}_1 + \cdots + d_p\mathbf{y}_p = \mathbf{0}$. Then, let $\mathbf{y}_{p+1}, \ldots, \mathbf{y}_k$ be the rest of the vectors of A, and notice that $d_1\mathbf{y}_1 + \cdots + d_p\mathbf{y}_p + 0\mathbf{y}_{p+1} + \cdots + 0\mathbf{y}_k = \mathbf{0}$ = a linear combination of all the vectors in A, possibly reordered. By our assumption, $d_1 = d_2 = \cdots = d_p = 0$. Now suppose the converse: that whenever $\mathbf{y}_1, \ldots, \mathbf{y}_p$ is a finite subset of A with $d_1\mathbf{y}_1 + \cdots + d_p\mathbf{y}_p = \mathbf{0}$, then $d_1 = \cdots = d_p = 0$. If, then $c_1\mathbf{x}_1 + \cdots + c_k\mathbf{x}_k = \mathbf{0}$ = a linear combination of all the vectors in A, then, since A is finite, this linear combination is a linear combination of finitely many vectors in A, so that, by assumption, $c_1 = c_2 = \cdots = c_k = 0$.

b. Now, suppose that A has the property that whenever $\mathbf{x}_1, \ldots, \mathbf{x}_p$ is a finite subset of A with $c_1\mathbf{x}_1 + \cdots + c_p\mathbf{x}_p = 0$, then $c_1 = \cdots = c_p = 0$. Suppose $\mathbf{x}$ is in $S[A]$, and that there are two finite subsets, say, $\mathbf{x}_1, \ldots, \mathbf{x}_p, \mathbf{y}_1, \ldots, \mathbf{y}_k$ and $\mathbf{x}_1, \ldots, \mathbf{x}_p, \mathbf{z}_1, \ldots \mathbf{z}_q$ with $\mathbf{x} = c_1\mathbf{x}_1 + \cdots + c_p\mathbf{x}_p + c_{p+1}\mathbf{y}_1 + \cdots + c_{p+k}\mathbf{y}_k = d_1\mathbf{x}_1 + \cdots + d_p\mathbf{x}_p + d_{p+1}\mathbf{z}_1 + \cdots + d_{p+q}\mathbf{z}_q$, where $\mathbf{x}_1, \ldots, \mathbf{x}_p$ are the vectors that are common to both finite sets of vectors, and the $\mathbf{y}_i$'s and $\mathbf{z}_i$'s are not in both sets. If we subtract the second from the first representation, then we get

$$\begin{aligned}\mathbf{0} &= c_1\mathbf{x}_1 + \cdots + c_p\mathbf{x}_p + c_{p+1}\mathbf{y}_1 + \cdots + c_{p+k}\mathbf{y}_k - d_1\mathbf{x}_1 - \cdots \\ &\quad - d_{p+1}\mathbf{z}_1 - \cdots - d_{p+q}\mathbf{z}_q \\ &= (c_1 - d_1)\mathbf{x}_1 + \cdots + (c_p - d_p)\mathbf{x}_p + c_{p+1}\mathbf{y}_1 + \cdots + c_{p+k}\mathbf{y}_k \\ &\quad + d_{p+1}\mathbf{z}_1 + \cdots + d_{p+q}\mathbf{z}_q\end{aligned}$$

Now, we have a linear combination of finitely many elements of A that is equal to the zero vector. Therefore, the coefficients are all 0, that is $(c_1 - d_1) = \cdots = (c - d_p) = c_{p+1} = \cdots = c_{p+k} = -d_{p+1} = \cdots = -d_{p+q} = 0$. So the "extra vectors" really were not there, and the coefficients of the $\mathbf{x}$'s are the same; $\mathbf{x} = c_1\mathbf{x}_1 + \cdots + c_p\mathbf{x}_p$, uniquely.

21. a. That V is a vector space is true, but tedious. For independence, use the definition of problem 19. Let B be a finite subset of A, and let $j(1), \ldots j(n)$ be nonnegative whole numbers chosen so that $B = \{t^{2j(1)}, \ldots, t^{2j(n)}\}$. If $c_1 t^{2j(1)} + \cdots + c_n t^{2j(n)} = \mathbf{0}$, then, since all the $j(i)$'s are different, $c_1 = c_2 = \cdots = c_n = 0$.

b. $S[A] = V$, since every polynomial with only even powers of t is a finite sum of scalars times even powers of t.

PROBLEM SET 2.5

1. Since $\begin{bmatrix} 2 & 1 & 0 \\ 6 & 1 & 0 \\ 0 & 0 & 0 \\ 1 & -1 & 1 \\ 3 & 4 & 1 \end{bmatrix}$ is row-equivalent to $\begin{bmatrix} 1 & 0 & 0 \\ 0 & 1 & 0 \\ 0 & 0 & 1 \\ 0 & 0 & 0 \\ 0 & 0 & 0 \end{bmatrix}$,

$A = \{(2,6,0,1,3), (1,1,0,-1,4), (0,0,0,1,1)\}$ is independent, and therefore is a basis for $S[A]$.

3. $\{(-4.5,-1,1,0), (-3,-1,0,1)\}$. If $\mathbf{x}$ is in K, then its components satisfy the two given homogeneous equations, so that $x_1 = 4.5x_3 - 3x_4$, and $x_2 = -x_3 - x_4$, so that $\mathbf{x} = x_3(-4.5,-1,1,0) + x_4(-3,-1,0,1)$. Since if $\mathbf{x} = 0$, $x_3 = x_4 = 0$, the vectors are independent and therefore form a basis for K. These two vectors together with $(1,0,0,0)$ and $(0,1,0,0)$ form a basis for $\mathbb{R}^4$.

5. $\{(1,0,0),(1,1,0),(1,1,1)\}$.

7. a. $\begin{bmatrix} 2 & 1 & 0 & : & 0 \\ 1 & 1 & -1 & : & 0 \end{bmatrix}$ is row equivalent to $\begin{bmatrix} 1 & 0 & 1 & : & 0 \\ 0 & 1 & -2 & : & 0 \end{bmatrix}$, so that $x_1 = -x_3$ and $x_2 = 2x_3$, and therefore $(x_1,x_2,x_3) = x_3(-1,2,1)$. All elements of K are therefore multiples of $(-1,2,1)$, and $\dim(K) = 1$.

b. $\{(-1,2,1),(1,0,0),(0,1,0)\}$ is a basis of $\mathbb{R}^3$. Any two of $(1,0,0)$, $(0,1,0)$ and $(0,0,1)$ will do for $\mathbf{y}_1$ and $\mathbf{y}_2$.

9. Since E_{ij} consists entirely of 0's except for a 1 in the i,jth spot, $M = \Sigma_{i=1}^{m}\Sigma_{j=1}^{n}c_{ij}E_{ij} = [c_{ij}]$. Therefore, if we are given M, we can write it as a linear combination of the E_{ij}, and if $M = 0_{mn}$ = then every c_{ij} must be 0. Therefore, the E_{ij} form a basis for M_{mn}.

11. $\{1,i\}$ is such a basis.

13. a. dim(scalar matrices) = 1 since they are all of the form cI_n.

b. dim(diagonal matrices) = n.

c. dim(upper triangular matrices) $= n + (n-1) + (n-2) + \cdots + 1 = n(n+1)/2$.

d. dim(lower triangular matrices) $= n(n+1)/2$.

15. $\begin{bmatrix} 1 & 1 & 0 \\ 0 & 1 & 1 \\ 0 & -1 & 1 \end{bmatrix}$ is row equivalent to I_3, so that A is a basis for $\mathbb{R}^3$.

17. No. There are continuous real-valued functions on $[0,1]$ that cannot be written as a finite linear combination of such exponentials, for example, the polynomials and trigonometric functions.

19. a. $\mathbf{x}$ is not $\mathbf{0}$, $\{\mathbf{e}_1,\mathbf{e}_2,\mathbf{e}_3\}$ is a basis for V, so that $\{\mathbf{x},\mathbf{e}_1,\mathbf{e}_2,\mathbf{e}_3\}$ is a spanning set for V. It must be dependent, and if $c\mathbf{x} + c_1\mathbf{e}_1 + c_2\mathbf{e}_2 + c_3\mathbf{e}_3 = \mathbf{0}$, then at least two of the coefficients must be nonzero, and one of them must be c (since the $\mathbf{e}_i$'s form a basis). Suppose, for the sake of notation, that c_1 is nonzero. Then $\mathbf{x} = -1/c(c_1\mathbf{e}_1 + c_2\mathbf{e}_2 + c_3\mathbf{e}_3)$ and, conversely, $\mathbf{e}_1 = -1/c_1(c\mathbf{x} + c_2\mathbf{e}_2 + c_3\mathbf{e}_3)$. Now, $\{\mathbf{x},\mathbf{e}_2,\mathbf{e}_3\}$ form a basis. Every element that can be written in terms of $\mathbf{e}_1,\mathbf{e}_2,\mathbf{e}_3$ can be written in terms of $\{\mathbf{x},\mathbf{e}_2,\mathbf{e}_3\}$, and it is easy to check that $\{\mathbf{x},\mathbf{e}_2,\mathbf{e}_3\}$ is independent.

b. If A is a linearly independent subset of V, then A has one, two, or three vectors. The case when A has one vector is part **a**. The case when A has three vectors is trivial since A is already a basis of V. The remaining case, when A consists of two vectors, is handled by adding to A any vector not in $S[A]$.

21. $p(t) = p_0 + p_1t + p_2t^2 + p_3t^3 + p_4t^4$, so $\int_0^1 p(t)\,dt = p_0t + (\frac{1}{2})p_1t^2 + (\frac{1}{3})p_2t^2 + (\frac{1}{4})p_3t^3 + (\frac{1}{5})p_4t^4|_0^1 = p_0 + (\frac{1}{2})p_1 + (\frac{1}{3})p_2 + (\frac{1}{4})p_3 + (\frac{1}{5})p_4 = 0$ and $p'(t) = p_1 + 2p_2t + 3p_3t^2 + 4p_4t^3$, so that $p'(2) = 0 = p_1 + 2p_2 + 12p_3 + 32p_4$. $\begin{bmatrix} 1 & \frac{1}{2} & \frac{1}{3} & \frac{1}{4} & \frac{1}{5} & : & 0 \\ 0 & 1 & 2 & 12 & 32 & : & 0 \end{bmatrix}$ is equivalent to $\begin{bmatrix} 1 & 0 & -\frac{5}{3} & -\frac{23}{4} & -\frac{79}{5} & : & 0 \\ 0 & 1 & 4 & 12 & 32 & : & 0 \end{bmatrix}$, so a basis for V is $\{\frac{5}{3} - 4t + t^2, \frac{23}{4} - 12t + t^3, \frac{79}{5} - 32t + t^4\}$.

23. No. $\begin{bmatrix} 1 & 0 \\ 0 & 1 \end{bmatrix} = \begin{bmatrix} 1 & 1 \\ 0 & 0 \end{bmatrix} - \begin{bmatrix} 0 & 1 \\ 1 & 0 \end{bmatrix} + \begin{bmatrix} 0 & 0 \\ 1 & 1 \end{bmatrix}$.

25. **a.** $-(0,1,1) - (0,2,3) + (0,3,4) = (0,0,0)$ and $-(1,0,0) + 9(0,1,1) - 4(0,2,3) + (1,-1,3) = (0,0,0)$.

b. 3.

c. For any $\mathbf{x}$ in A, $S[A] = S[A \backslash \{\mathbf{x}\}]$

d. Any three-element subset that contains at least one of $(1,0,0)$ and $(1,-1,3)$.

PROBLEM SET 2.6

1. **a.** $[1,-1,2]$. **b.** $[-\frac{3}{2},1,-\frac{1}{2}]$. **c.** $[0.45,0.025,-0.35]$.

3. **a.** $[6,-4]$. **b.** $[1,0]$. **c.** $[-4,-10]$. **d.** $[0,1]$.

5. **a.** See Section 2.5, problem 22.

b. $[\mathbf{x}]_F = [-\frac{11}{3},\frac{19}{3},-\frac{2}{3},\frac{1}{3}]$.

c. $\mathbf{x} = \begin{bmatrix} 1 & 0 \\ 1 & 1 \end{bmatrix}$.

7. **a.** $\begin{bmatrix} 1 & 2 \\ 6 & 3 \end{bmatrix}$ is row equivalent to I_2.

b. $P = \begin{bmatrix} 1 & 2 \\ 6 & 3 \end{bmatrix}$.

c. $Q = (-\frac{1}{9})\begin{bmatrix} 3 & -2 \\ -6 & 1 \end{bmatrix}$.

d. $PQ = QP = I_2$.

9. **a.** Both $\begin{bmatrix} -1 & 2 \\ 7 & -3 \end{bmatrix}$ and $\begin{bmatrix} 1 & 1 \\ 2 & 3 \end{bmatrix}$ are row equivalent to I_2.

b. $P = \begin{bmatrix} -10 & 9 \\ 9 & -7 \end{bmatrix}$.

11. If $A\mathbf{x} = \mathbf{0}$ for all $\mathbf{x}$, then $A\mathbf{e}_j = \mathbf{0} = [a_{1j} a_{2j} \cdots a_{mj}]^T =$ the jth column of A. Thus, all columns of A contain only 0's. $A\,\mathbf{x} = B\mathbf{x}$ for all $\mathbf{x}$ if and only if $(A - B)\mathbf{x} = \mathbf{0}$ for all $\mathbf{x}$, which happens if and only if $(A - B) = 0_{mn}$, if and only if $A = B$.

13. a. $\begin{bmatrix} 6 & 1 & 0 \\ 0 & -1 & 3 \\ 1 & -1 & 1 \end{bmatrix}$ is row equivalent to I_3.

b. $P = (\frac{1}{15}) \begin{bmatrix} 2 & -1 & 3 \\ 3 & 6 & -18 \\ 1 & 7 & -6 \end{bmatrix}$.

c. $[\mathbf{e}_1] = [\frac{2}{15}, \frac{1}{5}, \frac{1}{15}]$, $[\mathbf{e}_2] = [-\frac{1}{15}, \frac{2}{5}, \frac{7}{15}]$, and $[\mathbf{e}_3] = [\frac{1}{5}, -\frac{6}{5}, -\frac{2}{5}]$.

15. $P = \begin{bmatrix} 0 & 1 \\ 1 & 0 \end{bmatrix}$.

17. a. $[\mathbf{x}]_F = [2,-6,7,5]$, $[\mathbf{x}]_G = [\frac{2}{3}, \frac{26}{3}, -\frac{13}{3}, -\frac{7}{3}]$.

b. $P = (\frac{1}{3}) \begin{bmatrix} -2 & 1 & 1 & 1 \\ 1 & -2 & 1 & 1 \\ 1 & 1 & -2 & 1 \\ 1 & 1 & 1 & -2 \end{bmatrix}$.

c. $P([\mathbf{x}]_F^T) = [\frac{2}{3}, \frac{26}{3}, -\frac{13}{3}, -\frac{7}{3}]^T$.

d. $\mathbf{x} = \begin{bmatrix} 9 & 5 \\ 6 & 4 \end{bmatrix}$.

19. a. $\begin{bmatrix} 1 & 0 \\ 0 & 1 \end{bmatrix}$ is I_2, and $\begin{bmatrix} 2 & 1 \\ -3 & 6 \end{bmatrix}$ is row equivalent to I_2.

b. $[-6,3]$.
c. $[-2.6,-0.8]$.

21. $p = (\frac{1}{15}) \begin{bmatrix} 6 & -1 \\ 3 & 2 \end{bmatrix}$.

SUPPLEMENTARY EXERCISES

3b. $[x]_F = [0,-1,1]$

3c. $\begin{bmatrix} 0 & -\frac{1}{3} & \frac{1}{3} \\ -1 & \frac{1}{3} & \frac{2}{3} \\ 1 & \frac{1}{3} & -\frac{1}{3} \end{bmatrix}$

3d. The columns of P are the coordinates of the standard basis vectors with respect to the basis F.

5. Part **a** is a finite dimensional subspace while parts **b** and **c** are infinite dimensional subspaces; **d** and **e** are not subspaces.

7. Let $W = S[\mathbf{X}_1, \ldots, \mathbf{X}_m]$. Then $\dim(W) = m$, and $V = S[\mathbf{Y}_1, \ldots, \mathbf{Y}_m] \subseteq W$. The set $\{\mathbf{Y}_1, \ldots, \mathbf{Y}_m\}$ is linearly independent if and only if $\dim(V) = m$ or if and only if $V = W$. Now if A is an invertible matrix, then $\mathbf{X}_j = \Sigma_{k=1}^m b_{jk}\mathbf{Y}_k$, where $[b_{jk}]$ equals A^{-1}. This last equation implies that $W \subseteq V$, and hence that $V = W$. Thus $\{\mathbf{Y}_1, \ldots, \mathbf{Y}_m\}$ is a linearly independent set. Suppose that the matrix is not invertible. Then one of the rows of A can be written as a linear combination of the other $m-1$ rows. We may suppose that this is true for the first row. Thus, there are $m-1$ constants c_j such that

$$(a_{11}, \ . \ . \ . \ , a_{1m}) = \sum_{k=2}^{m} c_k(a_{k1}, \ . \ . \ . \ , a_{km})$$

Thus, $a_{1j} = \Sigma_{k=2}^{m} c_k a_{kj}$, for $j = 1, \ldots, m$. Hence $\mathbf{Y}_1 = \Sigma_{k=1}^{m} a_{1k}\mathbf{X}_k = \Sigma_{k=1}^{m}\Sigma_{j=2}^{m}(c_j a_{jk})\mathbf{X}_k = \Sigma_{j=2}^{m} c_j(\Sigma_{k=1}^{m} a_{jk}\mathbf{X}_k) = \Sigma_{j=2}^{m} c_j\mathbf{Y}_j$. Thus the set of vectors $\{\mathbf{Y}_1, \ldots, \mathbf{Y}_m\}$ is linearly dependent.

9. No.

11. Let $\mathbf{y}_j = \Sigma_{k=1}^{2} a_{jk}\mathbf{x}_k$, for $j = 1, 2, 3$. Then finding three constants c_j not all zero such that $\mathbf{0} = \Sigma_{j=1}^{3} c_j\mathbf{y}_j = \Sigma_{j=1}^{3} c_j(\Sigma_{k=1}^{2} a_{jk}\mathbf{x}_k) = \Sigma_{k=1}^{2}(\Sigma_{j=1}^{3} c_j a_{kj})\mathbf{x}_k$, reduces to finding a non-trivial solution to two homogeneous equations in three unknowns.

13. C consists of all 2×2 scalar matrices. Thus, a basis of C is $\{I_2\}$. If V is the set of $n \times n$ matrices, then C is still the set of scalar matrices and a basis for them is $\{I_n\}$.

PROBLEM SET 3.1

1. a. $L(\mathbf{x}+\mathbf{y}) = L(x_1+y_1, x_2+y_2, x_3+y_3) = (x_1+y_1)-(x_2+y_2)+(x_3+y_3) = x_1 - x_2 + x_3 + y_1 - y_2 + y_3 = L(\mathbf{x}) + L(\mathbf{y})$.
$L(c\mathbf{x}) = L(cx_1, cx_2, cx_3) = cx_1 - cx_2 + cx_3 = c(x_1 - x_2 + x_3) = cL(\mathbf{x})$.

b. $[1, -1, 1]$.

c. $L(\mathbf{e}_1) = 1$; $L(\mathbf{e}_2) = -1$; $L(\mathbf{e}_3) = 1$.

d. $\{(1,1,0), (-1,0,1)\}$.

3. a. $L(\mathbf{e}_1) = \frac{1}{3}(-1,1,-3,4,-1)$.
$L(\mathbf{e}_2) = \frac{1}{3}(5,-14,0,4,-1)$.
$L(\mathbf{e}_3) = \frac{1}{3}(1,2,3,6,1)$.

b. $L(x_1,x_2,x_3) = \frac{1}{3}(-x_1 + 5x_2 + x_3, x_1 - 14x_2 + 2x_3, -3x_1 + 3x_2, 4x_1 + 4x_2 + 6x_3, -x_1 - x_2 + x_3)$.

c. $A = \frac{1}{3}\begin{bmatrix} -1 & 5 & 1 \\ 1 & -14 & 2 \\ -3 & 0 & 3 \\ 4 & 4 & 6 \\ -1 & -1 & 1 \end{bmatrix}$.

5. Let $\mathbf{y}_1$ and $\mathbf{y}_2$ be in $L(K)$ and let a be a real number. Then there are vectors $\mathbf{x}_1$ and $\mathbf{x}_2$ in K with $L(\mathbf{x}_1) = \mathbf{y}_1$ and $L(\mathbf{x}_2) = \mathbf{y}_2$. Since K is a subspace, $\mathbf{x}_1 + \mathbf{x}_2$ and $a\mathbf{x}_1$ are in K; furthermore, $L(\mathbf{x}_1 + \mathbf{x}_2) = L(\mathbf{x}_1) + L(\mathbf{x}_2) = \mathbf{y}_1 + \mathbf{y}_2$, since L is linear. Similarly, $L(a\mathbf{x}_1) = aL(\mathbf{x}_1) = a\mathbf{y}_1$. Therefore, $\mathbf{y}_1 + \mathbf{y}_2$ and $a\mathbf{y}_1$ are back in $L(K)$.

7. $f(\mathbf{x}+\mathbf{y}) = f(x_1+y_1, x_2+y_2) = (x_1+y_1)(x_2+y_2) = x_1x_2 + y_1x_2 + x_1y_2 + y_1y_2$, whereas $f(\mathbf{x}) + f(\mathbf{y}) = f(x_1,x_2) + f(y_1,y_2) = x_1x_2 + y_1y_2$. Also, $f(c\mathbf{x}) = f(cx_1,cx_2) = c^2x_1x_2$, while $cf(\mathbf{x}) = cx_1x_2$.

9. Let $A = [a_{j1}]$ and let $\mathbf{e}_k = [\delta_{1k}]$. Then $A\mathbf{e}_k = [\Sigma_{1=1}^{n} a_{j1}\delta_{1k}] = [a_{jk}] =$ the kth column of A.

11. $\begin{bmatrix} d_1 & 0 \\ 0 & d_2 \end{bmatrix}$ multiplies vectors by d_1 in the $\mathbf{x}_1$ direction and by d_2 in the $\mathbf{x}_2$ direction, so, for example, $\begin{bmatrix} 2 & 0 \\ 0 & -0.5 \end{bmatrix}$ doubles all the x_1's and halves and reverses the sign of the x_2's.

13. V is P_n, and for W, take P_m, where $m \geq n+2$.
$L(p+q) = L[(p_0+q_0)+(p_1+p_2)t+\cdots+(p_n+q_n)t^n] = (p_0+q_0)t^2+(p_1+p_2)t^3+\cdots+(p_n+q_n)t^{n+2} = (p_0+\cdots+p_nt^n)t^2+(q_0+\cdots+q_nt^n)t^2 = L(p)+L(q)$.
$L(cp) = (cp_0+\cdots+cp_nt^n)t^2 = c(p_0+\cdots+p_nt^n)t^2 = cL(p)$.

15. For $p(t) = p_0+p_1t+\cdots+p_nt^n$, define $L(p) = p'(t) = p_1+2p_2t+\cdots+np_nt^{n-1}$; so that L maps into P_{n-1}. L is linear, for if p and q are polynomials in P_n and a is a real number, $L(p+q) = (p+q)' = p'+q' = L(p)+L(q)$, and $L(ap) = (ap)' = ap' = aL(p)$.

17. a. $L(\mathbf{e}_1) = \begin{bmatrix} 2 & -3 \\ 0 & 0 \end{bmatrix}$, $L(\mathbf{e}_2) = \begin{bmatrix} 1 & 5 \\ 0 & 0 \end{bmatrix}$, $L(\mathbf{e}_3) = \begin{bmatrix} 0 & 0 \\ 2 & -3 \end{bmatrix}$, and $L(\mathbf{e}_4) = \begin{bmatrix} 0 & 0 \\ 1 & 5 \end{bmatrix}$.

b. $L(a\mathbf{x}) = (a\mathbf{x})A = a(\mathbf{x}A) = aL(\mathbf{x})$, and $L(\mathbf{x}+\mathbf{y}) = (\mathbf{x}+\mathbf{y})A = \mathbf{x}A+\mathbf{y}A = L(\mathbf{x})+L(\mathbf{y})$.

c. For $L(\mathbf{x}) = A\mathbf{x}$: $L(\mathbf{e}_1) = \begin{bmatrix} 2 & 0 \\ 1 & 0 \end{bmatrix}$, $L(\mathbf{e}_2) = \begin{bmatrix} 0 & 2 \\ 0 & 1 \end{bmatrix}$, $L(\mathbf{e}_3) = \begin{bmatrix} -3 & 0 \\ 5 & 0 \end{bmatrix}$, and $L(\mathbf{e}_4) = \begin{bmatrix} 0 & -3 \\ 0 & 5 \end{bmatrix}$. For linearity, $L(a\mathbf{x}) = A(a\mathbf{x}) = a(A\mathbf{x}) = aL(\mathbf{x})$, and $L(\mathbf{x}+\mathbf{y}) = A(\mathbf{x}+\mathbf{y}) = A\mathbf{x}+A\mathbf{y} = L(\mathbf{x})+L(\mathbf{y})$.

PROBLEM SET 3.2

1. a. $\begin{bmatrix} 3 & 6 \\ -2 & 1 \end{bmatrix}$. **b.** $\frac{1}{7}\begin{bmatrix} 18 & -37 \\ 15 & 10 \end{bmatrix}$.

3. a. $A = \begin{bmatrix} a & b \\ c & d \end{bmatrix}$.

b. By induction on the power, n. $n = 1$. $L(\mathbf{x}) = A\mathbf{x}$. Inductive step: Suppose that $L^n(\mathbf{x}) = A^n\mathbf{x}$. Then $L^{n+1}(\mathbf{x}) = L^n[L(\mathbf{x})] = A^n[L(\mathbf{x})] = A^n(A\mathbf{x}) = A^{n+1}\mathbf{x}$.

c. It still works.

5. $A = cI_n$.

7. $(L_1+L_2)(\mathbf{x}) = L_1(\mathbf{x})+L_2(\mathbf{x}) = A_1\mathbf{x}+A_2\mathbf{x} = (A_1+A_2)\mathbf{x}$.

9. a. (0,2), (1,0), (2,0), and (3,−6).

b. $\begin{bmatrix} 0 & 1 & 2 & 3 \\ 2 & 0 & 0 & -6 \end{bmatrix}$.

c. $A\mathbf{x} = \begin{bmatrix} x_2+2x_3+3x_4 \\ 2x_1 \quad -6x_4 \end{bmatrix}$.

11. $L(\mathbf{f}_1) = -2\mathbf{f}_1$, and $L(\mathbf{f}_2) = 3\mathbf{f}_2$.

13. a. $(L_1 \circ L_2)(\mathbf{x}_1+\mathbf{x}_2) = L_1[L_2(\mathbf{x}_1+\mathbf{x}_2)] = L_1[L_2(\mathbf{x}_1)+L_2(\mathbf{x}_2)] = L_1[L_2(\mathbf{x}_1)]+L_1[L_2(\mathbf{x}_2)] = (L_1 \circ L_2)(\mathbf{x}_1)+(L_1 \circ L_2)(\mathbf{x}_2)$. $(L_1 \circ L_2)(a\mathbf{x}) = L_1[L_2(a\mathbf{x})] = L_1[aL_2(\mathbf{x})] = aL_1[L_2(\mathbf{x})] = a(L_1 \circ L_2)(\mathbf{x})$.

b. $(L_1 \circ L_2)(\mathbf{x}) = L_1[L_2(\mathbf{x})] = L_1(A_2\mathbf{x}) = A_1(A_2\mathbf{x}) = (A_1A_2)\mathbf{x}$.

15. $$\begin{bmatrix} 0 & \cdot & \cdot & \cdot & \cdot & \cdot & 0 \\ 1 & 0 & & & & & \cdot \\ 0 & \frac{1}{2} & 0 & \cdot & & & \cdot \\ \cdot & 0 & \frac{1}{3} & & \cdot & & \cdot \\ \cdot & & 0 & \cdot & & \cdot & \cdot \\ \cdot & & & \cdot & \cdot & & 0 \\ \cdot & & & & \cdot & \cdot & \\ 0 & \cdot & \cdot & \cdot & & 0 & 1/(n+1) \end{bmatrix}$$

The matrix is $(n+2) \times (n+1)$ and has all entries equal to 0 except for $1/n$'s down the diagonal right below the main diagonal.

17. $A = 0_{mn}$.

19. Since $L(\mathbf{x}) = \begin{bmatrix} a & b \\ c & d \end{bmatrix}\begin{bmatrix} -2 & 1 \\ 3 & 4 \end{bmatrix} = \begin{bmatrix} -2a+3b & a+4b \\ -2c+3d & c+4d \end{bmatrix}$,

$$A = \begin{bmatrix} -2 & 3 & 0 & 0 \\ 1 & 4 & 0 & 0 \\ 0 & 0 & -2 & 3 \\ 0 & 0 & 1 & 4 \end{bmatrix}$$

21. $$\begin{bmatrix} \frac{4}{3} & \frac{1}{3} & -2 & 1 & -\frac{7}{3} & \frac{5}{3} \\ \frac{1}{3} & \frac{4}{3} & 1 & -2 & \frac{5}{3} & -\frac{7}{3} \\ -\frac{5}{3} & \frac{1}{3} & 1 & 1 & \frac{2}{3} & \frac{5}{3} \\ \frac{1}{3} & -\frac{5}{3} & 1 & 1 & \frac{5}{3} & \frac{2}{3} \end{bmatrix}.$$

23. **a.** See Section 3.1, 14b. The vector space is different, but the mechanics of showing linearity are the same. It goes from V to V since $L(\cos jx) = (1/j)\sin jx$ and $L(\sin jx) = (-1/j)\cos jx$.

b. Basis = $\{\cos x, \sin x, \cos 2x, \sin 2x\}$. $A = \begin{bmatrix} 0 & -1 & 0 & 0 \\ 1 & 0 & 0 & 0 \\ 0 & 0 & 0 & -\frac{1}{2} \\ 0 & 0 & \frac{1}{2} & 0 \end{bmatrix}$.

PROBLEM SET 3.3

1. **a.** From $V = \mathbb{R}^3$ to $W = \mathbb{R}$. ker $= S[(1,0,-1), (0,1,0)]$, Rg $= R$.
b. From $V = \mathbb{R}$ to $W = \mathbb{R}^3$. ker $= \{\mathbf{0}\}$, Rg $= S[(1,0,1)]$.
c. From $V = \mathbb{R}^2$ to $W = \mathbb{R}^2$. ker $= \{\mathbf{0}\}$, Rg $= \mathbb{R}^2$.
d. From $V = \mathbb{R}^3$ to $W = \mathbb{R}^4$. ker $= \{\mathbf{0}\}$, Rg $= S[(1,0,1,0), (0,1,0,0), (0,0,0,1)]$.

3. $A = \begin{bmatrix} 2 & -5 & 3 \\ 1 & 3 & 0 \end{bmatrix}$, equivalent to $\begin{bmatrix} 1 & 0 & \frac{9}{11} \\ 0 & 1 & -\frac{3}{11} \end{bmatrix}$ so ker $= S[(-\frac{9}{11}, \frac{3}{11}, 1)]$ and $Rg = \mathbb{R}^2$, so, yes, there is a solution.

5. **a.** $\dim[\ker(L)] = 1$, $\dim[\text{Rg}(L)] = 2$, neither $1-1$ nor onto.
b. $\dim[\ker(L)] = 0$, $\dim[\text{Rg}(L)] = 4$, $1-1$, but not onto.

7. Let $\mathbf{y}$ be in Rg(L), so that there is an $\mathbf{x}$ in V with $L(\mathbf{x}) = \mathbf{y}$. If $\{\mathbf{x}_1, \mathbf{x}_2, \ldots, \mathbf{x}_n\}$ is any basis for V, then there are real numbers $c_1, c_2, \ldots, c_n$ with $\mathbf{x} =$

$c_1\mathbf{x}_1 + c_2\mathbf{x}_2 + \cdots + c_n\mathbf{x}_n$, so that $\mathbf{y} = L(\mathbf{x}) = L(c_1\mathbf{x}_1 + c_2\mathbf{x}_2 + \cdots + c_n\mathbf{x}_n) = L(c_1\mathbf{x}_1) + L(c_2\mathbf{x}_2) + \cdots + L(c_n\mathbf{x}_n) = c_1L(\mathbf{x}_1) + c_2L(\mathbf{x}_2) + \cdots + c_nL(\mathbf{x}_n)$ = a linear combination of $\{L(\mathbf{x}_1), \ldots, L(\mathbf{x}_n)\}$. Therefore, $\{L(\mathbf{x}_1), \ldots, L(\mathbf{x}_n)\}$ spans $\text{Rg}(L)$.

9. a. $\dim(V) = n = \dim[\ker(L)] + \dim[\text{Rg}(L)]$, so that $\dim[\text{Rg}(L)] \le n < m = \dim(W)$. Therefore, L cannot be onto.

b. No, L need not be onto. (Example: $A = 0_{mn}$.)

11. S is both $1-1$ and onto (with ker $= \{\mathbf{0}\}$ and Rg $= W$) if and only if c is nonzero. If $c = 0$, then $S = 0_{nn}$, and S is neither $1-1$ nor onto (ker $= V$ and Rg $= \{\mathbf{0}\}$.

13. Section 3.1, problem 13:

$$\ker = \{\mathbf{0}\},\ \text{Rg} = \{p(t) \text{ in } P_{n+2}: p_0 = p_1 = 0\}.$$

Section 3.1, problem 14:

a. $\ker = \{f \text{ in } C[0,1]: \int_0^1 f(t)\,dt = 0\}$, Rg $= R$.

b. $\ker = \{\mathbf{0}\}$, Rg = the functions on [0,1] with continuous first derivatives that equal 0 at 0.

Section 3.1, problem 15:

$$\ker = \{\text{constant polynomials}\},\ \text{Rg} = P_{n-1}.$$

15. a. $A^TA = \begin{bmatrix} 1 & 0 & 1 \\ 0 & 0 & 0 \\ 1 & 0 & 1 \end{bmatrix}$, $\ker A^TA = S[(1,0,-1), (0,1,0)] = \ker A$. A^TA is neither $1-1$ nor onto.

b. $A^TA = [2]$, $\ker A^TA = \{\mathbf{0}\} = \ker A$, Rg $= \mathbb{R}$: both $1-1$ and onto.

c. $A^TA = \begin{bmatrix} 1 & 1 \\ 1 & 2 \end{bmatrix}$, $\ker A^TA = \{\mathbf{0}\} = \ker A$, Rg $= \mathbb{R}^2$: both $1-1$ and onto.

d. $A^TA = 2\begin{bmatrix} 2 & 1 & 1 \\ 1 & 1 & 1 \\ 1 & 1 & 2 \end{bmatrix}$, $\ker A^TA = \{\mathbf{0}\} = \ker A$, Rg $= \mathbb{R}^3$: both $1-1$ and onto.

17. a. Suppose $\ker(L) = \{\mathbf{0}\}$. Then if $L(c_1\mathbf{x}_1 + c_2\mathbf{x}_2 + \cdots + c_n\mathbf{x}_n) = c_1L(\mathbf{x}_1) + c_2L(\mathbf{x}_2) + \cdots + c_nL(\mathbf{x}_n) = \mathbf{0}$, it follows that $c_1 = c_2 = \cdots = c_n = 0$. Therefore, $\{L(\mathbf{x}_1), L(\mathbf{x}_2), \ldots, L(\mathbf{x}_n)\}$ is independent. By problem 7, this set also spans $\text{Rg}(L)$ and must therefore be a basis for $\text{Rg}(L)$. Thus, the dimension of $\text{Rg}(L) = n = \dim(V)$, while $\dim[\ker(L)] = 0$. So, $\dim(V) = n = 0 + n = \dim[\ker(L)] + \dim[\text{Rg}(L)]$.

b. Suppose $\dim[\ker(L)] = n = \dim(V)$. Then by problem 7, the set $\{L(\mathbf{x}_1), L(\mathbf{x}_2), \ldots, L(\mathbf{x}_n)\}$ spans $\text{Rg}(L)$ whenever $\{\mathbf{x}_1, \ldots, \mathbf{x}_n\}$ is a basis for V. But $V = \ker(L)$ (otherwise, they could not have the same dimension), so that $L(\mathbf{x}_1) = L(\mathbf{x}_2) = \cdots = L(\mathbf{x}_n) = \mathbf{0}$. $\text{Rg}(L) = \{\mathbf{0}\}$, and we have $\dim(V) = n = n + 0 = \dim[\ker(L)] + \dim[\text{Rg}(L)]$.

19. $\ker(L) = S[1]$, $\text{Rg}(L) = S[1, t, t^2]$

$$A = \begin{bmatrix} 0 & 1 & 0 & 0 \\ 0 & 0 & 2 & 0 \\ 0 & 0 & 0 & 3 \\ 0 & 0 & 0 & 0 \end{bmatrix}.$$

PROBLEM SET 3.4

1. a. 1. **b.** 1. **c.** 2. **d.** 3.

3. a. From $\mathbb{R}^3$ to $\mathbb{R}^4$; dim(ker) = 1; dim(Rg) = 2.
b. From $\mathbb{R}^4$ to $\mathbb{R}^5$; dim(ker) = 0; dim(Rg) = 4.

5. a. dim(ker) = 0; dim(Rg) = 3; 1 − 1; not onto.
b. dim(ker) = 0; dim(Rg) = 4; 1 − 1; not onto.

7. a. Rank = 2. **b.** dim(ker) = 1, ker = $S[(3,-1,1)]$. **c.** dim(Rg) = 2, Rg = $\mathbb{R}^2 = S[(1,0),(0,1)]$. **d.** A is not 1 − 1. **e.** A is onto. **f.** Yes. $\{(-3,1,0) + c(3,-1,1)\}$.

9. a. Rank = 2. **b.** dim(ker) = 2, ker = $S[(-\frac{1}{2},0,1,0),(-\frac{1}{4},\frac{5}{2},0,1)]$. **c.** dim(Rg) = 2, Rg = $S[(0,1,2),(1,3,3)]$. **d.** Not 1 − 1. **e.** Not onto. **f.** Yes. $\{(0,-1,0,0)$ + any element of the kernel$\}$.

11. $A = \begin{bmatrix} 1 & 1 & 1 \\ 1 & 2 & 4 \end{bmatrix}$. Rank($A$) = 2. Infinitely many polynomials of degree ≤ 2 will pass through two given points.

13. a. $p(t) = 6$. **b.** $p(t) = \frac{8}{3} - (\frac{4}{3})t$. **c.** No solution in P_1.

15. a. $W = \mathbb{R}$; $L(\mathbf{f}+\mathbf{g}) = [(\mathbf{f}(0)+\mathbf{g}(0))]/2 + [\mathbf{f}(\frac{1}{2}) + \mathbf{g}(\frac{1}{2})] + [\mathbf{f}(1)+\mathbf{g}(1)]/2$
$= [\mathbf{f}(0)/2 + \mathbf{f}(\frac{1}{2}) + \mathbf{f}(1)/2]/2 + [\mathbf{g}(0)/2 + \mathbf{g}(\frac{1}{2}) + \mathbf{g}(1)/2]/2$
$= L(\mathbf{f}) + L(\mathbf{g})$, and $L(a\mathbf{f}) = (a\mathbf{f}(0)/2 + a\mathbf{f}(\frac{1}{2}) + a\mathbf{f}(1)/2]/2$
$= a[\mathbf{f}(0)/2 + \mathbf{f}(\frac{1}{2}) + \mathbf{f}(1)/2]/2 = aL(\mathbf{f})$.
b. $L_1(\mathbf{f}+\mathbf{g}) = L(\mathbf{f}+\mathbf{g}) - \int_0^1 \mathbf{f}(t) + \mathbf{g}(t)\,dt$
$= L(\mathbf{f}) + L(\mathbf{g}) - \int_0^1 \mathbf{f}(t)\,dt - \int_0^1 \mathbf{g}(t)\,dt$
$= L_1(\mathbf{f}) + L_1(\mathbf{g})$, and
$L_1(a\mathbf{f}) = L(a\mathbf{f}) - \int_0^1 a\mathbf{f}(t)\,dt = aL(\mathbf{f}) - a\int_0^1 \mathbf{f}(t)\,dt = aL_1(\mathbf{f})$.

17. Rowspace(A) = columnspace(A^T) = range(A^T).

19. If $\mathbf{p}(t) = p_0 + p_1 t + pt^2$, then $L(\mathbf{p}) = \begin{bmatrix} 1 & x_1 & x_1^2 \\ 1 & x_2 & x_2^2 \\ 1 & x_3 & x_3^2 \end{bmatrix} \begin{bmatrix} p_0 \\ p_1 \\ p_2 \end{bmatrix}$. If all of the x_i are distinct, then A is row equivalent to I_3, so that L is 1 − 1 and onto.

PROBLEM SET 3.5

1. a. $\begin{bmatrix} 2 & 0 \\ 0 & 2 \end{bmatrix}$. **b.** $P = (\frac{1}{7})\begin{bmatrix} 5 & -2 \\ 1 & 1 \end{bmatrix}$, $P^{-1} = \begin{bmatrix} 1 & 2 \\ -1 & 5 \end{bmatrix}$. **c.** $B = \begin{bmatrix} 2 & 0 \\ 0 & 2 \end{bmatrix}$.

3. $B = \begin{bmatrix} 2 & -2 & -\frac{4}{3} \\ 2 & -3 & -3 \\ -3 & 6 & 3 \end{bmatrix}$

5. a. $A = IAI^{-1}$. **b.** If $B = PAP^{-1}$, then $A = P^{-1}B(P^{-1})^{-1} = P^{-1}BP$. **c.** If $B = PAP^{-1}$ and $C = QBQ^{-1}$, than $C = QPAP^{-1}Q^{-1} = (QP)A(QP)^{-1}$.

7. With respect to F: $\begin{bmatrix} 0 & 1 \\ 3 & 0 \end{bmatrix}$ With respect to G: $\begin{bmatrix} 0 & 1 \\ 3 & 0 \end{bmatrix}$

P from F to G: $\begin{bmatrix} 1 & 1 \\ 3 & 1 \end{bmatrix}$. P^{-1} from G to F: $\begin{bmatrix} -\frac{1}{2} & \frac{1}{2} \\ \frac{3}{2} & -\frac{1}{2} \end{bmatrix}$.

9. $[L(\mathbf{x})]_G^T = A[\mathbf{x}]_F$ and $[L(\mathbf{x})]_{\mathscr{G}}^T = B[\mathbf{x}]_{\mathscr{F}}^T$, so that if $[\mathbf{x}]_{\mathscr{F}}^T = P[\mathbf{x}]_F^T$ and $[L(\mathbf{x})]_{\mathscr{G}}^T = Q[L(\mathbf{x})]_G^T$, then $A[\mathbf{x}]_F^T = [L(\mathbf{x})]_G^T = Q^{-1}[L(\mathbf{x})]_{\mathscr{G}}^T = Q^{-1}\{B([\mathbf{x}])_{\mathscr{F}}^T\} = Q^{-1}B(P([\mathbf{x}])_F^T = (Q^{-1}BP)[\mathbf{x}]_F^T$ for all $\mathbf{x}$. By problem 11 in problem set 2.6 $A = Q^{-1}BP$.

11. **a.** $\mathbf{x}$ is in ker(A) if and only if $P^{-1}\mathbf{x}$ is in ker(B), or $\mathbf{x}$ is in ker(B) if and only if $P\mathbf{x}$ is in ker(A).
    b. $\mathbf{y}$ is in Rg(A) if and only if $P^{-1}\mathbf{y}$ is in Rg(B), or $\mathbf{y}$ is in Rg(B) if and only if $P\mathbf{y}$ is in Rg(A).

SUPPLEMENTARY EXERCISES

3. **a.** rank = 3; dim(ker) = 0
   b. rank = 3; basis for kernel is {(2,7,29,−2)}
   c. rank = 2; basis for kernel is {(1,−2,0)(−1,0,1)}
5. The line $y = -x$ is mapped onto the line $x = 0$.
7. Let F, G, and H be the basis of V_1, V_2, and V_3, respectively, that are used to obtain the matrix representations of L_1 and L_2. Then $[L_2L_1(\mathbf{x})]_H = A_2[L_1(\mathbf{x})]_G = A_2A_1[\mathbf{x}]_F$. Thus, the matrix representation of L_2L_1 is A_2A_1.
9. $T(\mathbf{x}+\mathbf{y}) = L\mathbf{x} + L\mathbf{y} + \mathbf{a}$ and $T(\mathbf{x}) + T(\mathbf{y}) = L\mathbf{x} + \mathbf{a} + L\mathbf{y} + \mathbf{a}$. These are equal if and only if $\mathbf{a}$ equals the zero vector.
11. Parts **a** and **b** are straightforward. The kernel of the linear transformation is all of V, and its range is the zero vector.
13. Clearly $\{\mathbf{0}\}$ and V are invariant under any linear transformation. Suppose that $\mathbf{y}$ is in W_λ, then $L\mathbf{y} = \lambda\mathbf{y}$. Hence $L(L\mathbf{y}) = L(\lambda\mathbf{y}) = \lambda L(\mathbf{y})$. Thus, W_λ is an invariant subspace.

PROBLEM SET 4.1

1. **a.** −12. **b.** 0. **c.** 52. **d.** 336.
3. **a.** 0. **b.** 0. (After subtracting the first row from the second and the first row from the third, which does not change the determinant, row 3 is twice row 2, so the det is 0.)
5. det(E) = det(I) = 1.
7. **a.** For $n > 2$: subtract row 1 from rows two and three. Then, the third row is twice the first, and det = 0.
   b. Use part a with k = 1.
   c. If $n = 2$, then the det of part a is $a_1(a_2 + k) - a_2(a_1 + k) = a_1a_2 + a_1k - a_2a_1 - a_2k = k(a_1 - a_2)$.
9. **a.** $x_2 - x_1$. **b.** $(x_3 - x_2)(x_3 - x_1)(x_2 - x_1)$. **c.** Product $(x_j - x_i)$.

11. $\det(AA^{-1}) = \det(A)\det(A^{-1}) = \det(I) = 1$, so $\det(A^{-1}) = 1/\det(A)$.

13. $\det(A) = \det(PBP^{-1}) = \det(P)\det(B)\det(P^{-1}) = \det(P)\det(B)[1/\det(P)] = \det(B)$.

15. a. If $A = -A^T$, the $a_{ii} = -a_{ii}^T = -a_{ii}$. so $a_{ii} = 0$.

b. $\det A = \det A^T = \det(-A) = (-1)^3 \det A = -\det A$. Thus, $\det A = 0$.

c. If $A = \begin{bmatrix} 0 & a \\ -a & 0 \end{bmatrix}$, then $\det(A) = a^2$.

d. $\det(A) = \det[-(A)^T] = (-1)^n \det(A^T) = (-1)^n \det(A)$. If n is odd, then $\det(A) = 0$; if n is even, $\det(A)$ could be anything.

PROBLEM SET 4.2

1. a. $\operatorname{adj}(A) = \begin{bmatrix} 3 & 0 \\ 0 & 2 \end{bmatrix}$. **b.** $\operatorname{adj}(A) = \begin{bmatrix} b & 0 \\ 0 & a \end{bmatrix}$. **c.** $\operatorname{adj}(A) = \begin{bmatrix} 4 & -1 \\ -3 & 6 \end{bmatrix}$.

d. $\operatorname{adj}(A) = \begin{bmatrix} -2 & -4 \\ 1 & 2 \end{bmatrix}$.

3. a. $\begin{bmatrix} \frac{1}{2} & 0 \\ 0 & \frac{1}{3} \end{bmatrix}$. **b.** $\begin{bmatrix} 1/a & 0 \\ 0 & 1/b \end{bmatrix}$ as long as a,b are nonzero. $a = 0$ or $b = 0$: noninvertible. **c.** $\begin{bmatrix} \frac{4}{21} & -\frac{1}{21} \\ -\frac{3}{21} & \frac{6}{21} \end{bmatrix}$. **d.** Noninvertible.

5. a. $x_1 = 1.5, x_2 = 1, x_3 = 1, x_4 = 0.5$.

b. $A^{-1} = \begin{bmatrix} 0.5 & 0.5 & 0 & 0 \\ -0.5 & 0 & 0.5 & 0 \\ 0 & -0.5 & 0 & 0.5 \\ 0 & 0 & -0.5 & 0.5 \end{bmatrix}$. **c.** $A^{-1} \begin{bmatrix} 1 \\ 2 \\ 3 \\ 4 \end{bmatrix} = \begin{bmatrix} 1.5 \\ 1.0 \\ 1.0 \\ 0.5 \end{bmatrix}$.

7. a. Since $A\operatorname{adj}(A) = \det(A)I_n$, it follows that

$$\det(A)\det[\operatorname{adj}(A)] = \det(A)^n, \text{ or } \det[\operatorname{adj}(A)] = \det(A)^{n-1}, \text{ since } \det A \neq 0.$$

b. Suppose A is singular and $\operatorname{adj}(A)$ is invertible. Then from $A\operatorname{adj}(A) = (\det A)I$, we have $A = (\det A)B$, where $B = [\operatorname{adj}(A)]^{-1}$. But $\det A = 0$. Thus, $A = 0B = 0_{n,n}$. A contradiction, since $A = 0_{n,n}$ implies $\operatorname{adj}A = 0_{n,n}$.

PROBLEM SET 4.3

1. a. $x_1 = -\frac{13}{22}, x_2 = \frac{30}{11}$. **b.** $x_1 = \frac{14}{17}, x_2 = \frac{30}{17}$.

3. $x_1 = 1, x_2 = \frac{1}{7}, x_3 = \frac{6}{7}$.

5. a. $x_1 = -11, x_2 = \frac{5}{2}, x_3 = -\frac{23}{2}$.

b. $\begin{bmatrix} -1 & 3 & 1 & : & 7 \\ 1 & -1 & -1 & : & -2 \\ -1 & 6 & 2 & : & 3 \end{bmatrix} \begin{bmatrix} 1 & 0 & 0 & : & -11 \\ 0 & 1 & 0 & : & 2.5 \\ 0 & 0 & 1 & : & -11.5 \end{bmatrix}$.

c. $A^{-1} = \begin{bmatrix} -2 & 1 & 1 \\ \frac{1}{2} & \frac{1}{2} & 0 \\ -\frac{5}{2} & -\frac{3}{2} & 1 \end{bmatrix}$.

7. $x_1 = a, x_2 = 1, x_3 = -1$.

PROBLEM SET 4.4

1. a. $\frac{15}{2}$. **b.** $\frac{71}{2}$.

3. a. 15. **b.** 71.

5. The given equation is clearly that of a straight line (its linear in x_1 and x_2). Moreover the points (a_1, b_1) and (a_2, b_2) satisfy the equation. Thus, it is the equation for the straight line passing through the given points.

7. a. $\text{vol}(P) = 1$, $\text{vol}[L(P)] = \det(A)\text{vol}(P) = 5$.

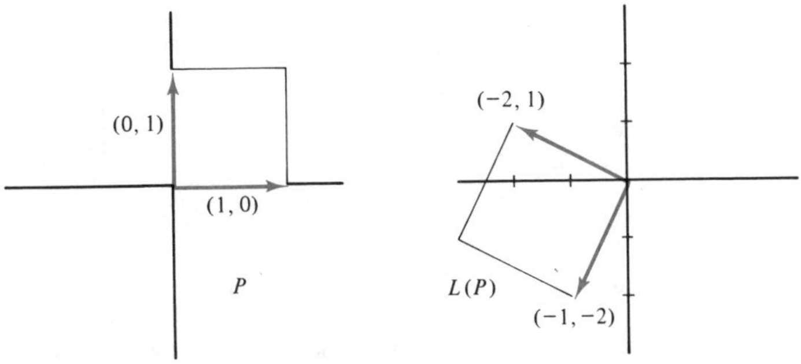

b. $\text{vol}(P) = 2$, $\text{vol}[L(P)] = 10$.

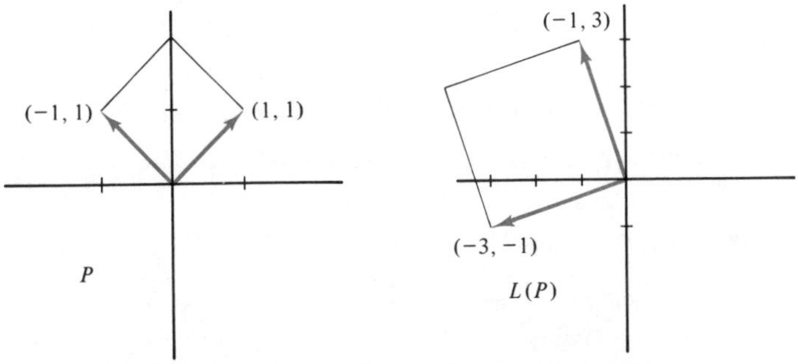

SUPPLEMENTARY EXERCISES

3. $x = -3, -2$

5. $(a(t)d(t) - b(t)c(t))' = (a'd + ad' - b'c - bc') = (a'd - b'c) + (ad' - bc')$.

7. Part **a** is linearly independent and part **b** is linearly dependent.

11. To see that the product of two invertible matrices is invertible, use the fact that a matrix is invertible if and only if its determinant is nonzero, and $\det(AB) = \det(A)\det(B)$. A similar argument is used to show that $SL(n)$ forms a group. To see that the set of all $n \times n$ matrices does not form a group under multiplication, observe that there are noninvertible matrices, O_{nn}, for example.

13a. If A and B are similar, there is a nonsingular matrix P such that $P^{-1}BP = A$. Then $\det(A) = \det(P^{-1}BP) = \det(P^{-1})\det(B)\det(P) = \det(B)$.

13b. $\det\begin{bmatrix} 2 & 1 \\ 1 & 1 \end{bmatrix} = \det\begin{bmatrix} 1 & 0 \\ 0 & 1 \end{bmatrix}$, yet the matrices are not similar.

PROBLEM SET 5.1

1. a. $\begin{bmatrix} 1-\lambda & 4 \\ 3 & 8-\lambda \end{bmatrix} = \lambda^2 - 9\lambda - 4.$

b. $\begin{bmatrix} 7-\lambda & 8 \\ 0 & 4-\lambda \end{bmatrix} = \lambda^2 - 11\lambda + 28.$

c. $\begin{bmatrix} 1-\lambda & 3 & 0 \\ 1 & 2-\lambda & 1 \\ 4 & -5 & 8-\lambda \end{bmatrix} = -\lambda^3 + 11\lambda^2 - 28\lambda + 9.$

3.

	Eigenvalue	Multiplicity	Dimension of eigenspace
a.	3	2	2
b.	3	2	1
c.	2	1	1
	4	1	1
d.	3	1	1
	2	1	1

5. Suppose that λ_0 is an eigenvalue, so that there is a nonzero vector $\mathbf{x}$ with $L(\mathbf{x}) = \lambda_0\mathbf{x}$; thus, the eigenspace of λ_0 must contain at least one nonzero vector. Now suppose that $\mathbf{x}_1$, $\mathbf{x}_2$, $\mathbf{x}$ are all in the eigenspace and that a is a real number. Then $L(\mathbf{x}_1 + \mathbf{x}_2) = L(\mathbf{x}_1) + L(\mathbf{x}_2) = \lambda_0\mathbf{x}_1 + \lambda_0\mathbf{x}_2 = \lambda_0(\mathbf{x}_1 + \mathbf{x}_2)$ and $L(a\mathbf{x}) = aL(\mathbf{x}) = a\lambda_0\mathbf{x} = \lambda_0(a\mathbf{x})$, so that both $\mathbf{x}_1 + \mathbf{x}_2$ and $a\mathbf{x}$ are back in the eigenspace. Thus the eigenspace is a subspace. Since it contains at least one nonzero vector, its dimension must be at least 1.

7. a. If λ_1 and λ_2 are the roots of the characteristic polynomial of $\begin{bmatrix} a_{11} & a_{12} \\ a_{21} & a_{22} \end{bmatrix}$, then

$\lambda^2 - c_1\lambda + c_2 = \lambda^2 - (\lambda_1 + \lambda_2)\lambda + \lambda_1\lambda_2 = \lambda^2 - (a_{11} + a_{22})\lambda + (a_{11}a_{22} - a_{12}a_{21}).$

b. $c_1 =$ the coefficient of λ^2 in $(\lambda - a_{11})(\lambda - a_{22})(\lambda - a_{33})$
$= (a_{11} + a_{22} + a_{33})$
$=$ the coefficient of λ^2 in $-(\lambda - \lambda_1)(\lambda - \lambda_2)(\lambda - \lambda_3)$
$= (\lambda_1 + \lambda_2 + \lambda_3)$, where $\lambda_1, \lambda_2, \lambda_3$ are the roots.
$c_3 =$ the constant term $= \lambda_1\lambda_2\lambda_3$
$a_{11}a_{22}a_{33} + a_{21}a_{32}a_{13} + a_{12}a_{23}a_{31} - a_{13}a_{22}a_{31} - a_{12}a_{21}a_{33} - a_{11}a_{32}a_{23} = \det(A).$

c. The coefficient of the t^{n-1} term is the sum of the eigenvalues and is also the trace (the sum of the elements along the main diagonal); the constant term is the det of A and is also the product of the eigenvalues.

9. All are 0_{22}.

11. Proof by induction on n. Assume $\mathbf{x}$ is an eigenvector with corresponding eigenvalue λ_0.
$n = 1$: $A\mathbf{x} = \lambda_0\mathbf{x}$.
Inductive step: Assume that $A^n\mathbf{x} = \lambda_0^n\mathbf{x}$. Then $A^{n+1}\mathbf{x} = AA^n\mathbf{x} = A(\lambda_0^n\mathbf{x}) = \lambda_0^n(A\mathbf{x}) = \lambda_0^n(\lambda_0\mathbf{x}) = \lambda_0^{n+1}\mathbf{x}$.

13.

	Eigenvalue	Eigenspace
a.	1	$S[(1,0,1,1)]$
	2	$S[(0,1,0,0)]$
	3	$S[(-1,0,1,0)]$
	4	$S[(1,1,0,1)]$
b.	0	$S[(-1,1,0,0),(0,0,1,0)]$
	1	$S[(-2,-1,6,1)]$

15. If $\det A$ is nonzero, then $p(A) = A^2 - (a_{11} + a_{22})A + \det(A)I_2 = 0$, and $A[A - (a_{11} + a_{22})I_2] = -\det(A)I_2$, or $A[-1/\det(A)][A - (a_{11} + a_{22})I_2] = I_2$, and $A^{-1} = [-1/\det(A)][A - (a_{11} + a_{22})I_2]$.

17. a. 6 and 9. **b.** $6 - c$ and $9 - c$. **c.** $6c$ and $9c$.

19. 0 is an eigenvalue if and only if $\dim[\ker(A)] > 0$, which happens if and only if A is noninvertible. Suppose that A is invertible with eigenvalues $\{\lambda_1, \ldots, \lambda_n\}$, so that the characteristic polynomial of A is $(-1)^n(\lambda - \lambda_1) \ldots (\lambda - \lambda_n)$. $\det(A^{-1} - \lambda I_n) = [1/\det(A)]\det(A)\det(A^{-1} - \lambda I_n) = [1/\det(A)]\det(I_n - \lambda A) = ((-1)^n\lambda^n/[\det(A)])\det[A - (1/\lambda)I_n] = [\lambda^n/\det(A)](1/\lambda - \lambda_1) \cdots (1/\lambda - \lambda_n)$. Thus, A^{-1} has eigenvalues $\{1/\lambda_1, \ldots, 1/\lambda_n\}$. If $A\mathbf{x} = \lambda_i\mathbf{x}$, then $\mathbf{x} = A^{-1}(\lambda_i\mathbf{x}) = \lambda_i A^{-1}(\mathbf{x})$, or $A^{-1}\mathbf{x} = (1/\lambda_i)\mathbf{x}$, so the eigenvectors are the same. That is, any eigenvector of A corresponding to eigenvalue λ_i is an eigenvector of A^{-1} corresponding to $1/\lambda_i$.

21. $\det(A^T - \lambda I_n) = \det[A^T - (\lambda I_n)^T] = \det[(A - \lambda I_n)^T] = \det(A - \lambda I_n)$.

23.

	Eigenvalue	Multiplicity	Dim of eigenspace
a.	1	4	3
b.	1	4	2
c.	1	4	1

25. If $A\mathbf{x} = \lambda_0\mathbf{x}$, and if B has all columns equal to $\mathbf{x}$, then $AB =$ the matrix that has all columns equal to $\lambda_0\mathbf{x} = \lambda_0 B$. Conversely, if $AB = \lambda_0 B$ and B is nonzero, then B has at least one nonzero column, say the ith one, $\mathbf{x}$. Then $A\mathbf{x} =$ the ith column of $AB =$ the ith column of $\lambda_0 B = \lambda_0\mathbf{x}$.

27.

Eigenvalue	Multiplicity	Dim of eigenspace
0	3	1

PROBLEM SET 5.2

1. a. $\{(-2,1),(0,1)\}$. **b.** $\{(2,1),(-1,1)\}$.

3. a. $\begin{bmatrix} (-1)^n & 0 \\ [(-1^{n+1} + 3^n]/2 & 3^n \end{bmatrix}$. **b.** $3^{n-1}\begin{bmatrix} 2(3^n) + 2^n & 2(3^n - 2^n) \\ 3^n - 2^n & 3^n + 2^{n+1} \end{bmatrix}$. **c.** $5^{n-1}\begin{bmatrix} 4 & 2 \\ 2 & 1 \end{bmatrix}$.

d. $\begin{bmatrix} 1 & 2 \\ 0 & 0 \end{bmatrix}$.

5. a.

Eigenvalue	Eigenvectors
1	(1,1,0,1)
2	(−1,0,1,0),(2,0,0,1)
3	(0,1,0,0)

b. $P^{-1} = \begin{bmatrix} 1 & -1 & 2 & 0 \\ 1 & 0 & 0 & 1 \\ 0 & 1 & 0 & 0 \\ 1 & 0 & 1 & 0 \end{bmatrix}$. **c.** $A^3 = \begin{bmatrix} 15 & 0 & 7 & -14 \\ 26 & 27 & 26 & -52 \\ 0 & 0 & 8 & 0 \\ 7 & 0 & 7 & -6 \end{bmatrix}$.

7. No limits exist except for the matrix in d. The limit there is $\begin{bmatrix} 1 & 2 \\ 0 & 0 \end{bmatrix}$.

9. $g(A) = \Sigma_{k=0}^{n} g_k A^k = \Sigma_{k=0}^{n} g_k \begin{bmatrix} \lambda_1^k & 0 \\ 0 & \lambda_2^k \end{bmatrix} = \begin{bmatrix} \Sigma_{k=0}^{n} g_k\lambda_1^k & 0 \\ 0 & \Sigma_{k=0}^{n} g_k\lambda_2^k \end{bmatrix} = \begin{bmatrix} g(\lambda)_1 & 0 \\ 0 & g(\lambda_2) \end{bmatrix}$.

In general, if A is diagonal with entries $\{\lambda_1, \ldots, \lambda_n\}$ down the main diagonal, then $g(A)$ is also diagonal with entries $\{g(\lambda_1), \ldots, g(\lambda_n)\}$ down the main diagonal.

11. a. Suppose that $D = [\lambda_i \delta_{ij}]$, so that $D^{1/2} = [\lambda_i^{1/2}\delta_{ij}]$. Then $(D^{1/2})^2 = D^{1/2}D^{1/2} = [\Sigma_{k=1}^{n} \lambda_i^{1/2}\delta_{ik}\lambda_k^{1/2}\delta_{kj}]$. If i does not equal j, then one of δ_{ik}, and δ_{kj} is always 0, so that the sum is 0. If $i = j$, then the sum is λ_i, so that $(D^{1/2})^2 = D$.

b. $(A^{1/2})^2 = PD^{1/2}P^{-1}PD^{1/2}P^{-1} = P(D^{1/2})^2P^{-1} = PDP^{-1} = A$.

13. a.
$$A^{-1/6} = \frac{1}{3}\begin{bmatrix} 2(9^{-1/6}) + 6^{-1/6} & 2(9^{-1/6}) - 2(6^{-1/6}) \\ 9^{-1/6} - 6^{-1/6} & 9^{-1/6} + 2(6^{-1/6}) \end{bmatrix}.$$
$$A^{2/3} = \frac{1}{3}\begin{bmatrix} 2(9^{2/3}) + 6^{2/3} & 2(9^{2/3}) - 2(6^{2/3}) \\ 9^{2/3} - 6^{2/3} & 9^{2/3} + 2(6^{2/3}) \end{bmatrix}.$$

b. $A^{-1/6} = 5^{-7/6}\begin{bmatrix} 4 & 2 \\ 2 & 1 \end{bmatrix}$, $A^{2/3} = 5^{-1/3}\begin{bmatrix} 4 & 2 \\ 2 & 1 \end{bmatrix}$.

c.
$$A^{-1/6} = \frac{1}{3}\begin{bmatrix} 2(3^{-1/6}) + 6^{-1/6} & -3^{-1/6} + 6^{-1/6} & 3^{-1/6} - 6^{-1/6} \\ -3^{-1/6} + 6^{-1/6} & 2(3^{-1/6}) + 6^{-1/6} & 3^{-1/6} - 6^{-1/6} \\ 3^{-1/6} - 6^{-1/6} & 3^{-1/6} - 6^{-1/6} & 2(3^{-1/6}) + 6^{-1/6} \end{bmatrix}.$$
$$A^{2/3} = \frac{1}{3}\begin{bmatrix} 2(3^{2/3}) + 6^{2/3} & -3^{2/3} + 6^{2/3} & 3^{2/3} - 6^{2/3} \\ -3^{2/3} + 6^{2/3} & 2(3^{2/3}) + 6^{2/3} & 3^{2/3} - 6^{2/3} \\ 3^{2/3} - 6^{2/3} & 3^{2/3} - 6^{2/3} & 2(3^{2/3}) + 6^{2/3} \end{bmatrix}.$$

PROBLEM SET 5.3

1. a. 50 percent. **b.** $x_{20} = 2^{20}10 = 10{,}485{,}760$.

3. $\begin{bmatrix} a_{n+1} \\ a_{n+2} \end{bmatrix} = \begin{bmatrix} 0 & 1 \\ \frac{1}{3} & \frac{2}{3} \end{bmatrix}\begin{bmatrix} a_n \\ a_{n+1} \end{bmatrix}$, $A^n\begin{bmatrix} a \\ b \end{bmatrix}$ approaches $\begin{bmatrix} a/4 + 3b/4 \\ a/4 + 3b/4 \end{bmatrix}$, as n goes to infinity.

5. a. $\begin{bmatrix} x_{k+1} \\ y_{k+1} \end{bmatrix} = \begin{bmatrix} (1-r) & r \\ r & (1-r) \end{bmatrix}\begin{bmatrix} x_k \\ y_k \end{bmatrix}$

b. As k goes to infinity, x_k and y_k both approach $(x_0 + y_0)/2$, where x_0 and y_0 are the original populations of the respective cities. (Unless $r = 1$, in which case, there is a limit only when $x_0 = y_0$.)

7. Yes. $\frac{3}{4} < r_1 < 1$ and $r_2 = \dfrac{25 - 23r_1}{28 - 27r_1}$.

SUPPLEMENTARY EXERCISES

3. See problem 13, supplementary exercises for Chapter 3.

5. $\lambda = 2$, multiplicity 3. Eigenspace $= S[(1,0,0)]$. The generalized eigenspace is all of R^3.

7. Suppose that $c_1\mathbf{x} + c_2L(\mathbf{x}) = \mathbf{0}$. Then $\mathbf{0} = L(\mathbf{0}) = L(c_1\mathbf{x} + c_2L(\mathbf{x})) = c_1L(\mathbf{x})$. Since $L(\mathbf{x})$ is not zero, we must have $c_1 = 0$, and then we may conclude that c_2 equals zero also. Thus, the set is linearly independent. Since every vector in R^2 can be written as a linear combination of two vectors for which L^2 equals zero, we must have L^2 identically zero.

9.
$$P = \begin{bmatrix} 1 & 0 & 0 & -2 \\ 0 & 1 & -3 & 0 \\ 0 & 0 & 1 & 0 \\ 0 & 0 & 0 & 1 \end{bmatrix} \quad A^{10} = \begin{bmatrix} 1 & 0 & 0 & 2^{11} - 2 \\ 0 & 1 & 3(2^{10} - 1) & 0 \\ 0 & 0 & 2^{10} & 0 \\ 0 & 0 & 0 & 2^{10} \end{bmatrix}$$

11. Eigenvalues are 1 and 4. Corresponding eigenvectors are $\sin t$ and $\sin 2t$.

13. $M = \left\{ \begin{bmatrix} a & b \\ 0 & c \end{bmatrix} : a, b, \text{ and } c \text{ arbitrary numbers} \right\}$. A basis for M is the set $\left\{ \begin{bmatrix} 1 & 0 \\ 0 & 0 \end{bmatrix}, \begin{bmatrix} 0 & 1 \\ 0 & 0 \end{bmatrix}, \begin{bmatrix} 0 & 0 \\ 0 & 1 \end{bmatrix} \right\}$.

PROBLEM SET 6.1

1. **a.** $\sqrt{5}$. **b.** $\sqrt{46}$. **c.** $\sqrt{70}$.

3. **a.** 0. **b.** $2ab$. **c.** -7.

5. **a.** $1/(5\sqrt{2})$. **b.** $-2/(\sqrt{697})$. **c.** $-1/(\sqrt{1022})$.

7. $\|\mathbf{x}+\mathbf{y}\|^2 = \|\mathbf{x}\|^2 + 2 < \mathbf{x}, \mathbf{y} > + \|\mathbf{y}\|^2 <= \|\mathbf{x}\|^2 + 2\|\mathbf{x}\|\,\|\mathbf{x}\| + \|\mathbf{y}\|^2 = (\|\mathbf{x}\| + \|\mathbf{y}\|)^2$, so $\|\mathbf{x}+\mathbf{y}\| \le \|\mathbf{x}\| + \|\mathbf{y}\|$.

9. **a.** $\|A\mathbf{x}\|^2 = \|\Sigma_{j=1}^n a_{ij}x_j\|^2 = \Sigma_{i=1}^m(\Sigma_{j=1}^n a_{ij}x_j)^2 \le \Sigma_{i=1}^m([\Sigma_{j=1}^n a_{ij}^2]^{1/2}[\Sigma_{j=1}^n x_j^2]^{1/2})^2 = \Sigma_{i=1}^m \Sigma_{j=1}^m a_{ij}^2(\Sigma_{j=1}^n x_j^2) = \|A\|^2\|\mathbf{x}\|^2$.
b. $\|A\mathbf{x}\|^2 = x^2 + 4xy + 4y^2 + 9x^2 - 6xy + y^2 \le x^2 + 2(x^2+y^2) + 4y^2 + 9x^2 + 3(x^2+y^2) + y^2 = 15x^2 + 10y^2 \le 15(x^2+y^2) = \|A\|^2\|\mathbf{x}\|^2$.

11. $x_j^2 \le x_1^2 + x_2^2 + x_3^2 + \cdots + x^2{}_n \le (|x_1| + |x_2| + \cdots + |x_n|)^2$.

13. $\|(\mathbf{x}_n - \mathbf{x})\| = \|(1/n, -2/n, 0)\| = \sqrt{5}/n$, which approaches 0 as n approaches infinity.

15. $|a_{jk} - a_{pjk}|^2 \le \Sigma_{j,k=1}^{m,n}(a_{jk} - a_{pjk})^2 = \|A - A_p\|^2 \le (\Sigma_{j,k=1}^{m,n}|a_{jk} - a_{pjk}|)^2$. Thus, every $|a_{jk} - a_{pjk}|$ goes to 0 iff $\|A - A_p\|$ does.

17. Since $A\mathbf{x} = [\Sigma_{j=1}^n a_{ij}x_j]$, each entry is the inner product of the ith row of A with $\mathbf{x}$, so the result follows.

19. Let $\mathbf{x}$ be the vector OA, $\mathbf{y}$ the vector OB. Then side AB has length $\|\mathbf{x}-\mathbf{y}\|$ which equals $\|\mathbf{x}\|$. Thus, $\|\mathbf{x}\|^2 = \|\mathbf{x}-\mathbf{y}\|^2 = \|\mathbf{x}\|^2 - 2\langle \mathbf{x},\mathbf{y}\rangle + \|\mathbf{y}\|^2$. Thus $2\langle \mathbf{x},\mathbf{y}\rangle = \|\mathbf{y}\|^2$. The vector from A to the midpoint of OB equals $\frac{1}{2}\mathbf{y} - \mathbf{x}$. Thus $\langle \mathbf{y}-\mathbf{x},\mathbf{y}\rangle = \frac{1}{2}\|\mathbf{y}\|^2 - \langle \mathbf{x},\mathbf{y}\rangle = 0$. Hence the two vectors are perpendicular.

PROBLEM SET 6.2

1. **a.** $(4.6, -9.2)$. **b.** $-\frac{10}{13}(2,3)$. **c.** $(7,0)$.

3. **a.** Trivial. The inner product is 0.
b. $-9/13(1,5)$ and $-20/13(-5,1)$.
c. $[-9/\sqrt{13}, -20/\sqrt{13}]_U$.

5. The vectors $(1,2,-1)$, $(-2,1,0)$, and $(1,2,5)$ are orthogonal. Hence any $\mathbf{x}$ in $\mathbb{R}^3$ can be written as $\mathbf{x} = c(1,2,-1) + c_2(-2,1,0) + c_3(1,2,5)$. If $\mathbf{x}$ is orthogonal to the first two vectors, then $c_1 = c_2 = 0$. Thus, $\mathbf{x} = c_3(1,2,5)$.

7. **a.** 45 degrees. **b.** 54.74 degrees. **c.** 60 degrees. **d.** 136.5 degrees.

9. U is orthogonal by problem 6. Each vector has length 1, so U is orthonormal.
a. $[2/\sqrt{14}, 2/\sqrt{38}, -9/\sqrt{133}]$.
b. $[-16/\sqrt{14}, 36/\sqrt{38}, 57/\sqrt{133}]$.
c. $[26/\sqrt{14}, 22/\sqrt{38}, -178/\sqrt{133}]$.

11. **a.** $2t$. **b.** t. **c.** $\sqrt{15}/4$.

13. For any two polynomials the formula in part **b** is valid. (Use integration by parts.) Thus, if $\mathbf{p}$ is to be orthogonal to its derivative we have $0 = \langle \mathbf{p},\mathbf{p}' \rangle = \mathbf{p}^2(1) - \mathbf{p}^2(0)$. Hence $\mathbf{p}$ is orthogonal to $\mathbf{p}'$ if and only if $\mathbf{p}^2(1) = \mathbf{p}^2(0)$ or $(a_0 + a_1 + a_2)^2 = a_0^2$ for $\mathbf{p}(t) = a_0 + a_1 t + a_2 t^2$.

15. a. The 1,1 and 3,3 entries equal $\frac{1}{2}$. The 1,3 and 3,1 entries equal $-\frac{1}{2}$, and all other entries are zero.

b. $\left\{\dfrac{\mathbf{e}_1 - \mathbf{e}_3}{\sqrt{2}}\right\}$.

c. $\left\{\dfrac{\mathbf{e}_1 + \mathbf{e}_3}{\sqrt{2}},\mathbf{e}_2,\mathbf{e}_4, \ldots\right\}$.

d. The 1,1 entry is one; all other entries are zero.

PROBLEM SET 6.3

1. a. $\{(1/\sqrt{2},-1/\sqrt{2},0),(0,0,1)\}$.
b. $\{1\sqrt{6}(1,-1,2),1/\sqrt{1074}(29,13,-8)\}$.

3. $(3,\frac{3}{2},3,\frac{3}{2})$.

5. $\frac{24}{7}$.

7. $\{\frac{1}{3}(0,1,1,1),1/\sqrt{15}(3,-2,1,1),1/\sqrt{35}(3,3,-4,1),1/\sqrt{7}(1,1,1,-2)\}$.

9. a. $\{1/\sqrt{5}(-2,1)\}$. **b.** $\left\{\dfrac{1}{\sqrt{14}}(-3,1,2)\right\}$. **c.** $\left\{1,\sqrt{11}(1,3,1,0), \dfrac{1}{\sqrt{11}}(0,-1,3,1)\right\}$.

11. Since P is orthogonal, $P^TP = [\Sigma_{k=1}^n p_{ki}p_{jk}] = [\delta_{ij}]$. Let $\mathbf{v}_j = \Sigma_{k=1}^n p_{kj}\mathbf{u}_k$. Then $\langle \mathbf{v}_i,\mathbf{v}_j \rangle = \Sigma_{1=1}^n \Sigma_{k=1}^n p_{1i}p_{kj}\langle \mathbf{u}_1,\mathbf{u}_k \rangle = \Sigma_{k=1}^n p_{ki}p_{kj} = \delta_{ij}$.

13. $\{1,\sqrt{3}(2t-1),\sqrt{5}(6t^2-6t+1)\}$.

15. a. $1/\sqrt{2}$.
b. $\int_0^1 \sin n\pi t\,dt = -1/n \cos n\pi t|_0^1 = 0$; $\int_0^1 \cos n\pi t\,dt = 1/n \sin n\pi t|_0^1 = 0$; $\int_0^1 \sin n\pi t \cos n\pi t\,dt = \frac{1}{2} \sin n\pi t|_0^1 = 0$. Assume now that n is not equal to m. $\int_0^1 \sin n\pi t \sin m\pi t\,dt = 1/((m^2 - n^2))(n \sin m\pi t \cos n\pi t - m \sin n\pi t \cos m\pi t)|_0^1 = 0$. The other two, $\int_0^1 \sin n\pi t \cos m\pi t\,dt$ and $\int_0^1 \cos n\pi t \cos m\pi t\,dt$, are similar.

17. a. $\langle c_1\mathbf{f}_1,c_2\mathbf{f}_2 \rangle = c_1c_2\langle \mathbf{f}_1,\mathbf{f}_2 \rangle = 0$.
b. $\langle c_1\mathbf{v}_1 + \cdots + c_k\mathbf{v}_k,c_{k+1}\mathbf{v}_{k+1} + \cdots + c_n\mathbf{v}_n \rangle = \Sigma_{i=1}^k \Sigma_{j=k+1}^n c_ic_j\langle \mathbf{v}_i,\mathbf{v}_j \rangle = 0$.

19. $\mathbf{x} = \text{Proj}_V\mathbf{x} + (\mathbf{x} - \text{Proj}_V\mathbf{x})$, where $\text{Proj}_V\mathbf{x}$ is in V and $\mathbf{x} - \text{Proj}_V\mathbf{x}$ is in $V^\perp$. Thus, we can write any $\mathbf{x}$ as a sum of two vectors, one in V and the other in $V^\perp$. To see that this is unique, suppose $\mathbf{x} = \mathbf{v}_1 + \mathbf{w}_1 = \mathbf{v}_2 + \mathbf{w}_2$, where $\mathbf{v}_i \in V$ and $\mathbf{w}_i \in V^\perp$. Then $\mathbf{0} = (\mathbf{v}_1 - \mathbf{v}_2) + (\mathbf{w}_1 - \mathbf{w}_2)$ and we have $\mathbf{v}_1 - \mathbf{v}_2 = \mathbf{w}_2 - \mathbf{w}_1$. But this equation says that $\mathbf{v}_1 - \mathbf{v}_2$ is in V and $V^\perp$. This implies that $\mathbf{v}_1 - \mathbf{v}_2 = \mathbf{0}$. Thus, $\mathbf{w}_1 - \mathbf{w}_2 = \mathbf{0}$ also.

21. See problem 19.

23. $A = \begin{bmatrix} \cos\theta & -\sin\theta \\ \sin\theta & \cos\theta \end{bmatrix}$, and $A^TA = I_2$.

PROBLEM SET 6.4

1. a. $\langle A\mathbf{x},\mathbf{y}\rangle = \langle \mathbf{x},A^T\mathbf{y}\rangle = x_1y_1 + 3x_1y_2 + 2x_2y_2$.
b. $\langle A\mathbf{x},\mathbf{y}\rangle = \langle \mathbf{x},A^T\mathbf{y}\rangle = 3x_1y_1 + x_1y_2 + 6x_2y_1 + 2x_2y_2 + 8x_3y_1 + 4x_3y_2$.
c. $\langle A\mathbf{x},\mathbf{y}\rangle = \langle \mathbf{x},A^T\mathbf{y}\rangle = 2x_1y_1 + 4x_2y_1 - x_1y_2 - x_2y_2 + 5x_1y_3 + 6x_2y_3$.

3. a. $\langle A\mathbf{x},\mathbf{x}\rangle = 2x_1^2 + (x_1 + x_2)^2 + x_2^2$.
b. $\langle A\mathbf{x},\mathbf{x}\rangle = ax_1^2 + 2bx_1x_2 + dx^2, = a\left(x_1 + \frac{b}{a}x_2\right)^2 + \frac{ad - b^2}{a}x_2^2$. Clearly A is positive definite if and only if $a > 0$ and $\frac{ad-b^2}{a} > 0$.

c. Let $A = \begin{bmatrix} a & b & c \\ b & d & e \\ c & e & f \end{bmatrix}$. $\langle A\mathbf{x},\mathbf{x}\rangle = ax_1^2 + 2bx_1x_2 + 2cx_1x_3 + dx_2^2 + 2ex_2x_3 + fx_3^2 = a\left(x_1 + \frac{b}{a}x_2 + \frac{c}{a}x_3\right)^2 + \frac{ad - b^2}{a}\left(x_2 + \frac{ae - bc}{ad - b^2}x_3\right)^2 + \left(\frac{a}{ad - b^2}\right)\det A\, x_3^2$. Hence, A is positive definite if and only if $a > 0$, $ad - b^2 > 0$, and $\det A > 0$.

5. a. The eigenvalues are $-1 + \sqrt{2}$ and $-1 - \sqrt{2}$. Basis: $\{(4 + 2\sqrt{2})^{-1/2}(1 + \sqrt{2},1), (4 - 2\sqrt{2})^{-1/2}(1 - \sqrt{2},1)\}$.
b. $\{\mathbf{e}_1,\mathbf{e}_2\}$.
c. The eigenvalues are $\frac{1}{2}(5 + \sqrt{5})$ and $\frac{1}{2}(5 - \sqrt{5})$. Basis:
$$\left\{\frac{(2\sqrt{5},5 - \sqrt{5})}{\sqrt{50 - 10\sqrt{5}}}, \frac{(2\sqrt{5},5 + \sqrt{5})}{\sqrt{50 + 10\sqrt{5}}}\right\}$$

7. a. $\{2^{-1/2}(1,1), 2^{-1/2}(-1,1)\}$.
b. $\left\{\frac{(-2,0,1)}{\sqrt{5}}, \frac{(1,-\sqrt{5},2)}{\sqrt{10}}, \frac{(1,\sqrt{5},2)}{\sqrt{10}}\right\}$.
c. Basis: $\{\mathbf{e}_1,\mathbf{e}_2,\mathbf{e}_3\}$.

9. a. $D = \begin{bmatrix} 6 & 0 & 0 \\ 0 & 9 & 0 \\ 0 & 0 & 9 \end{bmatrix}$ $P = \begin{bmatrix} 1/\sqrt{3} & 1/\sqrt{3} & -1/\sqrt{3} \\ 1/\sqrt{2} & 0 & 1/\sqrt{2} \\ -1/\sqrt{6} & 2/\sqrt{6} & 1/\sqrt{6} \end{bmatrix}$.

b. $D = \begin{vmatrix} 2 & 0 & 0 \\ 0 & 1 + 3\sqrt{2} & 0 \\ 0 & 0 & 1 - 3\sqrt{2} \end{vmatrix}$

$$P = \frac{1}{(5 + 2\sqrt{3})^{1/2}}\begin{bmatrix} 0 & 0 & 5 + 2\sqrt{3} \\ 1 & 1 + \sqrt{3} & 0 \\ 1 + \sqrt{3} & -1 & 0 \end{bmatrix}.$$

11. If A is an $n \times n$ symmetric matrix, the theorem in this section tells us an orthonormal basis of eigenvectors exists. Conversely suppose there is an orthonormal basis of eigenvectors of A. Then there is an orthogonal matrix P such that $PAP^T = D$, where D is a diagonal matrix. But then $A = P^TDP$ must be symmetric.

13. By 12, $\text{Rg}(A)$ is in $\ker(A^T)^\perp$ and $\text{Rg}(A^T)$ is in $\ker(A^\perp)$. $n = \dim(\ker A) + \dim(\text{Rg } A)$ and $m = \dim(\ker A^T) + \dim(\text{Rg } A^T)$. $\dim(\text{Rg } A) = \text{rank of } A = \dim(\text{Rg } A^T)$, so that $\dim(\mathrm{Rg}\, A) = \dim(\mathrm{Rg}\, A^T) = m - \dim(\ker A^T) = \dim[(\ker A^T)]^\perp$. Thus, $\mathrm{Rg}(A) = [\ker(A^T)]^\perp$.

15. a. Let $\{\mathbf{u}_1,\mathbf{u}_2\}$ be an orthonormal basis for $\mathbb{R}^2$, and set $\mathbf{y} = \langle \mathbf{x},L(\mathbf{u}_1)\rangle \mathbf{u}_1 + \langle \mathbf{x},L(\mathbf{u}_2)\rangle \mathbf{u}_2$. Then $\langle \mathbf{y},\mathbf{z}\rangle = \langle \langle \mathbf{x},L(\mathbf{u}_1)\rangle \mathbf{u}_1 + \langle \mathbf{x},L(\mathbf{u}_2)\rangle \mathbf{u}_2, c_1\mathbf{u}_1 + c_2\mathbf{u}_2\rangle = c_1\langle \mathbf{x},L(\mathbf{u}_1)\rangle + c_2\langle \mathbf{x},L(\mathbf{u}_2)\rangle = \langle \mathbf{x},L(c_1\mathbf{u}_1 + c_2\mathbf{u}_2)\rangle = \langle \mathbf{x},L(\mathbf{z})\rangle$.

b. If $L(\mathbf{z}) = A\mathbf{z}$, then $\langle L^T\mathbf{x},\mathbf{z}\rangle = \langle \mathbf{x},L(\mathbf{z})\rangle = \langle \mathbf{x},A\mathbf{z}\rangle = \langle A^T\mathbf{x},\mathbf{z}\rangle$ for all $\mathbf{x}$ and $\mathbf{z}$ in $\mathbb{R}^2$, so that $L^T(\mathbf{x}) = A^T\mathbf{x}$.

17. a. $L^T(p_0 + p_1t) = 3(2p_0 + p_1)(2t - 1)$.

b. $L^T(p_0 + p_1t) = \mathbf{0}$.

c. $L^T(p_0 + p_1t) = (3p_0 + 2p_1)(2t - 1)$.

PROBLEM SET 6.5

1. a. $\begin{bmatrix} 2 & 1 & : & 1 \\ 0 & 2 & : & 1 \\ 3 & 4 & : & 0 \end{bmatrix}$ is row equivalent to $\begin{bmatrix} 1 & 0 & : & 0.25 \\ 0 & 1 & : & 0.50 \\ 0 & 0 & : & -2.75 \end{bmatrix}$, so there is no solution. $\mathrm{Rg}(A) = \{(2c + d,2d,3c + 4d)\colon c,d \text{ real}\} = S[(2,0,3),(1,2,4)]$.

b. A^TA is row equivalent to I_2.

c. Solve: $\begin{bmatrix} 13 & 14 & : & 2 \\ 14 & 21 & : & 3 \end{bmatrix}$ to get $\begin{bmatrix} 1 & 0 & : & 0 \\ 0 & 1 & : & \frac{1}{7} \end{bmatrix}$, so that $\mathbf{x} = (0,\frac{1}{7})$.

d. $\|A\mathbf{x} - \mathbf{b}\| = \|(\frac{1}{7},\frac{2}{7},\frac{4}{7}) - (1,1,0)\| = (\frac{11}{7})^{1/2}$. Let $f(c,d) = \|A(c,d)^T - \mathbf{b}\|^2 = (2c + d - 1)^2 + (2d - 1)^2 + (3c + 4d)^2$. The first partials are $26c + 28d - 4$ and $28c + 42d - 6$. Set equal to 0 and solve to get $c = 0$ and $d = \frac{1}{7}$. Since $f_{cc} = 26 > 0$ and $f_{cc}f_{dd} - (f_{cd})(f_{cd}) = (26)(42) - (28)(28) > 0$, there is a min at $c = 0$, $d = \frac{1}{7}$. Argue that this min is absolute.

3. a. $13y = 78 - 26t$.

b. $62y = 41t + 345$.

5. a. $y = 9 + \frac{3}{2}t - \frac{1}{2}t^2$.

b. $y = \dfrac{-19}{33} - \dfrac{47}{616}t + \dfrac{83}{1848}t^2$.

SUPPLEMENTARY EXERCISES

3a. $6, 6/(\sqrt{5}\sqrt{41})$ **3b.** $19, 19/(\sqrt{62}\sqrt{45})$ **3c.** $37, 37/(\sqrt{39}\sqrt{66})$

5c. $t = c$ minimizes $f(t)$.

7a. Compare the two sides to show equality. We have $\|\mathbf{x} \times \mathbf{y}\|^2 = \{[(\|\mathbf{x}\|\,\|\mathbf{y}\|)]^2 - \langle \mathbf{x},\mathbf{y}\rangle^2\}^{1/2} = \|\mathbf{x}\|\,\|\mathbf{y}\|(1 - \cos^2\theta)^{1/2} = \|\mathbf{x}\|\,\|\mathbf{y}\|\sin\theta$.

7b. The term $\|\mathbf{y}\|\sin\theta$ equals the altitude of the parallelogram with base $\mathbf{x}$. Hence, the product $\|\mathbf{x}\|\,\|\mathbf{y}\|\sin\theta$ equals its area.

9a. Suppose that $PP^T = P^TP = I$. Then $\langle \mathbf{x},\mathbf{y}\rangle = \langle P^TP\mathbf{x},\mathbf{y}\rangle = \langle P\mathbf{x},P\mathbf{y}\rangle$.

9b. To obtain the converse statement, read the line in the proof of part **a** in reverse order to conclude that $\langle \mathbf{x},\mathbf{y}\rangle = \langle P^TP\mathbf{x},\mathbf{y}\rangle$ for every pair of vectors. Thus, $P^TP = I$. Similarly, $PP^T = I$.

11c. If $\mathbf{x}$ has constant length, then we have $\langle \mathbf{x},\mathbf{x} \rangle$ constant for all t. From part **b** we then conclude that $\langle \mathbf{x}',\mathbf{x} \rangle$ equals zero for all t. Thus, if $\mathbf{x}'$ is not zero, it is perpendicular to $\mathbf{x}$.

13a. $(I - P)^2 = I - 2P + P^2 = I - 2P + P = I - P$.

13b. (i) implies (ii). $\mathbf{x} = P\mathbf{y}$ implies $P\mathbf{x} = P^2\mathbf{y} = P\mathbf{y} = \mathbf{x}$. (ii) implies (iii). $(I - P)\mathbf{x} = \mathbf{x} - P\mathbf{x} = \mathbf{x} - \mathbf{x} = \mathbf{0}$. (iii) implies (i). $(I - P)\mathbf{x} = \mathbf{0}$ implies that $P\mathbf{x} = \mathbf{x}$. Hence $\mathbf{x}$ is in the range of P.

13c. Use part **b** and the fact that $I - P$ is also a projection.

13d. Suppose that $P = P^T$. Let $\mathbf{x}$ be in the kernel of P and $\mathbf{y}$ be in the range of P. Then $\langle \mathbf{x},\mathbf{y} \rangle = \langle \mathbf{x},P\mathbf{y} \rangle = \langle P^T\mathbf{x},\mathbf{y} \rangle = \langle P\mathbf{x},\mathbf{y} \rangle = 0$. Conversely, suppose that the kernel and range of P are orthogonal subspaces. Then $\langle P\mathbf{x},(I - P)\mathbf{y} \rangle = 0$ for every pair of vectors. This line also implies that $\langle (I - P^T)P\mathbf{x},\mathbf{y} \rangle = 0$, and that $\langle \mathbf{x},P^T(I - P)\mathbf{y} \rangle = 0$. Thus, both $P^T(I - P)$ and $(I - P^T)P$ equal the zero linear transformation. Equating these two, we have $P = P^TP = P^T$.

15b. The standard basis of M_{22} is also an orthonormal basis with respect to this inner product. An orthonormal basis of $S[E]$, obtained by using the Gram-Shmidt procedure, is

$$\left\{ \frac{1}{\sqrt{15}} \begin{bmatrix} 1 & 2 \\ -3 & 1 \end{bmatrix}, \quad \frac{-1}{\sqrt{165}} \begin{bmatrix} 8 & 1 \\ 6 & 8 \end{bmatrix}, \frac{1}{\sqrt{22}} \begin{bmatrix} 1 & -4 \\ -2 & 1 \end{bmatrix} \right\}$$

PROBLEM SET 7.1

1. **a.** 0.0001 and 9.9×10^{-5}.
b. 0.001 and 9.9×10^{-5}.
c. 0.01 and 9.9×10^{-5}.
d. 0.1 and 9.9×10^{-5}.

3. Suppose that $A = [a_{ij}]$ is $n \times m$ and $B = [b_{jk}]$ is $m \times p$; then $AB = [\Sigma_{j=1}^{n} a_{ij}b_{jk}]$ is $n \times p$. There are m multiplications per entry of AB and there are $m - 1$ additions per entry of AB. Since AB has np entries, the number of multiplications is nmp and the number of additions is $(m - 1)np$.

PROBLEM SET 7.2

1. **a.** $x_1 = -\frac{9}{88} = -0.0909$ (approx.) and $x_2 = \frac{19}{11} = 1.7273$ (approx.).
b. $x_1 = -\frac{30}{7} = 4.2857$, $x_2 = \frac{7}{12} = 1.7143$, $x_3 = \frac{53}{7} = 7.5714$.

3. Partial pivoting with four digits: $x_1 = 32.44$, $x_2 = -167.6$, and $x_3 = 157.1$. Exact answer: $x_1 = 9$, $x_2 = -36$, $x_3 = 30$.

PROBLEM SET 7.3

1.

Eigenvalue	Eigenvector
−10	(1,0)
1	$(\frac{1}{11},1)$

Power method for A:

n	$\mathbf{x}_n$	λ
0	(1,1)	—
1	(1,0.01099)	−10.11111
2	(1,−0.00110)	−9.98901
5	(1,0.00000)	−10.00001

Power method for $A - 10I$:

n	$\mathbf{x}_n$	λ	$\lambda - 10$
0	(1,1)	—	—
1	(0.090909,1)	11	1

3. a. Five iterations: $10^{-n} < 10^{-4}/\sqrt{2}$.
b. One iteration: $0\|(1,1)\| < 10$ automatically.

PROBLEM SET 8.1

1. a. $g(t) = 1/(2N+1)(1 + 2\Sigma_{k=1}^{n} \cos kt)$, so $g(t-c) = 1/(2N+1)[1 + 2\Sigma_{k=1}^{n}(\cos kt \cos kc - \sin kt \sin kc)] = (1/(2N+1)) + \Sigma_{k=1}^{n}(2 \sin kc)/(2N+1) \sin kt + \Sigma_{k=1}^{n}(2 \cos kc)/(2N+1) \cos kt$, which is in V.
b. $\Sigma_{j=-N}^{N} g_j(t) = 1$.
c. $g_1(t) + 1/(2N+1) + \Sigma_{k=1}^{n} - 2 \sin[2k/(2N+1)] \sin kt + \Sigma_{k=1}^{n} 2 \cos[2k/(2N+1)] \cos kt$.

3. See problem 15b, Section 6.3

5. a. $[f,f] = \Sigma_{j=-N}^{N} f(t_j)f(t_j) \geq 0$, with equality iff $f(t_j) = 0$ for all j, but for f in V, this means $f = 0$. Symmetry and linearity in the first argument are clear.
b. Since $g_j(t_k) = \delta_{jk}$, the result follows easily.
c. The kth coordinate of f with respect to the basis G is $f(t_k)$.

PROBLEM SET 8.2

1. a. $A^5 = (1/6^5)\begin{bmatrix} 108 & 99 & 36 \\ 108 & 108 & 27 \\ 81 & 108 & 54 \end{bmatrix}$.
b. Eigenvalues: $\frac{1}{2}, -\frac{1}{4} + i/4\sqrt{3}, -\frac{1}{4} - i/4\sqrt{3}$. The absolute value of the second two is $1/2\sqrt{3}$, which is $< \frac{1}{2}$.

3. $A^k/7^k = \frac{1}{7}\begin{bmatrix} 1 & 2 \\ 3 & 6 \end{bmatrix}$, so $A^k\mathbf{x}/7^k = (\frac{1}{7})(a+2b, 3a+6b) = (a+2b)/7(1,3)$. c can also be found by $c = \langle(a,b),(1,2)\rangle/\langle(1,3),(1,2)\rangle = (a+2b)/7$.

5. a. $A^5 = \begin{bmatrix} 2abc^2d & a^3c^2 + b^2cd^2 & a^2bc^2 \\ a^2c^3 & 2abc^2d & b^2c^2d \\ bc^2d^2 & a^2c^2d & abc^2d \end{bmatrix}$. Therefore, A has a largest positive eigenvalue with corresponding eigenvector with all positive components. The eigenvalue is simple and greater in absolute value than the others. As n goes to infinity, $A^n\mathbf{x}$ behaves like $\lambda^n c\mathbf{z}_0$.

b. The characteristic polynomial of A is $-\lambda^3 + bcd + ac\lambda$. 1 is an eigenvalue iff $c(a + bd) = 1$. In this case, the other eigenvalues are $\frac{1}{4} + \frac{1}{2}(1 - 4bcd)^{1/2}$, with absolute value $\leq (\frac{1}{2} - bcd)^{1/2} < 1$ if $bcd < \frac{1}{4}$ and absolute value equal to (bcd) if $bcd > \frac{1}{4}$. Note that $1 > 1 - ac = bcd > 0$.

PROBLEM SET 8.3

1. a. No—the second row does not sum to 1.
b. Yes.
c. No—the first row does not sum to 1.
d. Yes.
e. No—the second and third rows do not sum to 1.

3. -1 is an eigenvalue iff $1 + a + d + ad - bc = 0 = 1 + a + d + ad - (1 - a)(1 - d) = 2a + 2d = 0$ iff $a = d = 0$ and $b = c = 1$.

5. a. $p_{12} = \frac{4}{5}$ and $p_{21} = \frac{1}{3}$.

b.

	S_1	S_2
t_1	$\frac{4}{15}$	$\frac{11}{15}$
t_3	$\frac{991}{(15)^3}$	$\frac{2384}{(15)^3}$

c. Yes. (See problem 3.)

7. a. Obvious.

b. $P^2 = (\frac{1}{9})\begin{bmatrix} 3 & 2 & 2 & 2 \\ 2 & 3 & 2 & 2 \\ 2 & 2 & 3 & 2 \\ 2 & 2 & 2 & 3 \end{bmatrix}$.

9. a. P is clearly symmetric and positive. The sum of the entries in the ith row of $P = \Sigma_{j=1}^{n} p_{ij} =$ the sum of $(n - 1)$ terms equal to $(1 - a)/(n - 1)$ and one term equal to a. Hence p is stochastic.

b. $P(1,1, \ldots ,1)^T = [\Sigma_{j=1}^{n} p_{ij}] = (1,1, \ldots ,1)^T$, so $\mathbf{z}_0 = (1,1, \ldots ,1)$ is an eigenvector corresponding to eigenvalue 1. Since $c\langle(1,1, \ldots ,1), (1,1, \ldots ,1)\rangle = cn$, $\mathbf{p}_\infty = 1/n(1,1, \ldots ,1)$. This last statement follows from the fact that every entry of P is positive. Hence, we know that P is a regular stochastic matrix.

c. If $a = 1$, P is the $n \times n$ identity matrix I_n. In this case the limiting probability distribution equals the original probability distribution.

PROBLEM SET 8.4

1. a. $\begin{bmatrix} \exp(2) & 0 \\ 0 & \exp(-3) \end{bmatrix}$. **b.** $\begin{bmatrix} \exp(4) & 6(\exp(4) - \exp(3)) \\ 0 & \exp(3) \end{bmatrix}$. **c.** $\begin{bmatrix} \exp(2) & \exp(2) \\ 0 & \exp(2) \end{bmatrix}$.

3. a. $tA = tPBP^{-1} = P(tB)P^{-1}$.

b. $\exp(tA) = \Sigma_{n=0}^{\infty}(tA)^n/n! = \Sigma_{n=0}^{\infty}(PtBP^{-1})^n/n! = \Sigma_{n=0}^{\infty}P(tB)^nP^{-1}/n! = P(\Sigma_{n=0}^{\infty}(tB)^n/n!)P^{-1} = P(\exp(tB))P^{-1}$.

5. $x_1 = (\frac{1}{3})[2\exp(6t) + \exp(9t)]$
$x_2 = (\frac{1}{3})[2\exp(6t) + \exp(9t)]$
$x_3 = (\frac{1}{3})[-2\exp(6t) + 2\exp(9t)]$

PROBLEM SET 8.5

1. a. $K_0 = \{\mathbf{x}: \mathbf{x}$ is perpendicular to $\mathbf{a}\}$. If $\langle \mathbf{x},\mathbf{a}\rangle = c$, then $\mathbf{x} = \langle \mathbf{x},\mathbf{a}\rangle\mathbf{a}/(\langle \mathbf{a},\mathbf{a}\rangle) + \mathbf{x}_0$, where $\mathbf{x}_0$ is in $\{\mathbf{a}\}^\perp = K_0$, so $K_c = c\mathbf{a}/(\|\mathbf{a}\|^2) + K_0$.

b. dist $= |c_1 - c_2|/\|\mathbf{a}\|$.

3. a. $c = 0$.

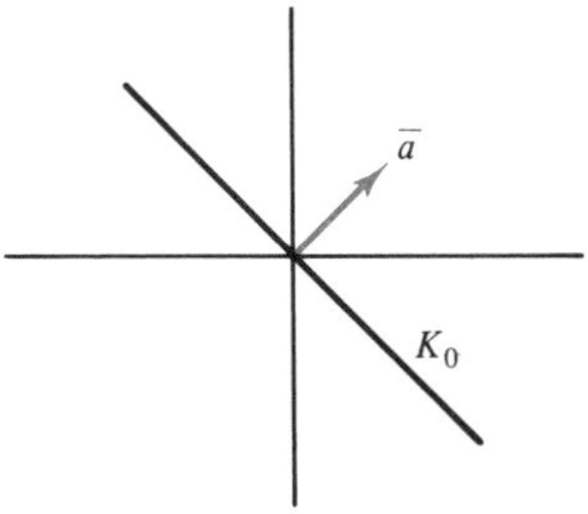

$c = 1$.

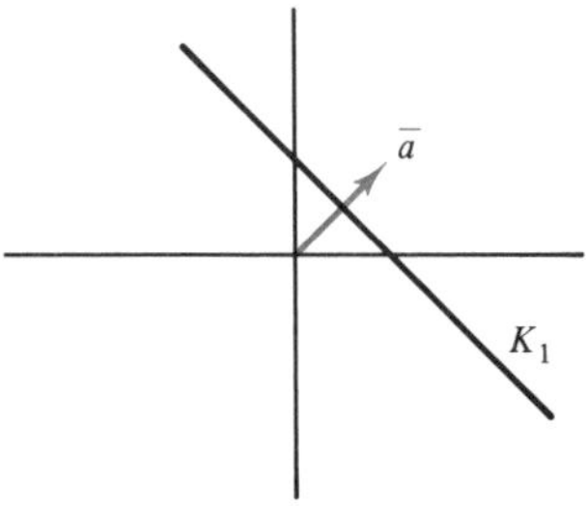

$c = -1$.

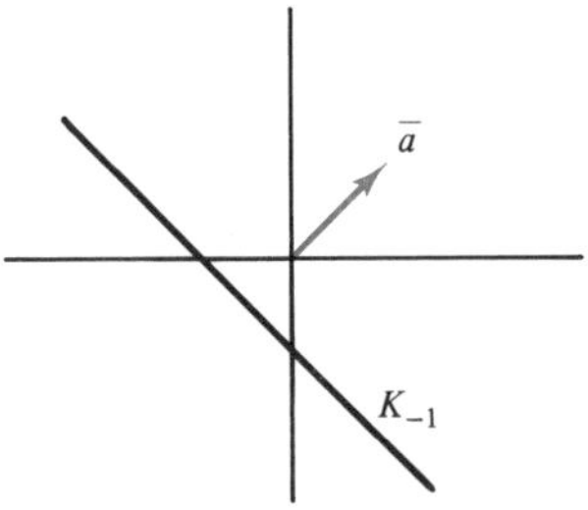

$c = 2$.

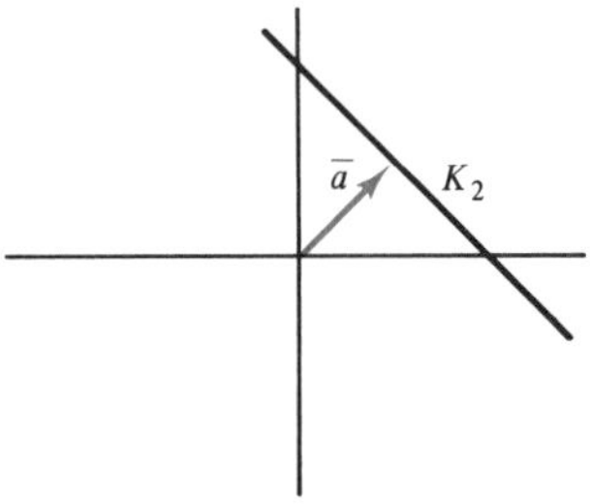

b. $A(K_1) = \{\mathbf{y}: \langle \mathbf{y}, \frac{1}{14}(-3,8)\rangle = 1\}$,

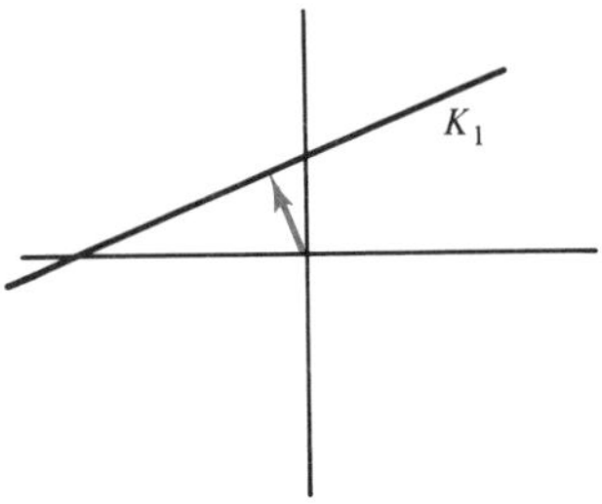

$A(K_{-1})$ is similar

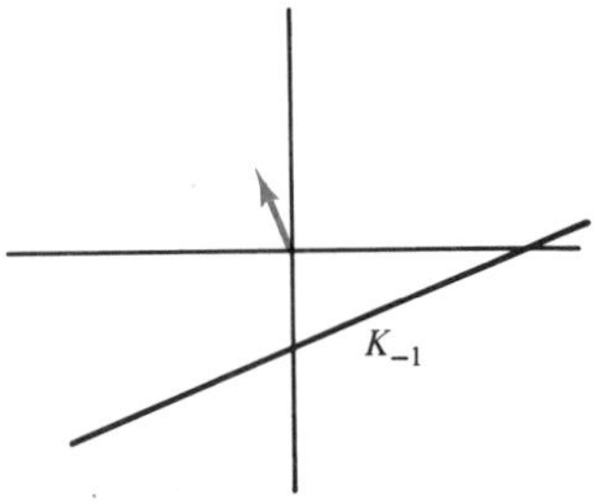

c. Distance $(AK_{c_1}, AK_{c_2}) = \dfrac{\sqrt{2}}{(\sqrt{73}/14)}\,\text{dist}(K_{c_1}, K_{c_2})$.

5. a. $(2+\sqrt{2})y_1^2 + (2\sqrt{2})y_2^2 = 1$.

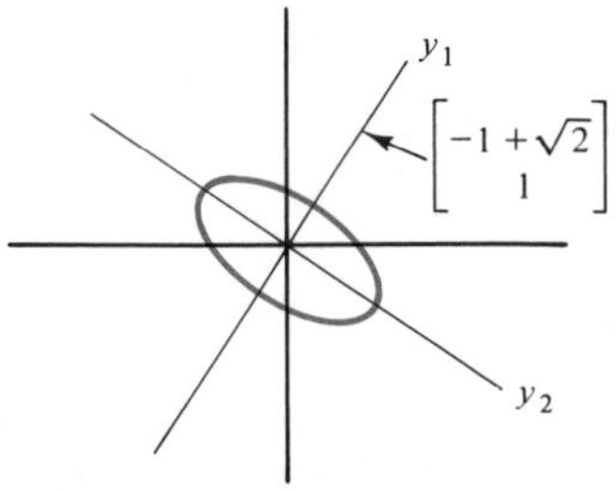

b. $(1+\sqrt{5})/2y_1^2 + (1-\sqrt{5})/2y_2^2 = 1$.

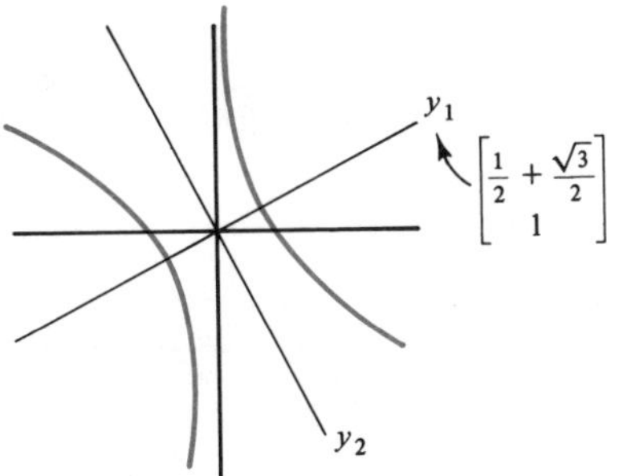

c. $y_1 = \frac{3}{4} - (y_2 - \frac{1}{2})^2$.

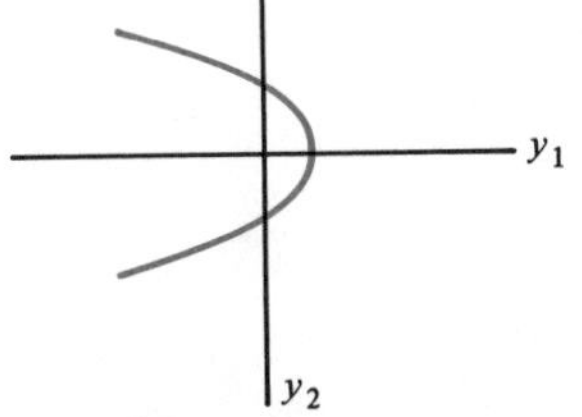

7. a. $\langle A\mathbf{x},\mathbf{x}\rangle = \langle (x_2, -x_1), (x_1, x_2)\rangle = x_2x_1 - x_1x_2 = 0$.

b. $\langle A\mathbf{x},\mathbf{x}\rangle = \langle [\Sigma_{j=1}^n a_{ij}x_j], [x_i]\rangle = \Sigma_{i=1}^n \Sigma_{j=1}^n a_{ij}x_jx_i = \Sigma_{i=1}^n a_{ii}x_i^2 + \Sigma_{i=1}^n \Sigma_{j=i+1}^n (a_{ij} + a_{ji})x_ix_j = 0$. The x_i's can be anything, so we have $a_{ii} = 0$ and $a_{ji} = -a_{ij}$ for all ij.

c. If $A = A^T = -A^T$, then $a_{ij} = a_{ji} = -a_{ji}$ for all ij, and $A = 0_{nn}$.

9. a. $(\frac{1}{2})(A + A^T)^T = \frac{1}{2}(A^T + A^{TT}) = \frac{1}{2}(A + A^T)$ and $\frac{1}{2}(A - A^T)^T = \frac{1}{2}(A^T - A^{TT}) = -\frac{1}{2}(A - A^T)$.

b. $A = B + C$, so $A^T = B^T + C^T = B - C$. Therefore, $B = \frac{1}{2}(A + A^T)$ and $C = (\frac{1}{2})(A - A^T)$.

INDEX